T0155944

Graduate Texts in Mathematics

Series Editors:

Sheldon Axler
San Francisco State University, San Francisco, CA, USA

Kenneth Ribet
University of California, Berkeley, CA, USA

Advisory Board:

Colin Adams, *Williams College, Williamstown, MA, USA*
Alejandro Adem, *University of British Columbia, Vancouver, BC, Canada*
Ruth Charney, *Brandeis University, Waltham, MA, USA*
Irene M. Gamba, *The University of Texas at Austin, Austin, TX, USA*
Roger E. Howe, *Yale University, New Haven, CT, USA*
David Jerison, *Massachusetts Institute of Technology, Cambridge, MA, USA*
Jeffrey C. Lagarias, *University of Michigan, Ann Arbor, MI, USA*
Jill Pipher, *Brown University, Providence, RI, USA*
Fadil Santosa, *University of Minnesota, Minneapolis, MN, USA*
Amie Wilkinson, *University of Chicago, Chicago, IL, USA*

Graduate Texts in Mathematics bridge the gap between passive study and creative understanding, offering graduate-level introductions to advanced topics in mathematics. The volumes are carefully written as teaching aids and highlight characteristic features of the theory. Although these books are frequently used as textbooks in graduate courses, they are also suitable for individual study.

For further volumes:
http://www.springer.com/series/136

Steven G. Krantz

Geometric Analysis of the Bergman Kernel and Metric

 Springer

Steven G. Krantz
Department of Mathematics
Washington University at St. Louis
St. Louis, MO, USA

ISSN 0072-5285
ISBN 978-1-4939-4429-3   ISBN 978-1-4614-7924-6 (eBook)
DOI 10.1007/978-1-4614-7924-6
Springer New York Heidelberg Dordrecht London

Mathematics Subject Classification (2010): 30C40, 32A25, 32A26, 32A36, 32H40, 46C05

© Springer Science+Business Media New York 2013
Softcover reprint of the hardcover 1st edition 2013
This work is subject to copyright. All rights are reserved by the Publisher, whether the whole or part of
the material is concerned, specifically the rights of translation, reprinting, reuse of illustrations, recitation,
broadcasting, reproduction on microfilms or in any other physical way, and transmission or information
storage and retrieval, electronic adaptation, computer software, or by similar or dissimilar methodology
now known or hereafter developed. Exempted from this legal reservation are brief excerpts in connection
with reviews or scholarly analysis or material supplied specifically for the purpose of being entered
and executed on a computer system, for exclusive use by the purchaser of the work. Duplication of
this publication or parts thereof is permitted only under the provisions of the Copyright Law of the
Publisher's location, in its current version, and permission for use must always be obtained from Springer.
Permissions for use may be obtained through RightsLink at the Copyright Clearance Center. Violations
are liable to prosecution under the respective Copyright Law.
The use of general descriptive names, registered names, trademarks, service marks, etc. in this publication
does not imply, even in the absence of a specific statement, that such names are exempt from the relevant
protective laws and regulations and therefore free for general use.
While the advice and information in this book are believed to be true and accurate at the date of
publication, neither the authors nor the editors nor the publisher can accept any legal responsibility for
any errors or omissions that may be made. The publisher makes no warranty, express or implied, with
respect to the material contained herein.

Printed on acid-free paper

Springer is part of Springer Science+Business Media (www.springer.com)

*Dedicated to the memory of Stefan Bergman,
an extraordinarily profound and original
mathematician*

Preface

The Bergman kernel and metric have been a seminal part of geometric analysis and partial differential equations since their invention by Stefan Bergman in 1922. Applications to holomorphic mappings, to function theory, to partial differential equations, and to differential geometry have kept the techniques plugged into the mainstream of mathematics for 90 years.

The Bergman kernel is based on a very simple idea: that the square-integrable holomorphic functions on a bounded domain in that complex space form a Hilbert space. Moreover, a simple formal argument shows that Hilbert space possesses a so-called reproducing kernel. This is an integration kernel which reproduces each element of the space. The kernel has wonderful invariance properties, leading to the Bergman metric. The Bergman kernel and metric have developed into powerful tools for function theory, analysis, differential geometry, and partial differential equations. The purpose of this book is to exposit this theory (particularly in the context of several complex variables), examine its key features, and bring the reader up to speed with some of the latest developments.

Bergman wrote several books about his kernel and contributed mightily to its development. The idea caught on widely, and the kernel became a standard device in the field. The Bergman metric was the first-ever Kähler metric, and that in turn spawned the vital subject of complex differential geometry. Ahlfors (in one variable), Chern (in several variables), Greene–Wu, and many others played a decisive role in this development.

Bergman's ideas received a major boost in the 1970s when Charles Fefferman did his Fields Medal-winning work on the boundary behavior of biholomorphic mappings. The key device in his analysis was the Bergman kernel and metric. Since then, a myriad of workers, from Bell and Ligocka to Webster to Greene–Krantz, Krantz–Li, Kim–Krantz, Isaev–Krantz, and many others, have developed and extended Bergman's theory. It is now part of the lingua franca of complex analysis, and the technique of reproducing kernels which it has spawned is part of every analyst's toolkit.

Fefferman's work inspired many others to examine the utility of the Bergman theory in the study of biholomorphic mappings. Bell's condition R, formulated in terms of regularity properties of the Bergman projection, has proved to be an influential and powerful weapon in the subject. In turn, condition R can be formulated in terms of the regularity theory of subelliptic partial differential equations, and this connection has had a key influence on the directions of research.

The Bergman metric was the first "universal" (in the sense that it can be constructed on virtually any domain) example of a metric that is invariant under biholomorphic mappings. [The Poincaré metric was of course the primordial example on the disc.] Today there are the Kobayashi–Royden metric, the Sibony metric, the Carathéodory metric, and many others. This is a useful tool in geometric analysis and function theory.

The connections of the Bergman kernel with partial differential equations, especially the extremal properties of the kernel and metric, are profound. Bergman himself explored applications of his theory to elliptic partial differential equations. Today we see the Bergman kernel as inextricably linked with the $\bar{\partial}$-Neumann problem. This link played a vital role in Fefferman's work and later proved crucial to Greene–Krantz and many of the other workers in the subject. Certainly Donald Spencer and J. J. Kohn were the pioneers of this symbiosis. Today the interaction is prospering. Another development is that the Bergman theory enjoys connections with the Monge–Ampère equation. That nonlinear partial differential equation contains important information about biholomorphic mappings and about the construction of geometries.

Connections with harmonic analysis are another exciting, and relatively new, aspect of the Bergman paradigm. Coifman–Rochberg–Weiss used the Bergman kernel in their proof of the H^1/BMO duality theorem on the ball, and Krantz–Li exploited it further in their study on strictly pseudoconvex domains. Many of the natural artifacts of harmonic analysis—including approach regions for Fatou theorems—are most propitiously formulated in terms of Bergman geometry or the boundary asymptotics of the kernel. The Bergman kernel is now a standard artifact of the harmonic analysis of several complex variables.

This text will in fact be a thoroughgoing treatment of all the basic analytic and geometric aspects of Bergman's theory. This will include

- Definitions of the Bergman kernel
- Definition and basic properties of the Bergman metric
- Calculation of the Bergman kernel and metric
- Invariance properties of the kernel and metric
- Boundary asymptotics of the kernel and metric
- Asymptotic expansions for the Bergman kernel
- Applications to function theory
- Applications to geometry
- Applications to partial differential equations
- Interpretations in terms of functional analysis
- The geometry of the Bergman metric
- Curvature of the Bergman metric
- The Bergman kernel and metric on manifolds

There are a few recent treatises on the Bergman kernel, notably those by Hedenmalm–Korenblum–Zhu and Duren–Schuster. But these books concentrate on the one-variable theory; they are also oriented towards the functional analysis aspects of the Bergman kernel. Our focus instead is the geometry of several complex variables and the contexts of real analysis, complex analysis, harmonic analysis, and differential geometry. This puts Bergman's ideas into a much broader arena and provides many more opportunities for applications and illustrations. We will certainly touch on the functional analysis properties of the Bergman projection, but these will not be our main focus. We shall also cover selected topics of the one complex variable theory. There is little overlap between this book and the two books cited above.

We would also be remiss not to mention the book of Ma and Marinescu on the Kähler geometry aspects of the Bergman theory. Certainly the Bergman metric was the very first Kähler metric, and this in turn has spawned the active and fruitful area of multivariable complex differential geometry. Various parts of the present book touch on this Kähler theory.

Lurking in the background behind Fefferman's biholomorphic mapping theorem were Bergman representative coordinates—yet another outgrowth of the Bergman kernel and metric. This is a much-underappreciated aspect of the theory and one that we shall treat in detail in the text. In fact there are many aspects of the Bergman theory that tend only to be known to experts and are not readily accessible in the literature. We intend to treat many of those. Several of the topics in this text appear here for the first time in book form.

We intend this to be a book for students as well as seasoned researchers. All needed background will be provided. The reader is only assumed to have had a solid course in complex variables and some basic background in real and functional analysis. A little exposure to geometry will be helpful, but is not a requirement. There are many illustrative examples and some useful figures. The book abounds with useful and instructive calculations, many of which cannot be found elsewhere. Every chapter ends with a selection of exercises, which should serve to help the reader get more directly involved in the subject matter. It will cause him/her to consult the literature, to calculate, and to learn by doing.

The book will help the novice reader to see how analysis is used in practice and how it can be evolved into a seminal tool for research. It is important for the student to see fundamental mathematics used in vitro in order to understand how research develops and grows.

It is a pleasure to thank E. M. Stein for introducing me to the Bergman kernel and Robert E. Greene for teaching me the geometric aspects. I have had many collaborators in my study of the kernel, and I offer them all my gratitude. I thank my editor, Elizabeth Loew, for her constant enthusiasm and support. And I thank the several referees for this book, who contributed a wealth of ideas and information.

St. Louis, MO, USA

Steven G. Krantz

Contents

Chapter 1
Introductory Ideas

In the early days of functional analysis—the early twentieth century—people did not
yet know what a Banach space was nor a Hilbert space. They frequently studied a
particular complete, infinite-dimensional space from a more abstract point of view.
The most common space to be studied in this regard was of course L^2. It was when
Stefan Bergman took a course from Erhard Schmidt on L^2 of the unit interval I that
he conceived of the idea of the Bergman space of square-integrable holomorphic
functions on the unit disc D. And the rest is history.

It is important for the Bergman theory that his space of holomorphic functions
has an inner product structure and that it is complete. The first of these properties
follows from the fact that it is a subspace of L^2; the second follows from a
fundamental inequality that we shall consider in the next section.

1.1 The Bergman Kernel

It is difficult to create an explicit integral formula, with holomorphic reproducing
kernel, for holomorphic functions on an arbitrary domain in $\mathbb{C}^n$.[1] Classical studies
which perform such constructions tend to concentrate on domains having a great
deal of symmetry (see, for instance, [HUA]). We now examine one of several
non-constructive approaches to this problem. This circle of ideas, due to Bergman
[BER1] and to Szegő [SZE] (some of the ideas presented here were anticipated by
the thesis of Bochner [BOC1]), will later be seen to have profound applications to
the boundary regularity of holomorphic mappings.

Bungart [BUN] and Gleason [GLE] have shown that *any* bounded domain in $\mathbb{C}^n$
will have a reproducing kernel for holomorphic functions such that the kernel itself
is holomorphic in the free variable. In other words, the formula has the form

---

[1] Here a *domain* is a connected, open set.

S.G. Krantz, *Geometric Analysis of the Bergman Kernel and Metric*,
Graduate Texts in Mathematics 268, DOI 10.1007/978-1-4614-7924-6_1,
© Springer Science+Business Media New York 2013

$$f(z) = \int_\Omega f(\zeta) K(z, \zeta) \, dV(\zeta),$$

and K is holomorphic in the z variable. Of course Bungart's and Gleason's proofs are highly nonconstructive, and one can say almost nothing about the actual form of the kernel K. The venerable Bochner–Martinelli kernel is easily constructed on any bounded domain with reasonable boundary (just as an application of Stokes's theorem) and the kernel is *explicit*—just like the Cauchy kernel in one complex variable. Also the kernel is the same for every domain. But the Bochner–Martinelli kernel is definitely *not* holomorphic in the free variable. On the other hand, Henkin [HEN], Kerzman [KER1], E. Ramirez [RAMI], and Grauert–Lieb [GRL] have given very explicit constructions of reproducing kernels on strictly pseudoconvex domains (see the definition below). And their kernels *are* holomorphic in the z variable. This matter is treated in [KRA1, Chap. 10].

In fact this last described result was considered to be quite a dramatic advance. For Henkin, Kerzman, Ramirez, and Grauert–Lieb provided us with a fairly explicit kernel, with an explicit and measurable singularity, that can not only reproduce but also create holomorphic functions. Such a kernel is very much like the Cauchy kernel in one complex variable. Thus at least on strictly pseudoconvex domains, we can perform many of the activities to which we are accustomed from the function theory of one complex variable. We can get formulas for derivatives of holomorphic functions, we can analyze power series, we can consider an analogue of the Cauchy transform, and (perhaps most importantly) we can write down solution operators for the $\bar\partial$ problem. People were optimistic that these new integral formulas would give a shot in the arm to the theory of function algebras—that they would now be able to study $H^\infty(\Omega)$ and $A(\Omega)$ on a variety of domains in $\mathbb{C}^n$ (see [GAM, Chap. II, IV] for the role model in $\mathbb{C}^1$). But this turned out to be too difficult.

The Bergman kernel is a canonical kernel that can be defined on any bounded domain. It has wonderful invariance properties and is a powerful tool for geometry and analysis. But it is difficult to calculate explicitly.

In this section we will see some of the invariance properties of the Bergman kernel. This will lead in later sections to the definition of the Bergman metric (in which all biholomorphic mappings become isometries) and to such other canonical constructions as representative coordinates. The Bergman kernel has certain extremal properties that make it a powerful tool in the theory of partial differential equations (see Bergman and Schiffer [BES]). Also the form of the singularity of the Bergman kernel (calculable for some interesting classes of domains) explains many phenomena of the function theory of several complex variables.

Let $\Omega \subseteq \mathbb{C}^n$ be a bounded domain (it is possible, but often tricky, to treat unbounded domains as well). Here a *domain* is a connected, open set. If the domain is smoothly bounded, then we may think of it as specified by a *defining function*:

$$\Omega = \{z \in \mathbb{C}^n : \rho(z) < 0\}.$$

It is customary to require that $\nabla \rho \neq 0$ on $\partial\Omega$. One can demonstrate the existence of a definiting function by using the implicit function theorem. See [KRPA2] for the latter and [KRA1] for a detailed consideration of defining functions.

Given a domain Ω as described in the last paragraph and a point $P \in \partial\Omega$, we say that w is a *complex tangent vector* at P and write $w \in \mathcal{T}_P(\partial\Omega)$ if

$$\sum_{j=1}^{n} \frac{\partial\rho}{\partial z_j}(P)w_j = 0.$$

The point P is said to be *strongly pseudoconvex* if

$$\sum_{j,k=1}^{n} \frac{\partial^2\rho}{\partial z_j \partial\bar{z}_k}(P)w_j\bar{w}_k > 0$$

for $0 \neq w \in \mathcal{T}_P(\partial\Omega)$. In fact a little elementary analysis shows that we can write the defining property of strong pseudoconvexity as

$$\sum_{j,k=1}^{n} \frac{\partial^2\rho}{\partial z_j \partial\bar{z}_k}(P)w_j\bar{w}_k \geq C|w|^2$$

and make the estimate uniform when P ranges over a compact, strongly pseudoconvex boundary neighborhood of Ω. Again, the book [KRA1, Chap. 3] has extensive discussion of the notion of strong pseudoconvexity.

Now let us return to the Bergman theory. Let dV denote the Lebesgue volume measure on Ω. Define the *Bergman space*

$$A^2(\Omega) = \left\{ f \text{ holomorphic on } \Omega : \int_{\Omega} |f(z)|^2 \, dV(z)^{1/2} \equiv \|f\|_{A^2(\Omega)} < \infty \right\}.$$

Of course we equip the Bergman space with the inner product

$$\langle f, g \rangle = \int_{\Omega} f(z)\bar{g}(z) \, dV(z).$$

Lemma 1.1.1. *Let $K \subseteq \Omega \subseteq \mathbb{C}^n$ be compact. There is a constant $C_K > 0$, depending on K and on n, such that*

$$\sup_{z \in K} |f(z)| \leq C_K \|f\|_{A^2(\Omega)}, \text{ all } f \in A^2(\Omega).$$

Proof. Since K is compact, there is an $r(K) = r > 0$ so that, for any $z \in K$, $B(z, r) \subseteq \Omega$. Here $B(z, r)$ is the usual Euclidean ball with center z and radius r.

Therefore for each $z \in K$ and $f \in A^2(\Omega)$, the mean-value property for holomorphic functions implies that

$$
\begin{aligned}
|f(z)| &= \left| \frac{1}{V(B(z,r))} \int_{B(z,r)} f(t)\, dV(t) \right| \\
&= \left| \frac{1}{V(B(z,r))} \int f(t)\chi_{B(z,r)}(t)\, dV(t) \right| \\
&\le (V(B(z,r)))^{-1/2} \|f\|_{L^2(B(z,r))} \\
&\le C(n)r^{-n}\|f\|_{A^2(\Omega)} \\
&\equiv C_K \|f\|_{A^2(\Omega)}. \qquad\qquad\qquad\square
\end{aligned}
$$

Lemma 1.1.2. *The space $A^2(\Omega)$ is a Hilbert space with the inner product $\langle f, g \rangle \equiv \int_\Omega f(z)\overline{g(z)}\, dV(z)$.*

Proof. Everything is clear except for completeness. Let $\{f_j\} \subseteq A^2$ be a sequence that is Cauchy in norm. Since L^2 is complete there is an L^2 limit function f. We need to see that f is holomorphic. But Lemma 1.1.1 yields that norm convergence implies normal convergence (i.e., uniform convergence on compact sets). Certainly holomorphic functions are closed under normal limits (just use the Cauchy theory of one complex variable). Therefore f is holomorphic and $A^2(\Omega)$ is complete. $\square$

Lemma 1.1.3. *For each fixed $z \in \Omega$, the functional*

$$
\Phi_z : f \mapsto f(z), \quad f \in A^2(\Omega)
$$

is a continuous linear functional on $A^2(\Omega)$.

Proof. This is immediate from Lemma 1.1.1 if we take K to be the singleton $\{z\}$. $\square$

We may now apply the Riesz representation theorem to see that there is an element $K_z \in A^2(\Omega)$ such that the linear functional Φ_z is represented by inner product with K_z : if $f \in A^2(\Omega)$, then, for all $z \in \Omega$, we have

$$
f(z) = \Phi_z(f) = \langle f, K_z \rangle.
$$

Definition 1.1.4. The *Bergman kernel* is the function $K(z, \zeta) = K_\Omega(z, \zeta) \equiv \overline{K_z(\zeta)}$, $z, \zeta \in \Omega$. It has the reproducing property

$$
f(z) = \int_\Omega K(z, \zeta) f(\zeta)\, dV(\zeta), \quad \forall f \in A^2(\Omega).
$$

Proposition 1.1.5. *The Bergman kernel $K(z, \zeta)$ is conjugate symmetric: $K(z, \zeta) = \overline{K(\zeta, z)}$.*

Proof. By its very definition, $\overline{K(\zeta, \cdot)} \in A^2(\Omega)$ for each fixed ζ. Therefore the reproducing property of the Bergman kernel gives

$$\int_{\Omega} K(z,t)\overline{K(\zeta,t)}\,dV(t) = \overline{K(\zeta,z)}.$$

On the other hand,

$$\int_{\Omega} K(z,t)\overline{K(\zeta,t)}\,dV(t) = \overline{\int K(\zeta,t)\overline{K(z,t)}\,dV(t)}$$

$$= \overline{\overline{K(z,\zeta)}} = K(z,\zeta). \qquad \Box$$

Proposition 1.1.6. *The Bergman kernel is uniquely determined by the properties that it is an element of $A^2(\Omega)$ in z, is conjugate symmetric, and reproduces $A^2(\Omega)$.*

Proof. Let $K'(z,\zeta)$ be another such kernel. Then

$$K(z,\zeta) = \overline{K(\zeta,z)} = \overline{\int K'(z,t)\overline{K(\zeta,t)}\,dV(t)}$$

$$= \overline{\int K(\zeta,t)\overline{K'(z,t)}\,dV(t)}$$

$$= \overline{\overline{K'(z,\zeta)}} = K'(z,\zeta). \qquad \Box$$

Since $L^2(\Omega)$ is a separable Hilbert space then so is its subspace $A^2(\Omega)$. Thus there is a countable, complete orthonormal basis $\{\phi_j\}_{j=1}^{\infty}$ for $A^2(\Omega)$.

Proposition 1.1.7. *Let L be a compact subset of Ω. Then the series*

$$\sum_{j=1}^{\infty} \phi_j(z)\overline{\phi_j(\zeta)}$$

sums uniformly on $L \times L$ to the Bergman kernel $K(z,\zeta)$.

Proof. By the Riesz–Fischer and Riesz representation theorems, we obtain

$$\sup_{z \in L}\left(\sum_{j=1}^{\infty}|\phi_j(z)|^2\right)^{1/2} = \sup_{z \in L}\left\|\{\phi_j(z)\}_{j=1}^{\infty}\right\|_{\ell^2}$$

$$= \sup_{\substack{\|\{a_j\}\|_{\ell^2}=1 \\ z \in L}}\left|\sum_{j=1}^{\infty}a_j\phi_j(z)\right|$$

$$= \sup_{\substack{\|f\|_{A^2}=1 \\ z \in L}}|f(z)|$$

$$\leq C_L. \qquad (1.1.7.1)$$

In the last inequality we have used Lemma 1.1.1. Therefore

$$\sum_{j=1}^{\infty} \left| \phi_j(z)\overline{\phi_j(\zeta)} \right| \le \left(\sum_{j=1}^{\infty} |\phi_j(z)|^2 \right)^{1/2} \left(\sum_{j=1}^{\infty} |\phi_j(\zeta)|^2 \right)^{1/2}$$

and the convergence is uniform over $z, \zeta \in L$. For fixed $z \in \Omega$, (1.1.7.1) shows that $\{\phi_j(z)\}_{j=1}^{\infty} \in \ell^2$. Hence we have that $\sum \phi_j(z)\overline{\phi_j(\zeta)} \in \overline{A^2(\Omega)}$ as a function of ζ. Let the sum of the series be denoted by $K'(z, \zeta)$. Notice that K' is conjugate symmetric by its very definition. Also, for $f \in A^2(\Omega)$, we have

$$\int K'(\cdot, \zeta) f(\zeta) \, dV(\zeta) = \sum \hat{f}(j)\phi_j(\cdot) = f(\cdot),$$

where convergence is in the Hilbert space topology. (Here $\hat{f}(j)$ is the jth Fourier coefficient of f with respect to the basis $\{\phi_j\}$.) But Hilbert space convergence dominates pointwise convergence (Lemma 1.1.1) so

$$f(z) = \int K'(z, \zeta) f(\zeta) \, dV(\zeta), \quad \text{all } f \in A^2(\Omega).$$

Therefore K' is the Bergman kernel.   □

Remark 1.1.8. It is worth noting explicitly that the proof of Proposition 1.1.7 shows that

$$\sum \phi_j(z)\overline{\phi_j(\zeta)}$$

equals the Bergman kernel $K(z, \zeta)$ *no matter what the choice* of complete orthonormal basis $\{\phi_j\}$ for $A^2(\Omega)$. This can be very useful information in practice.   □

Proposition 1.1.9. *If Ω is a bounded domain in $\mathbb{C}^n$, then the mapping*

$$P : f \mapsto \int_{\Omega} K(\cdot, \zeta) f(\zeta) \, dV(\zeta)$$

is the Hilbert space orthogonal projection of $L^2(\Omega, dV)$ onto $A^2(\Omega)$. We call P the Bergman projection.

Proof. Notice that P is idempotent and self-adjoint and that $A^2(\Omega)$ is precisely the set of elements of L^2 that are fixed by P.   □

Definition 1.1.10. Let $\Omega \subseteq \mathbb{C}^n$ be a domain and let $f : \Omega \to \mathbb{C}^n$ be a *holomorphic mapping*, that is, $f(z) = (f_1(z), \ldots, f_n(z))$ with $f_1, \ldots, f_n$ holomorphic on Ω. Let $w_j = f_j(z), j = 1, \ldots, n$. Then the *holomorphic Jacobian matrix* of f is the matrix

$$J_{\mathbb{C}} f = \frac{\partial(w_1, \ldots, w_n)}{\partial(z_1, \ldots, z_n)}.$$

Write $z_j = x_j + iy_j, w_k = \xi_k + i\eta_k, j, k = 1, \ldots, n$. Then the *real Jacobian matrix* of f is the matrix

$$J_{\mathbb{R}} f = \frac{\partial(\xi_1, \eta_1, \ldots, \xi_n, \eta_n)}{\partial(x_1, y_1, \ldots, x_n, y_n)}.$$

Proposition 1.1.11. *With notation as in the definition, we have*

$$\det J_{\mathbb{R}} f = |\det J_{\mathbb{C}} f|^2$$

whenever f is a holomorphic mapping.

Proof. We exploit the functoriality of the Jacobian. Let $w = (w_1, \ldots, w_n) = f(z) = (f_1(z), \ldots, f_n(z))$. Write $z_j = x_j + iy_j, w_j = \xi_j + i\eta_j, j = 1, \ldots, n$. Then, using the fact that f is holomorphic,

$$d\xi_1 \wedge d\eta_1 \wedge \cdots \wedge d\xi_n \wedge d\eta_n = (\det J_{\mathbb{R}} f(x, y)) dx_1 \wedge dy_1 \wedge \cdots \wedge dx_n \wedge dy_n. \quad (1.1.11.1)$$

On the other hand,

$$d\xi_1 \wedge d\eta_1 \wedge \cdots \wedge d\xi_n \wedge d\eta_n$$

$$= \frac{1}{(2i)^n} d\overline{w}_1 \wedge dw_1 \wedge \cdots \wedge d\overline{w}_n \wedge dw_n$$

$$= \frac{1}{(2i)^n} \overline{(\det J_{\mathbb{C}} f(z))} (\det J_{\mathbb{C}} f(z)) d\overline{z}_1 \wedge dz_1 \wedge \cdots \wedge d\overline{z}_n \wedge dz_n$$

$$= |\det J_{\mathbb{C}} f(z)|^2 dx_1 \wedge dy_1 \wedge \cdots \wedge dx_n \wedge dy_n. \quad (1.1.11.2)$$

Equating (1.1.11.1) and (1.1.11.2) gives the result. $\qquad \square$

Exercise for the Reader: Prove Proposition 1.1.11 using only matrix theory (no differential forms). This will give rise to a great appreciation for the theory of differential forms (see Bers [BERS, Chap. 7] for help).

Now we can prove the holomorphic implicit function theorem:

Theorem 1.1.12. *Let $f_j(w, z)$, $j = 1, \ldots, m$ be holomorphic functions of $(w, z) = ((w_1, \ldots, w_m), (z_1, \ldots, z_n)) \in \mathbb{C}^m \times \mathbb{C}^n$. Assume that*

$$f_j(w^0, z^0) = 0, \quad j = 1, \ldots, m,$$

and that

$$\det \left(\frac{\partial f_j}{\partial w_k} \right)_{j,k=1}^m \neq 0 \text{ at } (w^0, z^0).$$

Then the system of equations

$$f_j(w, z) = 0, \quad j = 1, \ldots, m,$$

has a unique holomorphic solution $w(z)$ in a neighborhood of z^0 that satisfies $w(z^0) = w^0$.

Proof. We rewrite the system of equations as

$$\operatorname{Re} f_j(w, z) = 0, \quad \operatorname{Im} f_j(w, z) = 0$$

for the $2m$ real variables $\operatorname{Re} w_k, \operatorname{Im} w_k, k = 1, \ldots, m$. By Proposition 1.1.11, the determinant of the Jacobian over $\mathbb{R}$ of this new system is the modulus squared of the determinant of the Jacobian over $\mathbb{C}$ of the old system. By our hypothesis, this number is nonvanishing at the point (w^0, z^0). Therefore the classical implicit function theorem (see Rudin [RUD1] or [KRPA2]) implies that there exist C^1 functions $w_k(z), k = 1, \ldots, m$, with $w(z^0) = w^0$ and that solve the system. Our job is to show that these functions are in fact holomorphic. When properly viewed, this is purely a problem of geometric algebra:

Applying exterior differentiation to the equations

$$0 = f_j(w(z), z), \quad j = 1, \ldots, m,$$

yields that

$$0 = df_j = \sum_{k=1}^m \frac{\partial f_j}{\partial w_k} dw_k + \sum_{k=1}^m \frac{\partial f_j}{\partial z_k} dz_k.$$

There are no $d\bar{z}_j$'s and no $d\bar{w}_k$'s because the f_j's are holomorphic.

The result now follows from linear algebra only: The hypothesis on the determinant of the matrix $(\partial f_j / \partial w_k)$ implies that we can solve for dw_k in terms of dz_j. Therefore w is a holomorphic function of z. $\qquad\square$

A holomorphic mapping $f : \Omega_1 \to \Omega_2$ of domains $\Omega_1 \subseteq \mathbb{C}^n, \Omega_2 \subseteq \mathbb{C}^n$ is said to be *biholomorphic* if it is one-to-one, onto, and $\det J_{\mathbb{C}} f(z) \neq 0$ for every $z \in \Omega_1$.

Exercise for the Reader: Use Theorem 1.1.12 to prove that a biholomorphic mapping has a holomorphic inverse (hence the name).

Remark 1.1.13. It is true, but not at all obvious, that the nonvanishing of the Jacobian determinant is a superfluous condition in the definition of "biholomorphic mapping"; that is, the nonvanishing of the Jacobian follows from the univalence of the mapping. A proof of this assertion is sketched in Exercise 37 at the end of [KRA1, Chap. 11]. □

In what follows we shall frequently denote the Bergman kernel for a given domain Ω by K_Ω.

Proposition 1.1.14. *Let Ω_1, Ω_2 be domains in $\mathbb{C}^n$. Let $f : \Omega_1 \to \Omega_2$ be biholomorphic. Then*

$$\det J_{\mathbb{C}} f(z) K_{\Omega_2}(f(z), f(\zeta)) \det \overline{J_{\mathbb{C}} f(\zeta)} = K_{\Omega_1}(z, \zeta).$$

Proof. Let $\phi \in A^2(\Omega_1)$. Then, by change of variable,

$$\int_{\Omega_1} \det J_{\mathbb{C}} f(z) K_{\Omega_2}(f(z), f(\zeta)) \det \overline{J_{\mathbb{C}} f(\zeta)} \phi(\zeta) \, dV(\zeta)$$

$$= \int_{\Omega_2} \det J_{\mathbb{C}} f(z) K_{\Omega_2}(f(z), \tilde{\zeta}) \det \overline{J_{\mathbb{C}} f(f^{-1}(\tilde{\zeta}))} \phi(f^{-1}(\tilde{\zeta}))$$

$$\times \det J_{\mathbb{R}} f^{-1}(\tilde{\zeta}) \, dV(\tilde{\zeta}).$$

By Proposition 1.1.11 this simplifies to

$$\det J_{\mathbb{C}} f(z) \int_{\Omega_2} K_{\Omega_2}(f(z), \tilde{\zeta}) \left\{ \left(\det J_{\mathbb{C}} f(f^{-1}(\tilde{\zeta})) \right)^{-1} \phi\left(f^{-1}(\tilde{\zeta})\right) \right\} \, dV(\tilde{\zeta}).$$

By change of variables, the expression in braces { } is an element of $A^2(\Omega_2)$. So the reproducing property of K_{Ω_2} applies and the last line equals

$$= \det J_{\mathbb{C}} f(z) \left(\det J_{\mathbb{C}} f(z) \right)^{-1} \phi\left(f^{-1}(f(z))\right) = \phi(z).$$

By the uniqueness of the Bergman kernel, the proposition follows. □

Proposition 1.1.15. *For $z \in \Omega \subset\subset \mathbb{C}^n$ it holds that $K_\Omega(z, z) > 0$.*

Proof. Now

$$K_\Omega(z, z) = \sum_{j=1}^{\infty} |\phi_j(z)|^2 \geq 0.$$

If in fact $K(z, z) = 0$ for some z, then $\phi_j(z) = 0$ for all j; hence, $f(z) = 0$ for every $f \in A^2(\Omega)$. This is absurd. $\qquad\square$

Definition 1.1.16. For any bounded domain $\Omega \subseteq \mathbb{C}^n$, we define a Hermitian metric on Ω by

$$g_{ij}(z) = \frac{\partial^2}{\partial z_i \, \partial \bar{z}_j} \log K(z, z), \quad z \in \Omega.$$

This means that the square of the length of a tangent vector $\xi = (\xi_1, \ldots, \xi_n)$ at a point $z \in \Omega$ is given by

$$|\xi|_{B,z}^2 = \sum_{i,j} g_{ij}(z)\xi_i \bar{\xi}_j.$$

The metric that we have defined is called the *Bergman metric*.

In a Hermitian metric $\{g_{ij}\}$, the length of a C^1 curve $\gamma : [0, 1] \to \Omega$ is given by

$$\ell(\gamma) = \int_0^1 \left(\sum_{i,j} g_{i,j}(\gamma(t))\gamma_i'(t)\overline{\gamma_j'(t)} \right)^{1/2} dt.$$

If P, Q are points of Ω, then their distance $d_\Omega(P, Q)$ in the metric is defined to be the infimum of the lengths of all piecewise C^1 curves connecting the two points.

Remark 1.1.17. It is not a priori obvious that the Bergman metric for a bounded domain Ω is given by a positive definite matrix at each point. We now outline a proof of this fact.

First we generate an orthonormal basis for the Bergman space. Fix $z_0 \in \Omega$. Let ϕ_0 be the (unique!) element of A^2 with $\phi_0(z_0)$ real, $\|\phi_0\| = 1$, and $\phi_0(z_0)$ maximal. (Why does such a ϕ_0 exist?) Let ϕ_1 be the (unique) element of A^2 with $\phi_1(z_0) = 0$, $(\partial\phi_1/\partial z_1)(z_0)$ real, $\|\phi_1\| = 1$, and $(\partial\phi_1/\partial z_1)(z_0)$ maximal. (Why does such a ϕ_1 exist?) Now ϕ_1 is orthogonal to ϕ_0, else ϕ_1 has nonzero projection on ϕ_0, leading to a contradiction. Continue this process to create an orthogonal system on Ω. Use Taylor series to see that it is complete. This circle of ideas comes from the elegant paper Kobayashi [KOB1].

Now let $\Omega \subseteq \mathbb{C}^n$ be a bounded domain and let (g_{ij}) be its Bergman metric. Use the ideas in the last paragraph to prove that the matrix $(g_{ij}(z))$ is positive definite, each $z \in \Omega$. [*Hint:* The crucial fact is that, for each $z \in \Omega$ and each j, there is an element $f \in A^2(\Omega)$ such that $\partial f/\partial z_j(z) \neq 0$.] $\qquad\square$

Proposition 1.1.18. *Let $\Omega_1, \Omega_2 \subseteq \mathbb{C}^n$ be domains and let $f : \Omega_1 \to \Omega_2$ be a biholomorphic mapping. Then f induces an isometry of Bergman metrics:*

$$|\xi|_{B,z} = |(J_{\mathbb{C}}f)\xi|_{B,f(z)}$$

for all $z \in \Omega_1, \xi \in \mathbb{C}^n$. Equivalently, f induces an isometry of Bergman distances in the sense that

$$d_{\Omega_2}(f(P), f(Q)) = d_{\Omega_1}(P, Q).$$

Proof. This is a formal exercise but we include it for completeness: From the definitions, it suffices to check that

$$\sum_{i,j} g_{i,j}^{\Omega_2}(f(z)) \, (J_{\mathbb{C}}f(z)w)_i \, \left(\overline{J_{\mathbb{C}}f(z)w}\right)_j = \sum g_{ij}^{\Omega_1}(z) w_i \overline{w}_j \qquad (1.1.18.1)$$

for all $z \in \Omega, w = (w_1, \ldots, w_n) \in \mathbb{C}^n$. But, by Proposition 1.1.14,

$$g_{ij}^{\Omega_1}(z) = \frac{\partial^2}{\partial z_i \overline{z}_j} \log K_{\Omega_1}(z, z)$$

$$= \frac{\partial^2}{\partial z_i \overline{z}_j} \log \left\{ |\det J_{\mathbb{C}}f(z)|^2 K_{\Omega_2}(f(z), f(z)) \right\}$$

$$= \frac{\partial^2}{\partial z_i \overline{z}_j} \log K_{\Omega_2}(f(z), f(z)) \qquad (1.1.18.2)$$

since $\log |\det J_{\mathbb{C}}f(z)|^2$ is locally

$$\log (\det J_{\mathbb{C}}f) + \log \left(\overline{\det J_{\mathbb{C}}f}\right) + C$$

hence is annihilated by the mixed second derivative. But line (1.1.18.2) is nothing other than

$$\sum_{\ell,m} g_{\ell,m}^{\Omega_2}(f(z)) \frac{\partial f_\ell(z)}{\partial z_i} \overline{\frac{\partial f_m(z)}{\partial \overline{z}_j}}$$

and (1.1.18.1) follows.

$\square$

Proposition 1.1.19. *Let $\Omega \subset\subset \mathbb{C}^n$ be a domain. Let $z \in \Omega$. Then*

$$K(z, z) = \sup_{f \in A^2(\Omega)} \frac{|f(z)|^2}{\|f\|_{A^2}^2} = \sup_{\|f\|_{A^2(\Omega)} = 1} |f(z)|^2.$$

Proof. Now

$$K(z,z) = \sum |\phi_j(z)|^2$$

$$= \left(\sup_{\|\{a_j\}\|_{\ell^2}=1} \left| \sum \phi_j(z)a_j \right| \right)^2$$

$$= \sup_{\|f\|_{A^2}=1} |f(z)|^2,$$

by the Riesz–Fischer theorem,

$$= \sup_{f \in A^2} \frac{|f(z)|^2}{\|f\|_{A^2}^2}. \qquad \square$$

We shall use this proposition in a moment. Meanwhile, we should like to briefly mention some open problems connected with the Bergman kernel:

The Lu Qi-Keng Conjecture

We have already noticed that $K_\Omega(z,z) > 0$, all $z \in \Omega$, any bounded Ω. It is reasonable to ask whether $K_\Omega(z,\zeta)$ is ever equal to zero. In fact various geometric constructions connected with the Bergman metric and associated biholomorphic invariants (which involve division by K) make it particularly desirable that K be nonvanishing.

If $\Omega = D$, the unit disc, then explicit calculation (which we perform below) shows that

$$K(z,\zeta) = \frac{1}{\pi} \frac{1}{(1 - z\bar{\zeta})^2},$$

hence, $K(z,\zeta)$ is nonvanishing on $D \times D$. Proposition 1.1.14 and the Riemann mapping theorem then show that the Bergman kernel for any proper simply connected subdomain of $\mathbb{C}$ is nonvanishing.

The Bergman kernel for the annulus was studied in Skwarczynski [SKW] and was seen to vanish at some points. It is shown in Suita and Yamada [SUY] that if $\Omega \subseteq \mathbb{C}$ is a multiply connected domain with smooth boundary, then K_Ω must vanish—this is proved by an analysis of differentials on the Riemann surface consisting of the double of Ω. By using the easy fact that the Bergman kernel for a product domain is the product of the Bergman kernels (exercise), we may conclude that any domain in $\mathbb{C}^2$ of the form $A \times \Omega$, where A is multiply connected, has a Bergman kernel with zeroes. The Lu Qi-Keng conjecture can be formulated as

Conjecture: A topologically trivial domain in $\mathbb{C}^n$ has nonvanishing Bergman kernel.

It is known (Greene and Krantz [GRK1, GRK2]) that a domain that is C^∞ sufficiently close to the ball in $\mathbb{C}^n$ has nonvanishing Bergman kernel. Also, if a domain Ω has Bergman kernel that is bounded from zero (and satisfies a modest geometric condition), then all "nearby" domains have Bergman kernel that is bounded from zero. Thus it came as a bit of a surprise when in Boas [BOA1] and [BOA2], it was shown that there exist topologically trivial domains—even ones with real analytic boundary and satisfying all reasonable additional geometric conditions—for which the Bergman kernel has zeroes. See also Wiegerinck [WIE], where interesting ideas contributing to the solution of this problem first arose. In Sect. 5.7 we treat the results of Boas.

Exercise for the Reader: The set of smoothly bounded domains for which the Lu Qi-Keng conjecture is true is closed in the Hausdorff topology on domains.

We shall say more about the Lu Qi-Keng conjecture in Sect. 5.7.

Smoothness to the Boundary of K_Ω

It is of interest to know whether K_Ω is smooth on $\overline{\Omega} \times \overline{\Omega}$. We can see from the formula above for the Bergman kernel of the disc that $K_D(z, z)$ blows up as $z \to 1^-$. In fact this property of blowing up prevails at any boundary point of a domain at which there is a peaking function (apply Proposition 1.1.19 to a high power of the peaking function). The reference Gamelin [GAM, p. 52 ff.] contains background information on peaking functions.

However, there is strong evidence that—as long as Ω is smoothly bounded—on compact subsets of

$$\overline{\Omega} \times \overline{\Omega} \setminus ((\partial\Omega \times \partial\Omega) \cap \{z = \zeta\})$$

the Bergman kernel will be smooth. For strictly pseudoconvex domains, this statement is true; its proof (see [KER2]) uses deep and powerful methods of partial differential equations. Unfortunately, on the Diederich–Fornæss worm domain (which is smoothly bounded, pseudoconvex but has many pathological properties), the Bergman kernel is *not* smooth as just indicated. See also [KRP1, KRP2] as well as [LIG1].

In what follows, a multi-index in $\mathbb{C}^n$ is an n-tuple $\alpha = (a_1, a_2, \ldots, a_n)$ of nonnegative integers. We write

$$z^\alpha \equiv z_1^{a_1} \cdot z_2^{a_2} \cdots \cdots z_n^{a_n}.$$

Also

$$\frac{\partial^\alpha}{\partial z^\alpha} \equiv \frac{\partial^{a_1}}{\partial z_1^{a_1}} \frac{\partial^{a_2}}{\partial z_2^{a_2}} \cdots \frac{\partial^{a_n}}{\partial z_n^{a_n}}$$

Perhaps the most central open problem in the function theory of several complex variables is to prove that a biholomorphic mapping of two smoothly bounded, pseudoconvex domains extends to a diffeomorphism of the closures. It is known (see Bell and Boas [BEB]) that a sufficient condition for this problem to have an affirmative answer on a smoothly bounded domain $\Omega \subseteq \mathbb{C}^n$ is that, for any multi-index α, there are constants $C = C_\alpha$ and $m = m_\alpha$ such that the Bergman kernel K_Ω satisfies

$$\sup_{z \in \Omega} \left| \frac{\partial^\alpha}{\partial z^\alpha} K_\Omega(z, \zeta) \right| \leq C \cdot \delta_\Omega(\zeta)^{-m}$$

for all $\zeta \in \Omega$. Here $\delta_\Omega(w)$ denotes the distance of the point $w \in \Omega$ to the boundary of the domain.

1.1.1   Calculating the Bergman Kernel

The Bergman kernel can almost never be calculated explicitly; unless the domain Ω has a great deal of symmetry—so that a useful orthonormal basis for $A^2(\Omega)$ can be determined—there are few techniques for determining K_Ω. Sometimes one can exploit the automorphism group of the domain (see [HUA] for an exemplary instance of this technique). We shall explore some of these ideas below.

In 1974 Fefferman [FEF1, Part I] introduced a new technique for obtaining an asymptotic expansion for the Bergman kernel on a large class of domains. (For an alternative approach, see Boutet de Monvel and Sjöstrand [BOS].) This work enabled rather explicit estimations of the Bergman metric and opened up an entire branch of analysis on domains in $\mathbb{C}^n$ (see, e.g., Fefferman [FEF2], Chern and Moser [CHM], Klembeck [KLE], and Greene and Krantz [GRK1, GRK2, GRK3, GRK4, GRK5, GRK6, GRK7, GRK8, GRK9, GRK10, GRK11]).

The Bergman theory that we have presented here would be a bit hollow if we did not at least calculate the kernel in a few instances. We complete the section by addressing that task.

Restrict attention to the ball $B \subseteq \mathbb{C}^n$. The functions z^α, α a multi-index, are each in $A^2(B)$ and are pairwise orthogonal by the symmetry of the ball. By the uniqueness of the power series expansion for an element of $A^2(B)$, the elements z^α form a complete orthonormal system on B (their closed linear span is $A^2(B)$). Setting

$$\gamma_\alpha = \int_B |z^\alpha|^2 \, dV(z),$$

we see that $\{z^\alpha / \sqrt{\gamma_\alpha}\}$ is a complete orthonormal system in $A^2(B)$. Thus by Proposition 1.1.7,

$$K(z, \zeta) = \sum_\alpha \frac{z^\alpha \overline{\zeta}^\alpha}{\gamma_\alpha}. \tag{1.1.1.1}$$

If we want to calculate the Bergman kernel for the ball in closed form, we need to calculate the γ_α's. This requires some lemmas from real analysis. These lemmas will be formulated and proved on $\mathbb{R}^N$ and $B_N = \{x \in \mathbb{R}^N : |x| < 1\}$.

Lemma 1.1.20. *We have that*

$$\int_{\mathbb{R}^N} e^{-\pi |x|^2} dx = 1.$$

Proof. The case $N = 1$ is familiar from calculus (or see [BKR, Sect. 6.6]). For the N-dimensional case, write

$$\int_{\mathbb{R}^N} e^{-\pi |x|^2} dx = \int_{\mathbb{R}} e^{-\pi x_1^2} dx_1 \cdots \int_{\mathbb{R}} e^{-\pi x_N^2} dx_N$$

and apply the one-dimensional result. $\square$

Let σ be the unique rotationally invariant area measure on $S_{N-1} = \partial B_N$ and let $\omega_{N-1} = \sigma(\partial B)$.

Lemma 1.1.21. *We have*

$$\omega_{N-1} = \frac{2\pi^{N/2}}{\Gamma(N/2)},$$

where

$$\Gamma(x) = \int_0^\infty t^{x-1} e^{-t} dt$$

is Euler's gamma function.

Proof. Introducing polar coordinates we have

$$1 = \int_{\mathbb{R}^N} e^{-\pi |x|^2} dx = \int_{S^{N-1}} d\sigma \int_0^\infty e^{-\pi r^2} r^{N-1} dr$$

or

$$\frac{1}{\omega_{N-1}} = \int_0^\infty e^{-\pi r^2} r^N \frac{dr}{r}.$$

Letting $s = r^2$ in this last integral and doing some obvious manipulations yields the result. $\square$

Now we return to $B \subseteq \mathbb{C}^n$. We set

$$\eta(k) = \int_{\partial B} |z_1|^{2k} \, d\sigma, \quad N(k) = \int_B |z_1|^{2k} \, dV(z), \quad k = 0, 1, \ldots.$$

Lemma 1.1.22. *We have*

$$\eta(k) = \pi^n \frac{2(k!)}{(k+n-1)!}, \quad N(k) = \pi^n \frac{k!}{(k+n)!}.$$

Proof. Polar coordinates show easily that $\eta(k) = 2(k+n)N(k)$. So it is enough to calculate $N(k)$. Let $z = (z_1, z_2, \ldots, z_n) = (z_1, z')$. We write

$$N(k) = \int_{|z|<1} |z_1|^{2k} \, dV(z)$$

$$= \int_{|z'|<1} \left(\int_{|z_1| \le \sqrt{1-|z'|^2}} |z_1|^{2k} \, dV(z_1) \right) dV(z')$$

$$= 2\pi \int_{|z'|<1} \int_0^{\sqrt{1-|z'|^2}} r^{2k} r \, dr \, dV(z')$$

$$= 2\pi \int_{|z'|<1} \frac{(1-|z'|^2)^{k+1}}{2k+2} \, dV(z')$$

$$= \frac{\pi}{k+1} \omega_{2n-3} \int_0^1 (1-r^2)^{k+1} r^{2n-3} \, dr$$

$$= \frac{\pi}{k+1} \omega_{2n-3} \int_0^1 (1-s)^{k+1} s^{n-1} \frac{ds}{2s}$$

$$= \frac{\pi}{2(k+1)} \omega_{2n-3} \beta(n-1, k+2),$$

where β is the classical beta function of special function theory (see Carrier et al. [CCP, p. 191] or Whittaker and Watson [WHW, pp. 235 ff.]). By a standard identity for the beta function we then have

$$N(k) = \frac{\pi}{2(k+1)}\omega_{2n-3}\frac{\Gamma(n-1)\Gamma(k+2)}{\Gamma(n+k+1)}$$

$$= \frac{\pi}{2(k+1)}\frac{2\pi^{n-1}}{\Gamma(n-1)}\frac{\Gamma(n-1)\Gamma(k+2)}{\Gamma(n+k+1)}$$

$$= \frac{\pi^n k!}{(k+n)!}.$$

This is the desired result. □

Lemma 1.1.23. *Let* $z \in B \subseteq \mathbb{C}^n$ *and* $0 < r < 1$. *The symbol* **1** *denotes the point* $(1, 0, \ldots, 0)$. *Then*

$$K_B(\mathbf{z}, r\mathbf{1}) = \frac{n!}{\pi^n}\frac{1}{(1 - rz_1)^{n+1}}.$$

Proof. Refer to formula (1.1.1.1) preceding Lemma 1.1.20. Then

$$K_B(\mathbf{z}, r\mathbf{1}) = \sum_\alpha \frac{\mathbf{z}^\alpha (r\mathbf{1})^\alpha}{\gamma_\alpha} = \sum_{k=0}^\infty \frac{z_1^k r^k}{N(k)}$$

$$= \frac{1}{\pi^n}\sum_{k=0}^\infty (rz_1)^k \cdot \frac{(k+n)!}{k!}$$

$$= \frac{n!}{\pi^n}\sum_{k=0}^\infty (rz_1)^k \binom{k+n}{n}$$

$$= \frac{n!}{\pi^n} \cdot \frac{1}{(1 - rz_1)^{n+1}}.$$

This is the desired result. □

Theorem 1.1.24. *If* $\mathbf{z}, \zeta \in B$, *then the Bergman kernel for the unit ball in* $\mathbb{C}^n$ *is*

$$K_B(\mathbf{z}, \zeta) = \frac{n!}{\pi^n}\frac{1}{(1 - \mathbf{z} \cdot \bar{\zeta})^{n+1}},$$

where $\mathbf{z} \cdot \bar{\zeta} = z_1\bar{\zeta}_1 + z_2\bar{\zeta}_2 + \cdots + z_n\bar{\zeta}_n$.

Proof. Let $\mathbf{z} = r\tilde{\mathbf{z}} \in B$, where $r = |\mathbf{z}|$ and $|\tilde{\mathbf{z}}| = 1$. Also, fix $\zeta \in B$. Choose a unitary rotation ρ such that $\rho\tilde{\mathbf{z}} = \mathbf{1}$. Then, by Proposition 1.1.14 and Lemma 1.1.23, we have

$$K_B(\mathbf{z}, \zeta) = K_B(r\tilde{\mathbf{z}}, \zeta) = K(r\rho^{-1}\mathbf{1}, \zeta)$$

$$= K(r\mathbf{1}, \rho\zeta) = \overline{K(\rho\zeta, r\mathbf{1})}$$

$$= \frac{n!}{\pi^n} \cdot \frac{1}{\left(1 - r(\overline{\rho\zeta})_1\right)^{n+1}}$$

$$= \frac{n!}{\pi^n} \cdot \frac{1}{\left(1 - (r\mathbf{1}) \cdot (\overline{\rho\zeta})\right)^{n+1}}$$

$$= \frac{n!}{\pi^n} \cdot \frac{1}{\left(1 - (r\rho^{-1}\mathbf{1}) \cdot \overline{\zeta}\right)^{n+1}}$$

$$= \frac{n!}{\pi^n} \cdot \frac{1}{(1 - \mathbf{z} \cdot \overline{\zeta})^{n+1}}. \qquad\qquad \square$$

Corollary 1.1.25. *The Bergman kernel for the unit disc in the complex plane is*

$$K_D(z, \zeta) = \frac{1}{\pi} \frac{1}{(1 - z\overline{\zeta})^2}.$$

Proposition 1.1.26. *The Bergman metric for the ball $B = B(0, 1) \subseteq \mathbb{C}^n$ is given by*

$$g_{ij}(z) = \frac{n+1}{(1 - |z|^2)^2} \left[(1 - |z|^2)\delta_{ij} + \overline{z}_i z_j\right].$$

Proof. Since $K(z, z) = n!/(\pi^n(1 - |z|^2)^{n+1})$, this is a routine computation that we leave to the reader. $\qquad\qquad \square$

Corollary 1.1.27. *The Bergman metric for the disc (i.e., the ball in dimension one) is*

$$g_{ij}(\zeta) = \frac{2}{(1 - |\zeta|^2)^2}, \quad i = j = 1,$$

This is the well-known Poincaré, or Poincaré-Bergman, metric.

Proposition 1.1.28. *The Bergman kernel for the polydisc $D^n(0, 1) \subseteq \mathbb{C}^n$ is the product*

$$K(z, \zeta) = \frac{1}{\pi^n} \prod_{j=1}^{n} \frac{1}{(1 - z_j\overline{\zeta}_j)^2}.$$

Proof. Exercise for the reader. Use the uniqueness property of the Bergman kernel.

$\square$

Exercise for the Reader: Calculate the Bergman metric for the polydisc.

1.1.2 The Poincaré-Bergman Distance on the Disc

If $D \subseteq \mathbb{C}$ is the unit disc, $\zeta \in D$, then Corollary 1.1.27 shows that

$$|w|_{B,z} = \left\{ \frac{2|w|^2}{(1 - |\zeta|^2)^2} \right\}^{1/2} = \frac{\sqrt{2}|w|}{1 - |\zeta|^2},$$

where the subscript B indicates that we are working in the Bergman metric. We now use this formula to derive an explicit expression for the Poincaré-Bergman distance from $0 \in D$ to $r + i0 \in D, 0 < r < 1$. Call this distance $d(0, r)$. Then

$$d(0, r) = \inf \left\{ \int_0^1 |\gamma'(t)|_{B,\gamma(t)} dt : \right.$$

$$\left. \gamma \text{ is a piecewise smooth curve in } D, \gamma(0) = 0, \gamma(1) = r + i0 \right\}.$$

Elementary comparisons show that, among curves of the form $\psi(t) = t + iw(t), 0 \le t \le 1$, the curve $\gamma(t) = tr + i0$ is the shortest in the Poincaré metric. Further elementary arguments show that a general curve of the form $\psi(t) = v(t) + iw(t)$ is always longer than some corresponding curve of the form $t + i\tilde{w}(t)$. We leave the details of these assertions to the reader. Thus

$$d(0, r) = \int_0^1 \frac{\sqrt{2}r}{(1 - (rt)^2)} dt$$

$$= \sqrt{2} \int_0^r \frac{1}{1 - t^2} dt$$

$$= \frac{1}{\sqrt{2}} \log \left(\frac{1 + r}{1 - r} \right).$$

Since rotations are conformal maps of the disc, we may next conclude that

$$d(0, re^{i\theta}) = \frac{1}{\sqrt{2}} \log \left(\frac{1 + r}{1 - r} \right).$$

Finally, if w_1, w_2 are arbitrary, then the Möbius transformation

$$\phi : z \mapsto \frac{z - w_1}{1 - \overline{w}_1 z}$$

satisfies $\phi(w_1) = 0, \phi(w_2) = (w_2 - w_1)/(1 - \overline{w}_1 w_2)$. Then Proposition 1.1.18 yields that

$$d(w_1, w_2) = d\left(0, \frac{w_2 - w_1}{1 - \overline{w}_1 w_2}\right)$$

$$= \frac{1}{\sqrt{2}} \left(\frac{1 + \left|\frac{w_2 - w_1}{1 - \overline{w}_1 w_2}\right|}{1 - \left|\frac{w_2 - w_1}{1 - \overline{w}_1 w_2}\right|}\right).$$

We note in passing that the expression $\rho(w_1, w_2) \equiv |(w_2 - w_1)/(1 - \overline{w}_1 w_2)|$ is called the *pseudohyperbolic distance*. It is also conformally invariant, but it does *not* arise from integrating an infinitesimal metric (i.e., lengths of tangent vectors at a point). A fuller discussion of both the Poincaré metric and the pseudohyperbolic metric on the disc may be found in [KRA9] and [GAR, Chap. 1].

1.1.3 Construction of the Bergman Kernel by Way of Differential Equations

It is actually possible to obtain the Bergman kernel of a domain in the plane from the Green's function for that domain (see [KRA5, Sect. 1.3.3]). Let us now summarize the key ideas. Unlike the first Bergman kernel construction, the present one will work for *any* domain with C^2 boundary. Thanks to work of Garabedian [GARA], one can say rather precisely what the Green's function of any planar domain is (see also [JAK]).

First, the fundamental solution for the Laplacian in the plane is the function

$$\Gamma(\zeta, z) = \frac{1}{2\pi} \log |\zeta - z| \, .$$

This means that $\Delta_\zeta \Gamma(\zeta, z) = \delta_z$ in the sense of distributions. (Observe that δ_z denotes the Dirac "delta mass" at z and Δ_ζ is the Laplacian in the ζ variable.) In more prosaic terms, the condition is that

$$\int \Gamma(\zeta, z) \cdot \Delta\varphi(\zeta) \, d\xi d\eta = \varphi(z)$$

for any C^∞ function φ with compact support. We write, as usual, $\zeta = \xi + i\eta$. (This topic is treated in detail in [KRA1, Chaps. 0, 1].)

Given a domain $\Omega \subseteq \mathbb{C}$, the *Green's function* is posited to be a function $G(\zeta, z)$ on $\Omega \times \Omega$ that satisfies

$$G(\zeta, z) = \Gamma(\zeta, z) - F_z(\zeta),$$

where $F_z(\zeta) = F(\zeta, z)$ is a particular harmonic function in the ζ variable (to be specified momentarily). Moreover, it is mandated that $G(\cdot, z)$ vanish on the boundary of Ω. One constructs the function $F(\cdot, z)$, for each fixed z, by solving a Dirichlet problem with boundary data $\Gamma(\cdot, z)$. Again, the reference [KRA1, p. 40] has all the particulars. It is worth noting, and this point is not completely obvious but is discussed in [KRA1, Chap. 1], that the Green's function is a symmetric function of its arguments.

The next proposition establishes a striking connection between the Bergman kernel and the classical Green's function.

Proposition 1.1.29. *Let $\Omega \subseteq \mathbb{C}$ be a bounded domain with C^2 boundary. Let $G(\zeta, z)$ be the Green's function for Ω and let $K(z, \zeta)$ be the Bergman kernel for Ω. Then*

$$K(z, \zeta) = 4 \cdot \overline{\frac{\partial^2}{\partial \zeta \partial \bar{z}} G(\zeta, z)}. \qquad (1.1.29.1)$$

Proof. Our proof will use a version of Stokes's theorem written in the notation of complex variables. It says that, if $u \in C^1(\overline{\Omega})$, then

$$\oint_{\partial U} u(\zeta) \, d\zeta = 2i \cdot \iint_U \frac{\partial u}{\partial \bar{\zeta}} \, d\xi \, d\eta, \qquad (1.1.29.2)$$

where again $\zeta = \xi + i\eta$. The reader is invited to convert this formula to an expression in ξ and η and to confirm that the result coincides with the standard real-variable version of Stokes's theorem that can be found in any calculus book (see, e.g., [THO, BLK]).

Now we already know that

$$G(\zeta, z) = \frac{1}{4\pi} \log(\zeta - z) + \frac{1}{4\pi} \log \overline{(\zeta - z)} + F(\zeta, z). \qquad (1.1.29.3)$$

Here we think of the logarithm as a multivalued holomorphic function; after we take a derivative, the ambiguity (which comes from an additive multiple of $2\pi i$) goes away.

Differentiating with respect to z (and using subscripts to denote derivatives), we find that

$$G_z(\zeta, z) = \frac{1}{4\pi} \frac{-1}{\zeta - z} + F_z(\zeta, z).$$

We may rearrange this formula to read

$$\frac{1}{\zeta - z} = -4\pi \cdot G_z(\zeta, z) + 4\pi F_z(\zeta, z).$$

We know that G, as a function of ζ, vanishes on $\partial\Omega$. Hence so does G_z. Let $f \in C^2(\overline{\Omega})$ be holomorphic on Ω. It follows that the Cauchy formula

$$f(z) = \frac{1}{2\pi i} \oint_{\partial\Omega} \frac{f(\zeta)}{\zeta - z} \, d\zeta$$

can be rewritten as

$$f(z) = -2i \oint_{\partial\Omega} f(\zeta) F_z(\zeta, z) \, d\zeta.$$

Now we apply Stokes's theorem (in the complex form) to rewrite this last as

$$f(z) = 4 \cdot \iint_{\Omega} (f(\zeta) F_z)_{\overline{\zeta}}(\zeta, z) \, d\xi \, d\eta,$$

where $\zeta = \xi + i\eta$. Since f is holomorphic and F is real valued, we may conveniently write this last formula as

$$f(z) = 4 \cdot \iint_{\Omega} f(\zeta) \overline{F_{\zeta\overline{z}}}(\zeta, z) \, d\xi \, d\eta.$$

Now formula (1.1.29.3) tells us that $F_{\zeta\overline{z}} = G_{\zeta\overline{z}}$. Therefore we have

$$f(z) = \iint_{\Omega} f(\zeta) 4\overline{G_{\zeta\overline{z}}}(\zeta, z) \, d\xi \, d\eta. \tag{1.1.29.4}$$

With a suitable limiting argument, we may extend this formula from functions f that are holomorphic and in $C^2(\overline{\Omega})$ to functions in $A^2(\Omega)$.

It is straightforward now to verify that $4\overline{G_{\zeta\overline{z}}}$ satisfies the first three characterizing properties of the Bergman kernel, just by examining our construction. The crucial reproducing property is of course formula (1.1.29.4). Then it follows that

$$K(z, \zeta) = 4 \cdot \overline{\frac{\partial^2}{\partial\zeta\partial\overline{z}}G(\zeta, z)}.$$

That is the desired result.   □

It is worth noting that the proposition we have just established gives a practical method for confirming the existence of the Bergman kernel—by relating it to the Green's function, whose existence is elementary. See [HAP1, HAP2] for a version of these techniques in the several complex variable contexts.

Now let us calculate. Of course the Green's function of the unit disc D is

$$G(\zeta, z) = \frac{1}{2\pi} \log |\zeta - z| - \frac{1}{2\pi} \log |1 - \zeta \bar{z}| ,$$

as a glance at any classical complex analysis text will tell us (see, e.g., [AHL] or [HIL]). Verify the defining properties of the Green's function for yourself.

With formula (1.1.29.1) in mind, we can make life a bit easier by writing

$$G(\zeta, z) = \frac{1}{4\pi} \log(\zeta - z) + \frac{1}{4\pi} \log(\overline{\zeta - z})$$
$$- \frac{1}{4\pi} \log (1 - \zeta \bar{z}) - \frac{1}{4\pi} \log \left(\overline{1 - \zeta \bar{z}} \right).$$

Here we think of the expression on the right as the concatenation of four multivalued functions, in view of the ambiguity of the logarithm function. This ambiguity is irrelevant for us because the derivative of the Green's function is still well defined (i.e., the derivative annihilates additive constants).

Now we readily calculate that

$$\frac{\partial G}{\partial \bar{z}} = \frac{1}{4\pi} \cdot \frac{-1}{\overline{\zeta - z}} + \frac{1}{4\pi} \cdot \frac{\zeta}{1 - \zeta \bar{z}}$$

and

$$\frac{\partial^2 G}{\partial \zeta \partial \bar{z}} = \frac{1}{4\pi} \cdot \frac{1}{(1 - \zeta \bar{z})^2} .$$

In conclusion, we may apply Proposition 1.1.29 to see that

$$K(z, \zeta) = \frac{1}{\pi} \cdot \frac{1}{(1 - z \cdot \bar{\zeta})^2} .$$

This result is consistent with that obtained in the other two calculations (Sects. 1.1.2 and 1.1.3). The Bergman metric, as before, is obtained by differentiation.

1.1.4 Construction of the Bergman Kernel by Way of Conformal Invariance

Let $D \subseteq \mathbb{C}$ be the unit disc. First we notice that if either $f \in A^2(D)$ or $\bar{f} \in A^2(D)$, then

$$f(0) = \frac{1}{\pi} \iint_D f(\zeta) \, dA(\zeta). \tag{1.1.4.1}$$

This is the standard, two-dimensional area form of the mean-value property for holomorphic or harmonic functions.

Of course the constant function $u(z) \equiv 1$ is in $A^2(D)$, so it is reproduced by integration against the Bergman kernel. Hence, for any $w \in D$,

$$1 = u(w) = \iint_D K(w, \zeta) u(\zeta) \, dA(\zeta) = \iint_D K(w, \zeta) \, dA(\zeta),$$

or

$$\frac{1}{\pi} = \frac{1}{\pi} \iint_D K(w, \zeta) \, dA(\zeta).$$

By (1.1.4.1), we may conclude that

$$\frac{1}{\pi} = K(w, 0)$$

for any $w \in D$.

Now, for $a \in D$ fixed, consider the Möbius transformation

$$h(z) = \frac{z - a}{1 - \bar{a}z}.$$

We know that

$$h'(z) = \frac{1 - |a|^2}{(1 - \bar{a}z)^2}.$$

We may thus apply Proposition 1.1.14 with $\phi = h$ to find that

$$\begin{aligned}
K(w, a) &= h'(w) \cdot K(h(w), h(a)) \cdot \overline{h'(a)} \\
&= \frac{1 - |a|^2}{(1 - \bar{a}w)^2} \cdot K(h(w), 0) \cdot \frac{1}{1 - |a|^2} \\
&= \frac{1}{(1 - \bar{a}w)^2} \cdot \frac{1}{\pi} \\
&= \frac{1}{\pi} \cdot \frac{1}{(1 - w\bar{a})^2}.
\end{aligned}$$

This is our formula for the Bergman kernel.

1.2 The Szegő and Poisson–Szegő Kernels

The basic theory of the Szegő kernel is similar to that for the Bergman kernel—they are both special cases of a general theory of "Hilbert spaces with reproducing kernel" (see [ARO]). Thus we only outline the basic steps here, leaving details to the reader. See Sect. 1.3 and also the paper [KRA11].

Let $\Omega \subseteq \mathbb{C}^n$ be a bounded domain with C^2 boundary. Let $A(\Omega)$ be those functions continuous on $\overline{\Omega}$ that are holomorphic on Ω. Let $H^2(\partial\Omega)$ be the space consisting of the closure in the $L^2(\partial\Omega, d\sigma)$ topology of the restrictions to $\partial\Omega$ of elements of $A(\Omega)$. Then $H^2(\partial\Omega)$ is a proper Hilbert subspace of $L^2(\partial\Omega)$. Each element $f \in H^2(\partial\Omega)$ has a natural holomorphic extension to Ω given by its Poisson integral Pf. It is a standard fact—see [KRA1, Chap. 8]—that

$$\lim_{\epsilon \to 0^+} f(\zeta - \epsilon \nu_\zeta) = f(\zeta)$$

for almost every ζ in the boundary of Ω. Here, as usual, ν_ζ is the unit outward normal to $\partial\Omega$ at the point ζ.

For each fixed $z \in \Omega$, the functional

$$\psi_z : H^2(\Omega) \ni f \mapsto Pf(z)$$

is continuous. (Why?—you may find the Bochner–Martinelli formula [KRA1] useful here. See also [KRA11]. We shall treat this matter in more detail in Example 1.2.1 below.) Let $\tilde{k}_z(\zeta)$ be the Hilbert space representative (coming from the Riesz representation theorem) for the functional ψ_z. Define the Szegő kernel $S(z, \zeta)$ by the formula

$$S(z, \zeta) = \overline{\tilde{k}_z(\zeta)} \ , z \in \Omega, \ \zeta \in \partial\Omega.$$

If $f \in H^2(\partial\Omega)$, then

$$f(z) = \int_{\partial\Omega} S(z, \zeta) f(\zeta) d\sigma(\zeta)$$

for all $z \in \Omega$. Here $d\sigma$ is $(2n - 1)$-dimensional Hausdorff measure on $\partial\Omega$. We shall not explicitly formulate and verify the various uniqueness and extremal properties for the Szegő kernel. The statements and proofs are exactly like those for the Bergman kernel. The reader is invited to consider these topics.

Example 1.2.1. We now want to describe the Szegő theory. In order to make this work, we need to present some preliminary results about the Bochner–Martinelli kernel and integral representation formula.

Definition 1.2.2. On $\mathbb{C}^n$ we let

$$\omega(z) \equiv dz_1 \wedge dz_2 \wedge \cdots \wedge dz_n$$

$$\eta(z) \equiv \sum_{j=1}^{n} (-1)^{j+1} z_j \, dz_1 \wedge \cdots \wedge dz_{j-1} \wedge dz_{j+1} \wedge \cdots \wedge dz_n.$$

The form η is sometimes called the *Leray form*. We shall often write $\omega(\bar{z})$ to mean $d\bar{z}_1 \wedge \cdots \wedge d\bar{z}_n$ and likewise $\eta(\bar{z})$ to mean $\sum_{j=1}^{n} (-1)^{j+1} \bar{z}_j \, d\bar{z}_1 \wedge \cdots \wedge d\bar{z}_{j-1} \wedge d\bar{z}_{j+1} \wedge \cdots \wedge d\bar{z}_n$.

The genesis of the Leray form is explained by the following lemma.

Lemma 1.2.3. *For any $z_0 \in \mathbb{C}^n$, any $\epsilon > 0$, we have*

$$\int_{\partial B(z^0,\epsilon)} \eta(\bar{z}) \wedge \omega(z) = n \int_{B(z^0,\epsilon)} \omega(\bar{z}) \wedge \omega(z).$$

Proof. Notice that $d\eta(\bar{z}) = \bar{\partial}\eta(\bar{z}) = n\omega(\bar{z})$. Therefore by Stokes's theorem,

$$\int_{\partial B(z^0,\epsilon)} \eta(\bar{z}) \wedge \omega(z) = \int_{B(z^0,\epsilon)} d\,[\eta(\bar{z}) \wedge \omega(z)].$$

Of course the expression in [ ] is saturated in dz's so, in the decomposition $d = \partial + \bar{\partial}$, only the term $\bar{\partial}$ will not die. Thus the last line equals

$$\int_{B(z^0,\epsilon)} [\bar{\partial}(\eta(\bar{z}))] \wedge \omega(z) = n \int_{B(z^0,\epsilon)} \omega(\bar{z}) \wedge \omega(z). \qquad \square$$

Remark 1.2.4. Notice that, by change of variables,

$$\int_{B(z^0,\epsilon)} \omega(\bar{z}) \wedge \omega(z) = \int_{B(0,\epsilon)} \omega(\bar{z}) \wedge \omega(z)$$

$$= \epsilon^{2n} \int_{B(0,1)} \omega(\bar{z}) \wedge \omega(z).$$

A straightforward calculation shows that

$$\int_{B(0,1)} \omega(\bar{z}) \wedge \omega(z)$$

$$= (-1)^{q(n)} \cdot (2i)^n \cdot (\text{volume of the unit ball in } \mathbb{C}^n \approx \mathbb{R}^{2n}),$$

where $q(n) = [n(n-1)]/2$. We denote the value of this integral by $W(n)$. $\qquad \square$

Theorem 1.2.5 (Bochner–Martinelli). *Let $\Omega \subseteq \mathbb{C}^n$ be a bounded domain with C^1 boundary. Let $f \in C^1(\overline{\Omega})$. Then, for any $z \in \Omega$, we have*

$$f(z) = \frac{1}{n W(n)} \int_{\partial\Omega} \frac{f(\zeta)\eta(\overline{\zeta} - \overline{z}) \wedge \omega(\zeta)}{|\zeta - z|^{2n}}$$

$$- \frac{1}{n W(n)} \int_{\Omega} \frac{\overline{\partial} f(\zeta)}{|\zeta - z|^{2n}} \wedge \eta(\overline{\zeta} - \overline{z}) \wedge \omega(\zeta).$$

Proof. Fix $z \in \Omega$. We apply Stokes's theorem to the form

$$M_z(\zeta) \equiv \frac{f(\zeta)\eta(\overline{\zeta} - \overline{z}) \wedge \omega(\zeta)}{|\zeta - z|^{2n}}$$

on the domain $\Omega_{z,\epsilon} \equiv \Omega \setminus \overline{B}(z, \epsilon)$, where $\epsilon > 0$ is chosen so small that $\overline{B}(z, \epsilon) \subseteq \Omega$. Note that Stokes's theorem does not apply to forms that have a singularity; thus we may not apply the theorem to L_z on any domain that contains the point z in either its interior or its boundary. This observation helps to dictate the form of the domain $\Omega_{z,\epsilon}$. As the proof develops, we shall see that it also helps to determine the outcome of our calculation.

Notice that

$$\partial(\Omega_{z,\epsilon}) = \partial\Omega \cup \partial B(z, \epsilon)$$

but that the two pieces are equipped with opposite orientations.
Thus by Stokes,

$$\int_{\partial\Omega} M_z(\zeta) - \int_{\partial B(z,\epsilon)} M_z(\zeta) = \int_{\partial\Omega_{z,\epsilon}} M_z(\zeta)$$

$$= \int_{\Omega_{z,\epsilon}} d_\zeta(M_z(\zeta)). \qquad (1.2.5.1)$$

Notice that we consider z to be fixed and ζ to be the variable. Now

$$d_\zeta M_z(\zeta) = \overline{\partial}_\zeta M_z(\zeta)$$

$$= \frac{\overline{\partial} f(\zeta) \wedge \eta(\overline{\zeta} - \overline{z}) \wedge \omega(\zeta)}{|\zeta - z|^{2n}}$$

$$+ f(\zeta) \cdot \left[\sum_{j=1}^{n} \frac{\partial}{\partial \overline{\zeta}_j} \left(\frac{\overline{\zeta}_j - \overline{z}_j}{|\zeta - z|^{2n}} \right) \right] \omega(\overline{\zeta}) \wedge \omega(\zeta). \qquad (1.2.5.2)$$

Observing that

$$\frac{\partial}{\partial \bar{\zeta}_j}\left(\frac{\bar{\zeta}_j - \bar{z}_j}{|\zeta - z|^{2n}}\right) = \frac{1}{|\zeta - z|^{2n}} - n\frac{|\bar{\zeta}_j - \bar{z}_j|^2}{|\zeta - z|^{2n+2}},$$

we find that the second term on the far right of (1.2.5.2) dies and we have

$$d_\zeta M_z(\zeta) = \frac{\overline{\partial} f(\zeta) \wedge \eta(\bar{\zeta} - \bar{z}) \wedge \omega(\zeta)}{|\zeta - z|^{2n}}.$$

Substituting this identity into (1.2.5.1) yields

$$\int_{\partial\Omega} M_z(\zeta) - \int_{\partial B(z,\epsilon)} M_z(\zeta) = \int_{\Omega_{z,\epsilon}} \frac{\overline{\partial} f(\zeta) \wedge \eta(\bar{\zeta} - \bar{z}) \wedge \omega(\zeta)}{|\zeta - z|^{2n}}. \qquad (1.2.5.3)$$

Next we remark that

$$\int_{\partial B(z,\epsilon)} M_z(\zeta) = f(z) \int_{\partial B(z,\epsilon)} \frac{\eta(\bar{\zeta} - \bar{z}) \wedge \omega(\zeta)}{|\zeta - z|^{2n}}$$

$$+ \int_{\partial B(z,\epsilon)} \frac{(f(\zeta) - f(z))\,\eta(\bar{\zeta} - \bar{z}) \wedge \omega(\zeta)}{|\zeta - z|^{2n}}$$

$$\equiv T_1 + T_2. \qquad (1.2.5.4)$$

Since $|f(\zeta) - f(z)| \leq C|\zeta - z|$ (and since each term of $\eta(\bar{\zeta} - \bar{z})$ has a factor of some $\bar{\zeta}_j - \bar{z}_j$), it follows that the integrand of T_2 is of size $O(|\zeta - z|)^{-2n+2} \approx \epsilon^{-2n+2}$. Since the surface over which the integration is performed has area $\approx \epsilon^{2n-1}$, it follows that $T_2 \to 0$ as $\epsilon \to 0^+$.

By Lemma 1.2.3 and the remark following, we also have

$$T_1 = \epsilon^{-2n} f(z) \int_{\partial B(z,\epsilon)} \eta(\bar{\zeta} - \bar{z}) \wedge \omega(\zeta)$$

$$= n\epsilon^{-2n} f(z) \int_{B(0,\epsilon)} \omega(\bar{\zeta}) \wedge \omega(\zeta)$$

$$= nW(n) f(z). \qquad (1.2.5.5)$$

Finally, (1.2.5.3)–(1.2.5.5) yield that

$$\left(\int_{\partial\Omega} M_z(\zeta)\right) - nW(n) f(z) + o(1) = \int_{\Omega_{z,\epsilon}} \overline{\partial} f(\zeta) \wedge \left[\frac{\eta(\bar{\zeta} - \bar{z})}{|\zeta - z|^{2n}}\right] \wedge \omega(\zeta).$$

Since

$$\left| \frac{\eta(\bar{\zeta} - \bar{z})}{|\zeta - z|^{2n}} \right| = O(|\zeta - z|^{-2n+1}),$$

the last integral is absolutely convergent as $\epsilon \to 0^+$ (remember that $\bar{\partial} f$ is bounded). Thus we finally have

$$f(z) = \frac{1}{nW(n)} \int_{\partial\Omega} L_z(\zeta) - \frac{1}{nW(n)} \int_{\Omega} \bar{\partial} f(\zeta) \wedge \frac{\eta(\bar{\zeta} - \bar{z})}{|\zeta - z|^{2n}} \wedge \omega(\zeta).$$

This is the Bochner–Martinelli formula.   □

Remark 1.2.6. We see that the Bochner–Martinelli formula is a quintessential example of a constructible integral formula. The kernel is quite explicit, and it is the same for all domains. For the Bergman kernel, and for other canonical kernels that we shall see below, this latter property does not hold.   □

We note that the classical Cauchy integral formula in one complex variable is an immediate consequence of our new Bochner–Martinelli formula.

Corollary 1.2.7. *If $\Omega \subseteq \mathbb{C}^n$ is bounded and has C^1 boundary and if $f \in C^1(\bar{\Omega})$ and $\bar{\partial} f = 0$ on Ω, then*

$$f(z) = \frac{1}{nW(n)} \int_{\partial\Omega} \frac{f(\zeta)\eta(\bar{\zeta} - \bar{z})}{|\zeta - z|^{2n}} \wedge \omega(\zeta). \tag{1.2.7.1}$$

Corollary 1.2.8. *In complex dimension 1, the last corollary says that*

$$f(z) = \frac{1}{2\pi i} \oint_{\partial\Omega} \frac{f(\zeta)}{\zeta - z} d\zeta.$$

Corollary 1.2.7 is particularly interesting. Like the classical Cauchy formula, it gives a constructible integral reproducing formula that is the same on all domains. Unlike the classical Cauchy formula, its kernel is *not* holomorphic in the free variable z. This makes the Bochner–Martinelli formula of limited utility in *constructing* holomorphic functions.

We note that Corollary 1.2.7 holds for broader classes of holomorphic functions—such as the Hardy classes. One sees this by a simple limiting argument. See our discussion of H^2 below.

Now we turn to the development of the Szegő theory. Let Ω be a bounded domain in $\mathbb{C}$ or $\mathbb{C}^n$ with C^1 boundary. Define $H^2(\Omega)$ as above. If $z \in \Omega$ is fixed, then, by inspection of the formula in Corollary 1.2.7 and the Schwarz inequality,

$$|f(z)| \le C \cdot \|f\|_{H^2(\Omega)} \, .$$

Thus we see that $H^2(\Omega)$ is a Hilbert space with reproducing kernel. A construction exactly like that for the Bergman kernel gives a new kernel called the Szegő kernel. We denote it by $S(z, \zeta)$.

Using the Szegő theory analogue of Proposition 1.1.7, we can actually calculate the Szegő kernel on the disc. We first note that $\{z^j\}_{j=0}^{\infty}$ forms a basis for the Hilbert space $H^2(D)$. This follows from the standard theory of power series for holomorphic functions on D. It is orthogonal by parity. It is complete by the uniqueness of the power series expansion. With a simple calculation, we can normalize the basis to the complete orthonormal basis $\{(1/\sqrt{2\pi})z^j\}_{j=0}^{\infty}$. Thus we see that

$$S(z, \zeta) = \sum_{j=0}^{\infty} \frac{1}{2\pi} z^j \overline{\zeta}^j = \frac{1}{2\pi} \cdot \frac{1}{1 - z \cdot \overline{\zeta}} \, .$$

Now it is instructive to write out the Szegő integral for a function in $A(D)$:

$$
\begin{aligned}
f(z) &= \int_0^{2\pi} f(e^{i\theta}) \cdot \frac{1}{2\pi} \cdot \frac{1}{1 - z \cdot e^{-i\theta}} \, d\theta \\
&= \frac{1}{2\pi i} \int_0^{2\pi} \frac{f(e^{i\theta})}{e^{i\theta} - z} \cdot i e^{i\theta} \, d\theta \\
&= \frac{1}{2\pi i} \oint_{\partial D} \frac{f(\zeta)}{\zeta - z} \, d\zeta .
\end{aligned}
$$

Thus we see that the canonical Szegő integral formula is in fact nothing other than the constructive Cauchy integral formula. But only on the disc!

We conclude this section by noting that the integral

$$Sg(z) = \int_{\partial \Omega} S(z, \zeta) g(\zeta) \, d\zeta$$

defines a projection from $L^2(\partial\Omega)$ to $H^2(\Omega)$. This is because the mapping is self-adjoint, idempotent, and fixes H^2. We call this mapping the *Szegő projection*. (Note that the Bergman projection is constructed similarly.)

Let $\{\phi_j\}_{j=1}^{\infty}$ be any complete, orthonormal basis for $H^2(\partial\Omega)$. Define

$$S'(z, \zeta) = \sum_{j=1}^{\infty} \phi_j(z) \overline{\phi_j(\zeta)} \, , \quad z, \zeta \in \Omega .$$

For convenience we tacitly identify here each function with its Poisson extension to the interior of the domain. Then, for $K \subseteq \Omega$ compact, the series defining S'

converges uniformly on $K \times K$. By a Riesz–Fischer argument, $S'(\cdot, \zeta)$ is the Poisson integral of an element of $H^2(\partial\Omega)$ and $S'(z, \cdot)$ is the conjugate of the Poisson integral of an element of $H^2(\partial\Omega)$. So S' extends to $(\overline{\Omega} \times \Omega) \cup (\Omega \times \overline{\Omega})$, where it is understood that all functions on the boundary are defined only almost everywhere. The kernel S' is conjugate symmetric. Also, by Riesz–Fischer theory, S' reproduces $H^2(\partial\Omega)$. Since the Szegő kernel is unique, it follows that $S = S'$.

The Poisson–Szegő kernel is obtained by a formal procedure from the Szegő kernel: This procedure manufactures a *positive* reproducing kernel from one that is not necessarily positive. The origin of this kernel may be found in [HUA, Chap. 3]. Note in passing that, just as we argued for the Bergman kernel in the last section, $S(z, z)$ is never 0 when $z \in \Omega$.

Proposition 1.2.9. *Define*

$$\mathcal{P}(z, \zeta) = \frac{|S(z, \zeta)|^2}{S(z, z)}, \quad z \in \Omega, \ \zeta \in \partial\Omega.$$

Then, for any $f \in A(\Omega)$ and $z \in \Omega$, it holds that

$$f(z) = \int_{\partial\Omega} f(\zeta)\mathcal{P}(z, \zeta)d\sigma(\zeta).$$

Proof. Fix $z \in \Omega$ and $f \in A(\Omega)$ and define

$$u(\zeta) = f(\zeta)\frac{\overline{S(z, \zeta)}}{S(z, z)}, \quad \zeta \in \partial\Omega.$$

Then $u \in H^2(\partial\Omega)$ hence

$$f(z) = u(z) = \int_{\partial\Omega} S(z, \zeta)u(\zeta)d\sigma(\zeta)$$

$$= \int_{\partial\Omega} \mathcal{P}(z, \zeta)f(\zeta)d\sigma(\zeta).$$

This is the desired formula. $\qquad\qquad\qquad\qquad\qquad\qquad\qquad\qquad\square$

Remark 1.2.10. In passing to the Poisson–Szegő kernel, we gain the advantage of positivity of the kernel (for more on this circle of ideas, see [KRA1, Chap. 8] and also [KAT, Chap. 1]). However, we lose something in that $\mathcal{P}(z, \zeta)$ is no longer holomorphic in the z variable nor conjugate holomorphic in the ζ variable. The literature on this kernel is rather sparse and there are many unresolved questions. The paper [KRA2] discusses some of the mapping properties of the Poisson–Szegő kernel. See also the more recent paper [KRA11]. It is an interesting historical fact that the Poisson–Szegő kernel was invented by Hua in [HUA], though he did *not* give the kernel this name. We say more about the Poisson–Szegő kernel in Sects. 3.2 and 3.3. $\square$

As an exercise, use the paradigm of Proposition 1.2.9 to construct a positive kernel from the Cauchy kernel on the disc (be sure to first change notation in the usual Cauchy formula so that it is written in terms of arc length measure on the boundary). What familiar kernel results?

Like the Bergman kernel, the Szegő and Poisson–Szegő kernels can almost never be explicitly computed. They can be calculated asymptotically in a number of important instances, however (see Fefferman [FEF1, Part I], Boutet de Monvel and Sjöstrand [BOS]). We will give explicit formulas for these kernels on the ball. The computations are similar in spirit to those in Sect. 1.1.2; fortunately, we may capitalize on much of the work done there.

Lemma 1.2.11. *The functions $\{z^\alpha\}$, where α ranges over multi-indices, are pairwise orthogonal and span $H^2(\partial B)$.*

Proof. The orthogonality follows from symmetry considerations. For the completeness, notice that it suffices to see that the span of $\{z^\alpha\}$ is dense in $A(B)$ in the uniform topology on the boundary. By the Stone–Weierstrass theorem, the closed algebra generated by $\{z^\alpha\}$ and $\{\bar{z}^\alpha\}$ is all of $C(\partial B)$. But the monomials $\bar{z}^\alpha, \alpha \neq 0$, are orthogonal to $A(B)$ (use the power series expansion about the origin to see this). The claimed density follows. $\square$

Lemma 1.2.12. *Let $1 = (1, 0, \dots, 0)$. Then*

$$S(z, 1) = \frac{(n-1)!}{2\pi^n} \frac{1}{(1-z_1)^n}.$$

Proof. We have that

$$S(z, 1) = \sum_\alpha \frac{z^\alpha \cdot 1^\alpha}{\|z_1^\alpha\|^2_{L^2(\partial B)}}$$

$$= \sum_{k=0}^\infty \frac{z_1^k}{\eta(k)}$$

$$= \frac{1}{2\pi^n} \sum_{k=0}^\infty \frac{z_1^k (k+n-1)!}{k!}$$

$$= \frac{(n-1)!}{2\pi^n} \sum_{k=0}^\infty \binom{k+n-1}{n-1} z_1^k$$

$$= \frac{(n-1)!}{2\pi^n} \frac{1}{(1-z_1)^n}.$$ $\square$

Lemma 1.2.13. *Let ρ be a unitary rotation on $\mathbb{C}^n$. For any $z \in \overline{B}, \zeta \in \partial B$, we have that $S(z, \zeta) = S(\rho z, \rho \zeta)$.*

Proof. This is a standard change of variables argument and we omit it. $\qquad\square$

Theorem 1.2.14. *The Szegő kernel for the ball is*

$$S(z, \zeta) = \frac{(n-1)!}{2\pi^n} \frac{1}{(1 - z \cdot \overline{\zeta})^n}.$$

Proof. Let $z \in B$ be arbitrary. Let ρ be the unique unitary rotation such that ρz is a multiple of $\mathbf{1}$. Then, by 1.2.13,

$$
\begin{aligned}
S(z, \zeta) &= S(\rho^{-1}\mathbf{1}, \zeta) \\
&= S(\mathbf{1}, \rho\zeta) = \overline{S(\rho\zeta, \mathbf{1})} \\
&= \frac{(n-1)!}{2\pi^n} \overline{\frac{1}{\left(1 - (\overline{\rho\zeta}) \cdot \mathbf{1}\right)^n}} \\
&= \frac{(n-1)!}{2\pi^n} \frac{1}{\left(1 - \overline{\zeta} \cdot (\rho^{-1}\mathbf{1})\right)^n} \\
&= \frac{(n-1)!}{2\pi^n} \frac{1}{(1 - z \cdot \overline{\zeta})^n}. \qquad\square
\end{aligned}
$$

Corollary 1.2.15. *The Poisson–Szegő kernel for the ball is*

$$\mathcal{P}(z, \zeta) = \frac{(n-1)!}{2\pi^n} \frac{(1 - |z|^2)^n}{|1 - z \cdot \overline{\zeta}|^{2n}}.$$

Exercise for the Reader: Calculate the Szegő and Poisson–Szegő kernel for the polydisc.

Let us now review some of our key ideas. We let Ω be a bounded domain in $\mathbb{C}$ or $\mathbb{C}^n$. Let $L^2(\Omega)$ be the square-integrable functions on Ω (with respect to the ordinary Lebesgue measure), and let $A^2(\Omega) \subseteq L^2(\Omega)$ be the subspace consisting of the holomorphic functions. This last space is known as the *Bergman space*. Then A^2 is a closed, Hilbert subspace of L^2. Thus we may consider the projection

$$P : L^2(\Omega) \to A^2(\Omega).$$

It is well known that this projection mapping is given by an integration kernel $K(z, \zeta)$ which is called the *Bergman kernel* (see [KRA1, Chap. 1] for details in this matter). As an instance, the Bergman kernel for Ω, the unit disc D in the complex plane is given by

$$K_D(z, \zeta) = \frac{1}{\pi} \cdot \frac{1}{(1 - z \cdot \overline{\zeta})^2}. \tag{1.2.16}$$

Of particular interest for us is the obvious fact that this K is smooth on $\overline{D} \times \overline{D} \setminus \mathcal{D}$, where $\mathcal{D}$ is the boundary diagonal.

We might also note that the Bergman kernel for the unit ball B in $\mathbb{C}^n$ is given by the formula

$$K_B(\zeta, \zeta) = \frac{n!}{\pi^n} \cdot \frac{1}{(1 - \zeta \cdot \overline{\zeta})^{n+1}}.$$

Again, one may see by inspection that K is smooth on $\overline{B} \times \overline{B} \setminus \mathcal{D}$.

Kerzman [KER2] has shown that a similar result is true on any smoothly bounded, strictly pseudoconvex domain: The Bergman kernel is smooth on the product of the closures less the boundary diagonal. In fact we now know (see the remark at the end of Kerzman's paper) that the result holds on any domain for which the $\overline{\partial}$-Neumann operator N is known to be pseudolocal (see [KER2] and [BEL3] for an explication of these ideas). This just means that $N\varphi$ is smooth wherever φ is smooth.

To see this, note that if $f \in A^2(\Omega)$ and $z \in \Omega$, then

$$\begin{aligned} f(z) &= \langle f, \delta_z \rangle \\ &= \langle Pf, \delta_z \rangle \\ &= \langle f, P\delta_z \rangle. \end{aligned}$$

Here, of course, δ_z is the Dirac delta mass at z. It follows that

$$K(z, \zeta) = \overline{P(\delta_z)}. \tag{1.2.17}$$

Here P is the Bergman projection. Of course $P(\delta_z)$ is an element of $A^2(\Omega)$, so it is certainly smooth on Ω. We are interested in the behavior of this function (of the variables z and ζ) as z and ζ tend to the boundary.

Now the well-known formula of Kohn, which we prove in Sect. 6.6, for the Bergman projection P is given by

$$P = I - \overline{\partial}^* N \overline{\partial}. \tag{1.2.18}$$

Here N is the $\overline{\partial}$-Neumann operator.

To be more specific, we can combine (1.2.17) and (1.2.18) to see that

$$K(z, \zeta) = \delta_\zeta - \overline{\partial}^* N \overline{\partial} \delta_\zeta. \tag{1.2.19}$$

Of course $\bar{\partial}$ and $\bar{\partial}^*$ are classical partial differential operators. So it is certainly the case that wherever a function φ is smooth, then also $\bar{\partial}\varphi$ and $\bar{\partial}^*\varphi$ are smooth. The pseudolocality of N means precisely that $N\psi$ is smooth wherever ψ is smooth. All in all then, line (1.2.19) tells us that if z is fixed inside Ω, then $K(z,\zeta)$ is smooth up to the boundary in the ζ variable (just because δ_z is). The estimates that come with the pseudolocality of N (see either [FOK] or [KRA4, Chap. 7] for the strictly pseudoconvex case) tell us further that, as $\zeta \to \partial\Omega$ and if z stays in a compact subset of the closure that is bounded from ζ, then K is still smooth.

With all this information in hand, it is a matter of interest to know for which domains the Bergman kernel will have a singularity when the two variables approach the same boundary point. In those circumstances the question of boundedness of the Bergman projection on various L^p and other spaces has a hope of being tractable, and the study of Condition R is accessible. But in fact Ligocka [LIG1] has shown that this contention fails on the Diederich–Fornaess worm domain (see [DIF1] as well as [CHS] and [KRP1]). Of course it is known (see [KRP2] or [CHS]) that the worm is bounded, pseudoconvex, and has smooth boundary. We treat the worm in Chap. 6.

We note in passing that the Bergman kernel for the bidisc does *not* satisfy the condition described in the last paragraph (of course the bidisc does not have smooth boundary). One might speculate that a similar failure occurs on a smoothly bounded, convex domain in $\mathbb{C}^2$ which has an analytic disc in the boundary.

Our purpose in the next sections is to show that the assertions being discussed here fail dramatically on the disc $D \subseteq \mathbb{C}$ when one considers the reproducing kernels for certain closed subspaces of $A^2(D)$.

1.3 Formal Ideas of Aronszajn

One of the first canonical integral formulas ever created was that of Bergman [BER1] and [BER2]. We shall present the Bergman idea in the context of a more general construction due to Nachman Aronszajn [ARO]. This is the idea of a Hilbert space with reproducing kernel. Fortunately Aronszajn's idea also entails the Szegő kernel and several other important reproducing kernels.

Definition 1.3.1. Let X be any set and let $\mathcal{H}$ be a Hilbert space of complex-valued functions on X. We say that $\mathcal{H}$ is a *Hilbert space with reproducing kernel* if, for each $x \in X$, the linear (point evaluation) map of the form

$$L_x : \mathcal{H} \longrightarrow \mathbb{C}$$
$$f \longmapsto f(x),$$

is continuous. We write this as

$$|f(x)| \leq C \cdot \|f\|_{\mathcal{H}}. \tag{1.3.1.1}$$

In this circumstance, the classical Riesz representation theorem (see [KRA14, Chap. 3]) tells us that, for each $x \in X$, there is a unique element $k_x \in \mathcal{H}$ such that

$$f(x) = \langle f, k_x \rangle \qquad \forall f \in \mathcal{H}. \tag{1.3.1.2}$$

We then define a function

$$K : X \times X \to \mathbb{C}$$

by the formula

$$K(x, y) \equiv \overline{k_x(y)}.$$

The function K is the *reproducing kernel* for the Hilbert space $\mathcal{H}$.

We see that K is uniquely determined by $\mathcal{H}$ because, again by the Riesz representation theorem, the element k_x for each $x \in \mathcal{H}$ is unique.

We know from our earlier discussion that, if $\{\varphi_j\}$ is a complete orthonormal basis for the Bergman space, then

$$K(x, y) = \sum_{j=1}^{\infty} \varphi_j(x)\overline{\varphi_j(y)}.$$

Here the convergence is in the Hilbert space topology in each variable. And in fact the fundamental property (1.3.1.1) of a Hilbert space with reproducing kernel shows that the convergence is uniform on compact subsets of $X \times X$.

1.4   A New Bergman Basis

We do our work in this section on the unit disc D in the complex plane $\mathbb{C}$.

Of course an orthogonal basis for the Bergman space of the disc D is given by $\{\zeta^j\}_{j=0}^{\infty}$. Calculating the L^2 norm of each of these elements, we find that an orthonormal basis for $A^2(D)$ is

$$\varphi_j(\zeta) = \frac{\sqrt{j+1}}{\sqrt{\pi}} \cdot \zeta^j, \quad j = 0, 1, 2, \ldots.$$

The basis is seen to be complete just by the theory of power series. Now it is a standard fact (see Sect. 1.1.2) that the Bergman kernel can be constructed from such a basis by the formula

$$K(z, \zeta) = \sum_{j=0}^{\infty} \varphi_j(z) \cdot \overline{\varphi_j(\zeta)}.$$

This series converges, uniformly on compact subsets of $D \times D$, to the kernel

$$K(z, \zeta) = \frac{1}{\pi} \cdot \frac{1}{(1 - z \cdot \overline{\zeta})^2}.$$

But now let us consider[2] the subspace $\mathcal{X}$ of $A^2(D)$ generated by the basis elements $\varphi_{2j}(\zeta)$, $j = 0, 1, 2, \ldots$. This is certainly a closed subspace of $A^2(D)$, and *its* Bergman kernel is given by

$$\sum_{j=0}^{\infty} \varphi_{2j}(z) \cdot \overline{\varphi_{2j}(\zeta)}. \tag{1.4.1}$$

Let us calculate explicitly the sum in (1.4.1). It is given by

$$\sum_{j=0}^{\infty} \frac{2j+1}{\pi} z^{2j} \cdot \overline{\zeta^{2j}}.$$

We may sum this series by examining the auxiliary expression

$$\sum_{j=0}^{\infty} \frac{2j+1}{\pi} \alpha^{2j} = \frac{d}{d\alpha} \left[\frac{1}{\pi} \cdot \sum_{j=0}^{\infty} \alpha^{2j+1} \right]$$

$$= \frac{d}{d\alpha} \left[\frac{1}{\pi} \cdot \alpha \cdot \frac{1}{1 - \alpha^2} \right]$$

$$= \frac{1}{\pi} \cdot \frac{\alpha^2 + 1}{(1 - \alpha^2)^2}.$$

We conclude that the Bergman kernel $K_{\mathcal{X}}$ for $\mathcal{X}$ is given by

$$K_{\mathcal{X}}(z, \zeta) = \frac{1}{\pi} \cdot \frac{(z\overline{\zeta})^2 + 1}{(1 - (z\overline{\zeta})^2)^2}. \tag{1.4.2}$$

The notable fact is that $K_{\mathcal{X}}$ blows up either when $z \cdot \overline{\zeta}$ tends to 1 or when $z \cdot \overline{\zeta}$ tends to -1. In other words, $K_{\mathcal{X}}$ blows up either when z and ζ tend to the *same* boundary point or when z and ζ tend to antipodal boundary points.

---

[2]The discussion here is based on unpublished work [BKP] of Boas, Krantz, and Peloso.

A companion result is obtained when one considers instead the space $\mathcal{Y}$ generated by the basis $\{\varphi_{2j+1}\}$, $j = 0, 1, 2, \ldots$. Its Bergman kernel is given by

$$\sum_{j=0}^{\infty} \varphi_{2j+1}(z) \cdot \overline{\varphi_{2j+1}(\zeta)}. \qquad (1.4.3)$$

Let us calculate explicitly the sum in (1.4.3). It is given by

$$\sum_{j=0}^{\infty} \frac{2j+2}{\pi} z^{2j+1} \cdot \overline{\zeta^{2j+1}}.$$

We may sum this series by examining the auxiliary expression

$$\sum_{j=0}^{\infty} \frac{2j+2}{\pi} \alpha^{2j+1} = \frac{d}{d\alpha} \left[\frac{1}{\pi} \cdot \sum_{j=0}^{\infty} \alpha^{2j+2} \right]$$

$$= \frac{d}{d\alpha} \left[\frac{1}{\pi} \cdot \alpha^2 \cdot \frac{1}{1-\alpha^2} \right]$$

$$= \frac{1}{\pi} \cdot \frac{2\alpha}{(1-\alpha^2)^2}.$$

We conclude that the Bergman kernel $K_{\mathcal{Y}}$ for $\mathcal{Y}$ is given by

$$K_{\mathcal{Y}}(z, \zeta) = \frac{1}{\pi} \cdot \frac{2z\overline{\zeta}}{(1-(z\overline{\zeta})^2)^2}. \qquad (1.4.4)$$

Again, this new Bergman kernel has boundary singularities either when z and ζ tend to the *same* boundary point or when z and ζ tend to antipodal boundary points.

It is notable that

$$K_{\mathcal{X}}(z, \zeta) + K_{\mathcal{Y}}(z, \zeta) = \frac{1}{\pi} \cdot \frac{1}{(1-z \cdot \overline{\zeta})^2} = K_D(z, \zeta).$$

It may be noted that the even part (in the z variable) of the classical Bergman kernel for the disc D is

$$\frac{1}{\pi} \cdot \frac{(z\overline{\zeta})^2 + 1}{(1-(z\overline{\zeta})^2)^2}.$$

This is precisely the kernel that we found for the Bergman space $\mathcal{X}$ generated by the basis of monomials with even index. Likewise the odd part (in the z variable) of the classical Bergman kernel for the disc D is

$$\frac{1}{\pi} \cdot \frac{2z\bar{\zeta}}{(1 - (z\bar{\zeta})^2)^2} \cdot$$

This is precisely the kernel that we found for the Bergman space $\mathcal{Y}$ generated by the basis of monomials with odd index.

1.5 Further Examples

Of course one is by no means restricted to doing analysis modulo 2. One could examine the Bergman space generated by φ_0, φ_3, φ_6, etc., and calculate the corresponding Bergman kernel. We shall not do so here, but simply note that the resulting kernel has boundary singularities at the *third* roots of unity. In other words, there are three boundary singularities.

A similar result obtains if one considers the Bergman kernel for the space generated by φ_{jm}, where m is any fixed positive integer. In that case, there are boundary singularities at the mth roots of unity.

A rather more dramatic example is obtained when we consider the space $\mathcal{Z}$ generated by the basis $\{\varphi_{2^j}\}$, $j = 0, 1, 2, \ldots$. The corresponding Bergman kernel is

$$K_{\mathcal{Z}}(z, \zeta) = \sum_{j=0}^{\infty} \frac{2^j + 1}{\pi} z^{2^j} \cdot \overline{\zeta^{2^j}} = \sum_{j=0}^{\infty} \frac{2^j + 1}{\pi} (z \cdot \bar{\zeta})^{2^j} \cdot \cdot$$

We shall not sum this series explicitly. But we instead analyze the auxiliary holomorphic function

$$\Phi(\alpha) = \sum_{j=0}^{\infty} \frac{2^j + 1}{\pi} \alpha^{2^j}. \tag{1.5.1}$$

The sequence of exponents is lacunary, and the radius of convergence of the series is 1. Thus the Hadamard gap theorem applies, and we see that the series in (1.5.1) defines a holomorphic function with a singularity at every boundary point of the disc. In other words, the holomorphic functions Φ does not analytically continue past any boundary point. From this we conclude that the kernel $K_{\mathcal{Z}}$ has boundary singularities at *every* boundary point of the unit disc D.

If we consider the complementary basis φ_1, φ_3, φ_5, φ_6, φ_7, etc., and the Bergman kernel $K'_{\mathcal{Z}}$ associated to the space it generates, then we must conclude that it, too, has boundary singularities at every boundary point. This is true because

$$K_{\mathcal{Z}} + K'_{\mathcal{Z}} = K_D.$$

What is particularly interesting is that the first basis in this example (the one with a lacunary sequence of indices) is quite sparse, while the second is not.

We conclude this discussion by briefly treating an example in which the selected basis and also its complementary basis are lacunary in a certain sense but have similar density properties. Namely, let

$$\mathcal{J}_1 = \left\{ 0\,,\,2\,,\quad 5,6,7,8\,,\quad 17,18,\ldots,32\,,\quad 65,66,\ldots,\ldots,128,\quad \ldots \right\}$$

and

$$\mathcal{J}_2 = \left\{ 1\,,\,3,4\,,\quad 9,10,\ldots,16\,,\quad 33,34,\ldots,64\,,\quad \ldots \right\}.$$

We see that $\mathcal{J}_1$ and $\mathcal{J}_2$ are disjoint and their union is all the nonnegative integers.

Now the Bergman kernel for the space generated by $\mathcal{J}_1$ is given by

$$\sum_{j=1}^{\infty} \sum_{\ell=2^{2j}+1}^{2^{2j+1}} \frac{\ell+1}{\pi} z^{\ell} \bar{\zeta}^{\ell}.$$

The inner sum (we omit the details), with $\alpha = z \cdot \bar{\zeta}$, may be calculated to be

$$\frac{1}{\pi} \cdot \frac{(2^{2j+1}+2)\alpha^{2^{2j+1}+2} - (2^{2j}+2)\alpha^{2^{2j}+2} - (2^{2j+1}+2)\alpha^{2^{2j+1}+1} + (2^{2j}+2)\alpha^{2^{2j}+1} - \alpha^{2^{2j+1}+2} + \alpha^{2^{2j}+2}}{(\alpha-1)^2}.$$

Hence, the kernel is

$$\sum_{j=1}^{\infty} \frac{1}{\pi} \cdot \frac{(2^{2j+1}+2)(z\bar{\zeta})^{2^{2j+1}+2} - (2^{2j}+2)(z\bar{\zeta})^{2^{2j}+2} - (2^{2j+1}+2)(z\bar{\zeta})^{2^{2j+1}+1}}{((z\bar{\zeta})-1)^2}$$

$$+ \frac{(2^{2j}+2)(z\bar{\zeta})^{2^{2j}+1} - (z\bar{\zeta})^{2^{2j+1}+2} + (z\bar{\zeta})^{2^{2j}+2}}{((z\bar{\zeta})-1)^2}.$$

The problem of calculating the sum of this series appears to be intractable. But there is reason to believe that the only singularity is at $1 \in \partial D$. The analysis for the kernel for the basis $\mathcal{J}_2$ would be similar.

1.6   A Real Bergman Space

The space $\mathbf{h}^2(D)$ of square-integrable *harmonic* functions on the disc D is a Hilbert space with reproducing kernel in the sense of Aronszajn (see [ARO]). In particular, if $K \subseteq D$ is compact, then there is a constant $C = C(K)$ so that if $f \in \mathbf{h}^2(D)$, then

$$\sup_{z \in K} |f(z)| \leq C \cdot \|f\|_{\mathbf{h}^2}.$$

As a result, there is a "Bergman kernel" for this space.

It may be noted that a basis for $\mathbf{h}^2(D)$ consists of

$$\{\zeta^j\}_{j=0}^{\infty} \bigcup \{\bar{\zeta}^j\}_{j=1}^{\infty}.$$

The Bergman kernel for $\mathbf{h}^2$ is easily calculated; it is essentially a derivative of the usual Poisson kernel.

If we instead calculate the Bergman kernel for the space generated just by $\{\zeta^j\}$ or for the space generated just by $\{\bar{\zeta}^j\}$, then we obtain the usual Bergman kernel or (up to an additive constant) the conjugate of the usual Bergman kernel.

On the other hand, one might consider the space $\mathbf{h}^2(D)$ of harmonic functions u on the disc that satisfy

$$\sup_{0<r<1} \int_0^{2\pi} |u(re^{i\theta})|^2 \, d\theta < \infty.$$

This is analogous to the Hardy space, but now we are focusing on *harmonic* functions. It is easy to verify that this is a Hilbert space with reproducing kernel. Calculating the kernel, one finds that it is the classical Poisson kernel. Given Stokes's theorem, it is no surprise that the kernel for $\mathbf{h}^2$ is a derivative of the kernel for $\mathbf{h}^2$.

One should note that, in this example, the Bergman kernel for a subspace does *not* have extra singularities. All the kernels being discussed here have just the classical singularity at 1.

1.7 The Behavior of the Singularity in a General Setting

In this section we build (at least philosophically) on the earlier material and prove the next result. We note that, in this theorem, "boundary singularity" for the Bergman kernel means a boundary point of the unit disc $D \subseteq \mathbb{C}$ which has no neighborhood to which the Bergman kernel directly analytically continues as z, ζ both approach the point. In particular, "boundary singularity" does not necessarily mean that the kernel is blowing up at the indicated point.

Theorem 1.7.1. *Let $\mathbb{Z}^+$ denote the collection of nonnegative integers. Write*

$$\mathbb{Z}^+ = \mathcal{I}_1 \cup \mathcal{I}_2,$$

where $\mathcal{I}_1 \cap \mathcal{I}_2 = \emptyset$ and each of $\mathcal{I}_1, \mathcal{I}_2$ is an infinite set.

Let B_1 be the Bergman space on the unit disc D generated by the basis $\mathcal{I}_1$ and B_2 the Bergman space on D generated by the basis $\mathcal{I}_2$. Correspondingly, let K_1 be the Bergman kernel for B_1 and K_2 the Bergman kernel for B_2. Then K_1, K_2 each have more than one boundary singularity.

Proof. Let $\{\varphi_j\}_{j=1}^{\infty}$ be a basis for B_1. Then the corresponding Bergman kernel is

$$K_1(z, \zeta) = \sum_{j=1}^{\infty} \varphi_j(z) \cdot \overline{\varphi}_j(\zeta).$$

In particular, we see immediately that the kernel depends on $z \cdot \overline{\zeta}$. Since we are considering $\{\varphi_j\}$ as a subset of the particular basis $\{\sqrt{(j+1)/\pi}\zeta^j\}$, we know that the Bergman kernel for B_1 is a sum of *some* of the terms $(j+1)(z\overline{\zeta})^j$ (for convenience we omit the factor of $1/\pi$). Thus it suffices for us to study the single-variable power series

$$\sum_j (j+1)t^j.$$

Here the sum is taken over *some*, but not *all*, of the j. Also t is a complex variable. Certainly a series of this form has radius of convergence 1 and has a singularity at the boundary point 1. This corresponds to the singularity of the Bergman kernel on the boundary diagonal. The question that we must study is whether this one-variable power series has any other singular points on the boundary of D.

Note that the derivative or the integral of the series has the very same singularities. So we may as well study the series

$$\sum_j t^j,$$

where the sum is taken over *some* of the j, but not *all* of the j.

Now a classical result of Szegő (see [SZE]) comes into play. This result says that a power series with finitely many distinct coefficients (in our case the coefficients are all either 0 or 1) has either the boundary circle as its natural boundary (meaning that *every* boundary point is singular) or else the series sums to a rational function (whose Maclaurin series coefficients form a sequence that is eventually periodic). We know from our earlier calculations that, in the second case, there are polynomials p and q such that the series represents a rational function of the form

$$p(t) + \frac{q(t)}{1 - t^m}.$$

Here m is the period of the coefficient sequence.

Translating these ideas back into the language of z and ζ, we find that we have the following two cases for our Bergman kernel:

(1) The subspace B_1 of the Bergman space has a basis of monomials with an eventually periodic structure. In this case, the kernel has, for each fixed $z \in \partial D$, singularities at finitely many values (more than one) of $\zeta \in \partial D$.
(2) Our subspace of the Bergman space has a basis of monomials that lacks an eventually periodic structure. In this case, the kernel function is singular when z, ζ approach two arbitrary (and, in general, distinct) boundary points. □

The second case here holds in particular for either $\mathcal{I}_1$ or $\mathcal{I}_2$ when the set $\mathcal{I}_1$ is lacunary as in Sect. 1.4. We conclude by noting that the results of this section and the last are unpublished work of Boas, Krantz, and Peloso.

1.8 The Annulus

Let us first consider the Bergman kernel on the domain

$$D' = \{\zeta \in \mathbb{C} : |\zeta| > 1\}.$$

One may either utilize the transformation formula

$$K_{\Omega_1}(z, \zeta) = \Phi'(z) \cdot K_{\Omega_2}(\Phi(z), \Phi(\zeta)) \cdot \overline{\Phi'(\zeta)}$$

for a conformal mapping $\Phi : \Omega_1 \to \Omega_2$ or else exploit the fact that if $\{\psi_j\}_{j=1}^{\infty}$ is a complete orthonormal basis for the Bergman space on Ω_2, then the set $\{(\psi_j \circ \Phi) \cdot \Phi'\}_{j=1}^{\infty}$ is a complete orthonormal basis for the Bergman space on Ω_1. By either means, with the mapping $\Phi(\zeta) = 1/\zeta$ from ${}^c\overline{D}$ to D, we find that the Bergman kernel for the complement of the closed unit disc D' is

$$K_{D'}(z, \zeta) = \frac{1}{\pi} \frac{1}{(1 - z\bar{\zeta})^2}.$$

We note that the indicated calculations show that a complete orthonormal basis for the Bergman space on D' is $\{1/\zeta^j\}_{j=2}^{\infty}$. Now calculations just like those in Sect. 1.5 show that if we let $\mathcal{I}_1 = \{0, 2, 4, 6, \dots\}$ and $\mathcal{I}_2$ the complementary set, then the Bergman kernels corresponding to these two bases each have two singularities in the boundary (at 1 and -1).

Now we may consider the situation on the annulus

$$A = \{\zeta \in \mathbb{C} : 1/2 < |\zeta| < 2\}.$$

Bergman [BER2] has shown that an explicit formula for the Bergman kernel on A would entail elliptic functions. But we can derive an approximate formula that is good enough for our purposes as follows:

Note that $\{\zeta^j\}_{j=-\infty}^{\infty}$ is an orthogonal basis for $A^2(A)$. Moreover, a straightforward calculation shows that

$$\|\zeta^j\|_{A^2(A)} = \sqrt{\frac{\pi}{j+1}} \cdot \sqrt{\frac{2^{4j+4}-1}{2^{2j+2}}} \quad , \quad j \neq -1$$

and

$$\|\zeta^{-1}\|_{A^2(A)} = \sqrt{2\pi} \cdot \sqrt{\log 4}.$$

Thus the Bergman kernel for the annulus A is

$$K_A(z,\zeta) = \frac{1}{2\pi \cdot \log 4} z^{-1}\overline{\zeta}^{-1} + \sum_{j \neq -1} \frac{j+1}{\pi} \frac{2^{2j+2}}{2^{4j+4}-1} z^j \overline{\zeta}^j.$$

The usual error analysis shows then that

$$\sum_{j=0}^{\infty} \frac{j+1}{\pi} \frac{2^{2j+2}}{2^{4j+4}-1} z^j \overline{\zeta}^j = \sum_{j=0}^{\infty} \frac{j+1}{\pi} \frac{2^{2j+2}}{2^{4j+4}} z^j \overline{\zeta}^j + \mathcal{E}(z,\zeta),$$

where $\mathcal{E}$ is a bounded error term with bounded derivatives of all orders.

An analysis of the terms with index less than or equal to -2 gives that

$$\sum_{j=-\infty}^{-2} \frac{j+1}{\pi} \cdot \frac{2^{2j+2}}{2^{4j+4}-1} z^j \overline{\zeta}^j = \sum_{j=2}^{\infty} \frac{j-1}{\pi} \cdot \frac{2^{2j-2}}{2^{4j-4}-1} z^{-j} \overline{\zeta}^{-j} + \mathcal{F},$$

where $\mathcal{F}$ is an error term as usual.

These sums are straightforward to calculate, and we find that the Bergman kernel for the annulus A is given by

$$K_A(z,\zeta) = \frac{1}{2\pi \log 4} z^{-1}\overline{\zeta}^{-1} + \frac{4}{\pi} \cdot \frac{1}{(4 - z \cdot \overline{\zeta})^2} + \frac{4}{\pi} \cdot \frac{1}{(1 - 4z \cdot \overline{\zeta})^2} + \mathcal{G}(z,\zeta),$$

where $\mathcal{G}$ is an error term. Notice that the second term reflects the outer boundary of the annulus and the third term reflects the inner boundary.

It is easy to see from these calculations that, if we were to consider the Bergman space on the annulus corresponding to just the basis elements with even index y, then the resulting kernel would have two singularities on the outer boundary of the annulus and two singularities on the inner boundary. Refer to [KRA3] for more on these phenomena.

1.9   A Direct Connection Between the Bergman and Szegő Kernels

1.9.1   Introduction

Two of the most classical and well-established reproducing formulas in complex analysis are those of S. Bergman and G. Szegő. The first of these is a formula for the Bergman space, and the associated integral lives on the interior of the domain in question. The latter of these is a formula for the Hardy space, and the associated integral lives on the boundary of the domain. For formal reasons, the Bergman integral gives rise to a projection from $L^2(\Omega)$ to $A^2(\Omega)$ (the Bergman space); likewise, the Szegő integral gives rise to a projection from $L^2(\partial\Omega)$ to $H^2(\partial\Omega)$ (the Hardy space).

Since both of the artifacts in question here are canonical, it is natural to suspect that there is some relationship between the two integral formulas. After all, they both reproduce functions that are continuous on the closure of the domain and holomorphic on the interior. In the present section we establish such a connection—very explicitly—on a variety of domains in $\mathbb{C}^1$ and $\mathbb{C}^n$. This is done by way of a moderately subtle calculation using Stokes's theorem. The calculation itself has some intrinsic interest, but the main point is the relationship between the canonical integrals and the associated projections. See [KRA15] for the details of these ideas.

In separate calculations we treat the situation on the disc D, the unit ball B, and on a strongly pseudoconvex domain Ω. While the first two calculations are very explicit, in some sense it is the third of these calculations which is most natural and most satisfying.

It is a pleasure to thank Jürgen Leiterer, who contributed a number of useful ideas to these calculations.

1.9.2   The Case of the Disc

Let D be the unit disc in $\mathbb{C}$. In this context, the Szegő kernel is

$$S(z, \zeta) = \frac{1}{2\pi} \cdot \frac{1}{1 - z \cdot \bar{\zeta}}$$

and the Bergman kernel is

$$K(z, \zeta) = \frac{1}{\pi} \cdot \frac{1}{(1 - z \cdot \bar{\zeta})^2}.$$

Take f to be real analytic on a neighborhood of $\overline{D}$. Now we can calculate

$$\frac{1}{2\pi} \int_{\partial D} f(\zeta) S(z,\zeta) d\sigma(\zeta) = \frac{1}{2\pi} \int_{\partial D} f(\zeta) \cdot \frac{1}{1 - z\bar{\zeta}} \frac{\left[\bar{\zeta} d\zeta - \zeta d\bar{\zeta}\right]}{2i}$$

$$= \frac{1}{4\pi i} \int_{\partial D} \frac{f(\zeta)\bar{\zeta}}{1 - z \cdot \bar{\zeta}} \, d\zeta - \frac{1}{4\pi i} \int_{\partial D} \frac{f(\zeta)\zeta}{1 - z \cdot \bar{\zeta}} \, d\bar{\zeta}$$

$$\overset{\text{(Stokes)}}{=} \frac{1}{4\pi i} \iint_D \frac{f(\zeta)}{1 - z \cdot \bar{\zeta}} \, d\bar{\zeta} \wedge d\zeta + \frac{1}{4\pi i} \iint_D \frac{f(\zeta)\bar{\zeta}z}{(1 - z \cdot \bar{\zeta})^2} \, d\bar{\zeta} \wedge d\zeta$$

$$- \frac{1}{4\pi i} \iint_D \frac{\partial(f \cdot \zeta)/\partial\zeta}{1 - z \cdot \bar{\zeta}} \, d\zeta \wedge d\bar{\zeta} + \frac{1}{4\pi i} \iint_D \frac{\partial f/\partial\bar{\zeta} \cdot \bar{\zeta}}{1 - z \cdot \bar{\zeta}} \, d\bar{\zeta} \wedge d\zeta$$

$$= \frac{1}{4\pi i} \iint_D \frac{f(\zeta)}{(1 - z \cdot \bar{\zeta})^2} \, d\bar{\zeta} \wedge d\zeta - \frac{1}{4\pi i} \iint_D \frac{f(\zeta)z\bar{\zeta}}{(1 - z \cdot \bar{\zeta})^2} \, d\bar{\zeta} \wedge d\zeta$$

$$+ \frac{1}{4\pi i} \iint_D \frac{f(\zeta)\bar{\zeta}z}{(1 - z \cdot \bar{\zeta})^2} \, d\bar{\zeta} \wedge d\zeta - \frac{1}{4\pi i} \iint_D \frac{\partial(f \cdot \zeta)/\partial\zeta}{1 - z \cdot \bar{\zeta}} \, d\zeta \wedge d\bar{\zeta}$$

$$+ \frac{1}{4\pi i} \iint_D \frac{\partial f/\partial\bar{\zeta} \cdot \bar{\zeta}}{1 - z \cdot \bar{\zeta}} \, d\bar{\zeta} \wedge d\zeta$$

$$= \frac{1}{4\pi i} \iint_D \frac{f(\zeta)}{(1 - z \cdot \bar{\zeta})^2} \, d\bar{\zeta} \wedge d\zeta - \frac{1}{4\pi i} \iint_D \frac{\partial(f \cdot \zeta)/\partial\zeta}{1 - z \cdot \bar{\zeta}} \, d\zeta \wedge d\bar{\zeta}$$

$$+ \frac{1}{4\pi i} \iint_D \frac{\partial f/\partial\bar{\zeta} \cdot \bar{\zeta}}{1 - z \cdot \bar{\zeta}} \, d\bar{\zeta} \wedge d\zeta$$

$$= \frac{1}{4\pi i} \iint_D \frac{f(\zeta)}{(1 - z \cdot \bar{\zeta})^2} \, d\bar{\zeta} \wedge d\zeta - \frac{1}{4\pi i} \iint_D \frac{\partial f/\partial\zeta \cdot \zeta}{1 - z \cdot \bar{\zeta}} \, d\zeta \wedge d\bar{\zeta}$$

$$- \frac{1}{4\pi i} \iint_D \frac{f(\zeta)}{1 - z \cdot \bar{\zeta}} \, d\zeta \wedge d\bar{\zeta} + \frac{1}{4\pi i} \iint_D \frac{\partial f/\partial\bar{\zeta} \cdot \bar{\zeta}}{1 - z \cdot \bar{\zeta}} \, d\bar{\zeta} \wedge d\zeta$$

$$= \frac{1}{4\pi i} \iint_D \frac{f(\zeta)}{(1 - z \cdot \bar{\zeta})^2} \, d\bar{\zeta} \wedge d\zeta - \frac{1}{4\pi i} \iint_D \frac{\partial f/\partial\zeta \cdot \zeta}{1 - z \cdot \bar{\zeta}} \, d\zeta \wedge d\bar{\zeta}$$

$$- \frac{1}{4\pi i} \iint_D \frac{f(\zeta)}{(1 - z \cdot \bar{\zeta})^2} \, d\zeta \wedge d\bar{\zeta} + \frac{1}{4\pi i} \iint_D \frac{f(\zeta)z\bar{\zeta}}{(1 - z \cdot \bar{\zeta})^2} \, d\zeta \wedge d\bar{\zeta}$$

$$+ \frac{1}{4\pi i} \iint\limits_{D} \frac{\partial f/\partial \overline{\zeta} \cdot \overline{\zeta}}{1 - z \cdot \overline{\zeta}} \, d\overline{\zeta} \wedge d\zeta$$

$$= \frac{1}{2\pi i} \iint\limits_{D} \frac{f(\zeta)}{(1 - z \cdot \overline{\zeta})^2} \, d\overline{\zeta} \wedge d\zeta - \frac{1}{4\pi i} \iint\limits_{D} \frac{\partial f/\partial \zeta \cdot \zeta}{1 - z \cdot \overline{\zeta}} \, d\zeta \wedge d\overline{\zeta}$$

$$+ \frac{1}{4\pi i} \iint\limits_{D} \frac{f(\zeta)z\overline{\zeta}}{(1 - z \cdot \overline{\zeta})^2} \, d\zeta \wedge d\overline{\zeta} + \frac{1}{4\pi i} \iint\limits_{D} \frac{\partial f/\partial \overline{\zeta} \cdot \overline{\zeta}}{1 - z \cdot \overline{\zeta}} \, d\overline{\zeta} \wedge d\zeta$$

$$= A - B + C + D.$$

Certainly $A = \int_D f(\zeta) K(z, \zeta) \, dA(\zeta)$, where K is the Bergman kernel of the disc. So this is the Bergman projection. Now we shall analyze the terms $-B + C + D$ to see how the Szegő kernel differs from the Bergman kernel. We shall determine that, for certain monomials in a basis for the Bergman (Szegő) space, the sum of these three terms is zero. While for others it is not. In the latter case the difference will be controllable.

First we consider a monomial $f(\zeta) = \zeta^k \overline{\zeta}^m$ with $k = 0, 1, \ldots, m = 1, 2, \ldots,$ and $k < m$. In this case,

$$Sf(z) = \int_{\partial D} f(\zeta) S(z, \zeta) \, d\sigma(\zeta)$$

$$= \int_{\partial D} \zeta^k \overline{\zeta}^m \frac{1}{1 - z\overline{\zeta}} \, d\sigma(\zeta).$$

Now we expand the Szegő kernel in a Neumann series, and, counting powers of ζ and powers of $\overline{\zeta}$, we see immediately by parity that the integral is zero.

A similar calculation shows that, for these values of m and k, $Bf(z) \equiv 0$.

But matters are different when $k \geq m$. In this case, for the Szegő integral, we may write $\zeta^k \overline{\zeta}^m = \zeta^{k-m} \equiv h(z)$ for $|\zeta| = 1$. Hence,

$$Sf(z) = Sh(z)$$

$$= h(z)$$

$$= z^{k-m}.$$

By contrast,

$$Bf(z) = \frac{1}{2\pi i} \int_D \frac{\zeta^k \overline{\zeta}^m}{(1 - z\overline{\zeta})^2} \, d\overline{\zeta} \wedge d\zeta$$

$$= \frac{1}{2\pi i} \int_D \zeta^k \overline{\zeta}^m \left(\sum_{j=0}^{\infty} z^j \overline{\zeta}^j \right)^2 d\overline{\zeta} \wedge d\zeta$$

$$= \frac{1}{2\pi i} \sum_{j,\ell=0}^{\infty} z^{j+\ell} \int_D \zeta^k \overline{\zeta}^{m+j+\ell} \, d\overline{\zeta} \wedge d\zeta$$

$$= \frac{1}{2\pi i} \sum_{\substack{0 \le j,\ell \le k-m \\ j+\ell=k-m}} z^{k-m} \int_D |\zeta|^{2k} \, d\overline{\zeta} \wedge d\zeta$$

$$= \frac{1}{2\pi i} \sum_{\substack{0 \le j,\ell \le k-m \\ j+\ell=k-m}} z^{k-m} \frac{2\pi i}{k+1}$$

$$= \frac{k-m+1}{k+1} z^{k-m}$$

$$= \left(1 - \frac{m}{k+1} \right) z^{k-m}.$$

Thus we see that

$$S f(z) - B f(z) = \frac{m}{k+1} z^{k-m}.$$

As a consequence, if now f is a function that is real analytic in a neighborhood of $\overline{D}$, then we may write

$$f(\zeta) = \sum_{k \ge 0, m > 0} \alpha_{k,m} \zeta^k \overline{\zeta}^m.$$

As a consequence of our calculations above, we then find that

$$S f(z) - B f(z) = \sum_{k \ge 0, k \ge m} \frac{m}{k+1} z^{k-m}.$$

This series of course converges uniformly on compact subsets of the open unit disc D.

We treat the case of the Bergman and Szegő projections on the ball below.

Given the Fefferman's asymptotic expansion for the Bergman kernel [FEF1] and Boutet de Monvel–Sjöstrand's asymptotic expansion for the Szegő kernel [BOS], one would expect a like calculation (up to a controllable error term) on a smoothly bounded, strongly pseudoconvex domain. Unfortunately we do not know enough about the canonical kernels on domains of finite type to be able to predict what will happen there. We explore the strongly pseudoconvex case below.

In a more recent work, Chen and Fu [CHF] have explored some new comparisons of the Bergman and Szegö kernels. A sample theorem is this:

Theorem 1.9.1. *Let $\Omega \subseteq \mathbb{C}^n$ be a pseudoconvex domain with C^2 boundary. Then*

(1) For any $0 < a < 1$, there exists a constant $C > 0$ such that

$$\frac{S(z,z)}{K(z,z)} \leq C\delta(z)|\log \delta(z)|^{n/a}.$$

(2) If there is a neighborhood U of $\partial\Omega$, a bounded, continuous plurisubharmonic function φ on $U \cap \Omega$, and a defining function ρ of Ω satisfying $i\partial\bar{\partial}\varphi \geq i\rho^{-1}\partial\bar{\partial}\rho$ on $U \cap \Omega$ as currents, then there exists constants $0 < a < 1$ and $C > 0$ such that

$$\frac{S(z,z)}{K(z,z)} \geq C\delta(z)|\log \delta(z)|^{-1/a}.$$

These authors further show that, for a C^2-bounded *convex* domain, the quotient S/K is comparable to δ without any logarithmic factor.

The techniques used in the work of Chen and Fu are weighted estimates for the $\bar{\partial}$ operator (in the spirit of Hörmander's work [HOR1]) and also an innovative use of the Diederich–Fornæss index (see [DIF3]). We can say no more about the details here.

We turn next to an examination of the situation on the unit ball B in $\mathbb{C}^n$.

1.9.3   The Unit Ball in $\mathbb{C}^n$

For simplicity we shall in fact restrict attention to complex dimension 2. In that situation, the area measure $d\sigma$ on the boundary is given by

$$d\sigma = \frac{1}{16}\left[\zeta_1 d\zeta_2 \wedge d\bar{\zeta}_1 \wedge d\bar{\zeta}_2 - \zeta_2 d\zeta_1 \wedge d\bar{\zeta}_1 \wedge d\bar{\zeta}_2 + \bar{\zeta}_1 d\bar{\zeta}_2 \wedge d\zeta_1 \wedge d\zeta_2 - \bar{\zeta}_2 d\bar{\zeta}_1 \wedge d\zeta_1 \wedge d\zeta_2\right].$$

As a result, we have

$$\int_{\partial B} f(\zeta) S(z, \zeta) \, d\sigma(\zeta)$$

$$= \frac{1}{2\pi^2} \cdot \frac{1}{16} \iiint_{\partial B} f(\zeta) \frac{1}{(1 - z \cdot \bar{\zeta})^2} \left[\zeta_1 d\zeta_2 \wedge d\bar{\zeta}_1 \wedge d\bar{\zeta}_2 - \zeta_2 d\zeta_1 \wedge d\bar{\zeta}_1 \wedge d\bar{\zeta}_2 \right.$$

$$\left. + \bar{\zeta}_1 d\bar{\zeta}_2 \wedge d\zeta_1 \wedge d\zeta_2 - \bar{\zeta}_2 d\bar{\zeta}_1 \wedge d\zeta_1 \wedge d\zeta_2 \right]$$

$$= \frac{1}{32\pi^2} \iiiint_B \frac{\partial f / \partial \zeta_1}{(1 - z \cdot \bar{\zeta})^2} \cdot \zeta_1 \, d\zeta_1 \wedge d\zeta_2 \wedge d\bar{\zeta}_1 \wedge d\bar{\zeta}_2$$

$$+ \frac{1}{32\pi^2} \iiiint_B \frac{f}{(1 - z \cdot \bar{\zeta})^2} \, d\zeta_1 \wedge d\zeta_2 \wedge d\bar{\zeta}_1 \wedge d\bar{\zeta}_2$$

$$- \frac{1}{32\pi^2} \iiiint_B \frac{\partial f / \partial \zeta_2}{(1 - z \cdot \bar{\zeta})^2} \cdot \zeta_2 \, d\zeta_2 \wedge d\zeta_1 \wedge d\bar{\zeta}_1 \wedge d\bar{\zeta}_2$$

$$- \frac{1}{32\pi^2} \iiiint_B \frac{f}{(1 - z \cdot \bar{\zeta})^2} \, d\zeta_2 \wedge d\zeta_1 \wedge d\bar{\zeta}_1 \wedge d\bar{\zeta}_2$$

$$+ \frac{1}{32\pi^2} \iiiint_B \frac{\partial f / \partial \bar{\zeta}_1}{(1 - z \cdot \bar{\zeta})^2} \cdot \bar{\zeta}_1 \, d\bar{\zeta}_1 \wedge d\bar{\zeta}_2 \wedge d\zeta_1 \wedge d\zeta_2$$

$$+ \frac{1}{32\pi^2} \iiiint_B \frac{f}{(1 - z \cdot \bar{\zeta})^2} \, d\bar{\zeta}_1 \wedge d\bar{\zeta}_2 \wedge d\zeta_1 \wedge d\zeta_2$$

$$+ \frac{1}{32\pi^2} \iiiint_B \frac{2f \cdot \bar{\zeta}_1 z_1}{(1 - z \cdot \bar{\zeta})^3} \, d\bar{\zeta}_1 \wedge d\bar{\zeta}_2 \wedge d\zeta_1 \wedge d\zeta_2$$

$$- \frac{1}{32\pi^2} \iiiint_B \frac{\partial f / \partial \bar{\zeta}_2}{(1 - z \cdot \bar{\zeta})^2} \cdot \bar{\zeta}_2 \, d\bar{\zeta}_2 \wedge d\bar{\zeta}_1 \wedge d\zeta_1 \wedge d\zeta_2$$

$$- \frac{1}{32\pi^2} \iiiint_B \frac{f}{(1 - z \cdot \bar{\zeta})^2} \, d\bar{\zeta}_2 \wedge d\bar{\zeta}_1 \wedge d\zeta_1 \wedge d\zeta_2$$

$$- \frac{1}{32\pi^2} \iiiint_B \frac{2f \cdot \bar{\zeta}_2 z_2}{(1 - z \cdot \bar{\zeta})^3} \, d\bar{\zeta}_2 \wedge d\bar{\zeta}_1 \wedge d\zeta_1 \wedge d\zeta_2.$$

Now we may group together like terms to obtain

$$= -\frac{1}{8\pi^2} \iiint\limits_B \frac{f(\zeta)}{(1 - z \cdot \bar\zeta)^3} \, d\bar\zeta_1 \wedge d\zeta_1 \wedge d\bar\zeta_2 \wedge d\zeta_2$$

$$+ \frac{3}{16\pi^2} \iiint\limits_B \frac{f(\zeta) \cdot (z \cdot \bar\zeta)}{(1 - z \cdot \bar\zeta)^3} \, d\bar\zeta_1 \wedge d\zeta_1 \wedge d\bar\zeta_2 \wedge d\zeta_2$$

$$+ \frac{1}{32\pi^2} \iiint\limits_B \frac{\partial f/\partial \zeta_1}{(1 - z \cdot \bar\zeta)^2} \cdot \zeta_1 \, d\zeta_1 \wedge d\zeta_2 \wedge d\bar\zeta_1 \wedge d\bar\zeta_2$$

$$- \frac{1}{32\pi^2} \iiint\limits_B \frac{\partial f/\partial \zeta_2}{(1 - z \cdot \bar\zeta)^2} \cdot \zeta_2 \, d\zeta_2 \wedge d\zeta_1 \wedge d\bar\zeta_1 \wedge d\bar\zeta_2$$

$$+ \frac{1}{32\pi^2} \iiint\limits_B \frac{\partial f/\partial \bar\zeta_1}{(1 - z \cdot \bar\zeta)^2} \cdot \bar\zeta_1 \, d\bar\zeta_1 \wedge d\bar\zeta_2 \wedge d\zeta_1 \wedge d\zeta_2$$

$$- \frac{1}{32\pi^2} \iiint\limits_B \frac{\partial f/\partial \bar\zeta_2}{(1 - z \cdot \bar\zeta)^2} \cdot \bar\zeta_2 \, d\bar\zeta_2 \wedge d\bar\zeta_1 \wedge d\zeta_1 \wedge d\zeta_2$$

$$= -A + B + C - D + E - F.$$

Now $-A$ is just the usual Bergman integral on the ball B in $\mathbb{C}^2$. We can analyze the other terms just as we did for the disc. As an instance, let us examine B. We take $f(\zeta) = \zeta^\alpha \bar\zeta^\beta$, where α and β are multi-indices. Then we may write (ignoring dimensionality constants which are of no interest)

$$B = \iiint\limits_B \frac{f(\zeta) \cdot (z \cdot \bar\zeta)}{(1 - z \cdot \bar\zeta)^3} \, dV(\zeta)$$

$$= \iiint\limits_B (\zeta^\alpha \bar\zeta^\beta)(z \cdot \bar\zeta) \left(\sum_{j=0}^\infty (z \cdot \bar\zeta)^j \right)^3 \, dV(\zeta).$$

And now we see by inspection that if $\beta \geq \alpha$, then the integral equals 0.

However, if $\beta < \alpha$, then the integral does not vanish. Take, for instance, the case when $f(\zeta) = \zeta_1^2 \bar\zeta_1$. Then our integral becomes

$$\iiint\limits_B (\zeta_1^2 \bar\zeta_1)(z \cdot \bar\zeta) \left(\sum_{j=0}^\infty (z \cdot \bar\zeta)^j \right)^3 \, dV = \iiint\limits_B (\zeta_1^2 \bar\zeta_1)(z_1 \bar\zeta_1) \cdot 1 \, dV,$$

where we have used parity to see that most of the terms vanish. Now this last is

$$z_1 \iiint_B |\zeta_1|^4 \, dV(\zeta) = \frac{\pi^2}{12} z_1 \, .$$

The terms C, D, E, F are calculated similarly, and we find that the sum of these error terms is a constant times z_1. A similar calculation reveals that the Szegő projection of this same f is a different constant times z_1. So the situation is completely analogous to the disc case.

1.9.4 Strongly Pseudoconvex Domains

We again, for simplicity, restrict attention to $\mathbb{C}^2$. In the seminal paper [FEF1] and [FEF2], Fefferman shows that, near a strongly pseudoconvex boundary point, the Bergman kernel may be written (in suitable local coordinates) as

$$\frac{2}{\pi^2} \cdot \frac{1}{(1 - z \cdot \overline{\zeta})^3} + \mathcal{E}(z, \zeta) \, ,$$

where $\mathcal{E}$ is an error term of strictly lower order (in the sense of pseudodifferential operators) than the Bergman kernel.

In the important paper [BOS], Boutet de Monvel and Sjöstrand show that, near a strongly pseudoconvex boundary point, the Szegő kernel may be written (in suitable local coordinates) as

$$\frac{1}{2\pi^2} \cdot \frac{1}{(1 - z \cdot \overline{\zeta})^2} + \mathcal{F}(z, \zeta) \, ,$$

where $\mathcal{F}$ is an error term of strictly lower order (in the sense of pseudodifferential or Fourier integral operators) than the Szegő kernel.

We now take advantage of these two asymptotic expansions to say something about the relationship between the Bergman and Szegő projections on a smoothly bounded strongly pseudoconvex domain.

Now fix a smoothly bounded, strongly pseudoconvex domain Ω with defining function ρ (see [KRA1] for this notion). Let U be a tubular neighborhood of $\partial\Omega$, and let V be a relatively compact subdomain of U that is also a tubular neighborhood of $\partial\Omega$. Let φ_j be a partition of unity that is supported in U and sums to be identically 1 on V. We assume that each φ_j has support so small that both the Fefferman and Boutet de Monvel–Sjöstrand expansions are valid on the support of φ_j. Then we write

$$\int_{\partial\Omega} f(\zeta) S(z, \zeta) \, d\sigma(\zeta) = \iiint_{\partial\Omega} f(\zeta) S(z, \zeta) \omega(\zeta)$$

$$= \sum_j \iiint_{\partial\Omega} \varphi_j(\zeta) f(\zeta) S(z, \zeta) \omega(\zeta),$$

where ω is the differential form that is equivalent to area measure on the boundary. And now, using Boutet de Monvel–Sjöstrand and using the notable lemma of Fefferman [FEF1] and [FEF2] that says that a strongly pseudoconvex boundary point is the ball up to fourth order, one can write each term of this last sum as

$$\frac{1}{2\pi} \iiint\limits_{\partial B} \bar{\varphi}_j(\zeta) f(\zeta) \cdot \frac{1}{(1 - z \cdot \bar{\zeta})^2} \Big[\zeta_1 d\zeta_2 \wedge d\bar{\zeta}_1 \wedge d\bar{\zeta}_2$$

$$- \zeta_2 d\zeta_1 \wedge d\bar{\zeta}_1 \wedge d\bar{\zeta}_2 + \bar{\zeta}_1 d\bar{\zeta}_2 \wedge d\zeta_1 \wedge d\zeta_2 - \bar{\zeta}_2 d\bar{\zeta}_1 \wedge d\zeta_1 \wedge d\zeta_2 \Big] + \mathcal{G},$$

where the error term $\mathcal{G}$ arises from approximating $\partial\Omega$ by ∂B, from approximating the Szegő kernel S by the kernel for the ball, by applying a change of variable to φ_j, and also by approximating ω by the differential form that we used on the ball.

Now we may carry out the calculations using Stokes's theorem just as in the last subsection to finally arrive at the assertion that the last integral equals

$$\iiiint\limits_{B} \bar{\tilde{\varphi}}_j(\zeta) \frac{f(\zeta)}{(1 - z \cdot \bar{\zeta})^3} \, dV + \mathcal{H}.$$

We cannot make the error term $\mathcal{H}$ disappear, but it is smoothly bounded hence negligible. Finally, we can use the Fefferman asymptotic expansion to relate this last integral to the Bergman projection integral on the strongly pseudoconvex domain Ω.

In summary, we have used Stokes's theorem to relate the Szegő projection integral on a smoothly bounded, strongly pseudoconvex domain to the Bergman projection integral on that domain. In this context, we do not get a literal equality. Instead we get an equality up to a controllable error term.

1.9.5 Concluding Remarks

Certainly one of the fundamental problems of the function theory of several complex variables is to understand the canonical kernels in as much detail as possible. This discussion is a contribution to that program.

1.10 Multiply Connected Domains

Now what about multiply connected domains? A useful result in [KRA3] shows the following. Let Ω be smoothly bounded with connectivity k, and let $S_1, S_2, \ldots, S_k$ be the boundary curves of Ω. Suppose that S_k bounds the unbounded component of the complement. Let Ω_k be the bounded region bounded by S_k and let Ω_j, $j = 1, 2, \ldots, k - 1$ be the unbounded region bounded by S_j. Then, for $z, \zeta \in \Omega$,

$$K_\Omega(z, \zeta) = K_{\Omega_1}(z, \zeta) + K_{\Omega_2}(z, \zeta) + \cdots + K_{\Omega_k}(z, \zeta) + \mathcal{E}(z, \zeta),$$

where $\mathcal{E}$ is a bounded function with bounded derivatives. See Sect. 1.14 below. Thus K can be written as the sum of a Bergman kernel for a domain (namely, Ω_k) that is conformally equivalent to the disc plus Bergman kernels for domains which are conformally equivalent to the complement of the closure of the disc (namely, Ω_1, $\ldots, \Omega_{k-1}$). We already understand the Bergman kernels for those domains.

Meanwhile, we shall explore a slightly different direction. Let $\Omega \subseteq \mathbb{C}$ be a domain—multiply connected or not. Let λ be a nontrivial automorphism of that domain. This means that λ is a one-to-one, onto, holomorphic mapping of the domain to itself. Let $\{\varphi_j\}$ be a complete orthonormal basis for the Bergman space on Ω. Let $\mathcal{I}_0$ denote the collection of those basis elements φ_j such that $\varphi_j \circ \lambda = \varphi_j$ (as an example, think of the even basis elements on the disc, with the automorphism being $\zeta \mapsto -\zeta$). Assume that $\mathcal{I}_0$ is a proper subset of the full basis. Let $X = X_{\mathcal{I}_0}$ be the subspace of the full Bergman space generated by $\mathcal{I}_0$.

Now let us consider the Bergman kernel $K_{\mathcal{I}_0}$ for X. Certainly this kernel will have the boundary diagonal $\mathcal{D} = \{(z, \zeta) \in \partial\Omega \times \partial\Omega : z = \zeta\}$ as usual as a singular set. But it will also have the image $\mathcal{S} = \{(z, \zeta) : \zeta \in \partial\Omega, z = \lambda(\zeta)\}$ of $\mathcal{D}$ under λ as a singular set. Thus we now have a fairly general criterion for recognizing multiple singular sets.

See Sect. 6.8 for more on multiply connected domains.

1.11 The Bergman Kernel for a Sobolev Space

We may define the Bergman kernel on the disc for the Sobolev space W^1 and it appears to be (up to a bounded error term)

$$\frac{1}{\pi} \log(1 - z\bar{\zeta}).$$

Specifically, set $\varphi_j(\zeta) = \zeta^j$. We calculate that

$$\iint_D |\varphi_j(\zeta)|^2 \, dA = \iint_D |\zeta^j|^2 \, dA = \frac{\pi}{j + 1}$$

and

$$\iint_D |\varphi_j'(\zeta)|^2 \, dA = \iint_D |j\zeta^{j-1}|^2 \, dA = j\pi.$$

Thus

$$\|\varphi_j\|_{W^1} = \sqrt{\pi} \cdot \sqrt{\frac{j^2 + j + 1}{j + 1}}.$$

Thus the full Bergman kernel for W^1 is given by

$$\sum_{j=0}^{\infty} \frac{1}{\pi} \cdot \frac{j+1}{j^2+j+1} \cdot z^j \bar\zeta^j = \frac{1}{\pi} + \sum_{j=1}^{\infty} \frac{1}{\pi} \cdot \frac{j+1}{j^2+j+1} \cdot z^j \bar\zeta^j = \frac{1}{\pi} + \sum_{j=1}^{\infty} \frac{1}{\pi} \cdot \frac{1}{j} \cdot z^j \bar\zeta^j + \mathcal{E},$$

where $\mathcal{E}$ is an error term which is bounded and has one bounded derivative. So $\mathcal{E}$ is negligible from the point of view of determining where the kernel has singularities (i.e., where it blows up).

We look at

$$\frac{1}{\pi} + \frac{1}{\pi} \sum_{j=1}^{\infty} \frac{1}{j} \alpha^j = \frac{1}{\pi} + \frac{1}{\pi} \sum_{j=1}^{\infty} \int \alpha^{j-1}$$

$$= \frac{1}{\pi} + \frac{1}{\pi} \int \sum_{j=1}^{\infty} \alpha^{j-1}$$

$$= \frac{1}{\pi} + \frac{1}{\pi} \int \frac{1}{\alpha} \sum_{j=1}^{\infty} \alpha^j$$

$$= \frac{1}{\pi} + \frac{1}{\pi} \int \frac{1}{\alpha} \left[\sum_{j=0}^{\infty} \alpha^j - 1 \right]$$

$$= \frac{1}{\pi} + \frac{1}{\pi} \int \frac{1}{\alpha} \left[\frac{1}{1-\alpha} - 1 \right]$$

$$= \frac{1}{\pi} + \frac{1}{\pi} \int \frac{1}{1-\alpha}$$

$$= \frac{1}{\pi} - \frac{1}{\pi} \log(1 - \alpha).$$

Thus the Bergman kernel for the order 1 Sobolev space is given by

$$K(z, \zeta) = \frac{1}{\pi} - \frac{1}{\pi} \log(1 - z\bar\zeta).$$

Also the kernel for the space generated just by the monomials with even index seems to be given by (up to a bounded error term)

$$\frac{1}{\pi} \left(\log(z\bar\zeta) + \frac{1}{2} \log(1 - z\bar\zeta) - \frac{1}{2} \log(1 + z\bar\zeta) \right).$$

To see this, we look at

$$\sum_{j=0}^{\infty} \frac{1}{\pi} \cdot \frac{2j+1}{(2j)^2 + 2j + 1} z^{2j} \overline{\zeta}^{2j} = \frac{1}{\pi} + \frac{1}{\pi} \sum_{j=1}^{\infty} \frac{1}{2j} z^{2j} \overline{\zeta}^{2j} + \mathcal{F}.$$

Here, as in the first calculation, $\mathcal{F}$ is a bounded term with one bounded derivative. So it is negligible from the point of view of our calculation.

Thus we wish to calculate

$$\frac{1}{\pi} + \frac{1}{\pi} \sum_{j=1}^{\infty} \frac{1}{2j} \alpha^{2j} = \frac{1}{\pi} + \frac{1}{\pi} \sum_{j=1}^{\infty} \int \alpha^{2j-1}$$

$$= \frac{1}{\pi} + \frac{1}{\pi} \int \frac{1}{\alpha} \sum_{j=1}^{\infty} \alpha^{2j}$$

$$= \frac{1}{\pi} + \frac{1}{\pi} \int \frac{1}{\alpha} \left[\sum_{j=0}^{\infty} \alpha^{2j} - 1 \right]$$

$$= \frac{1}{\pi} + \frac{1}{\pi} \int \frac{1}{\alpha} \left[\frac{1}{1 - \alpha^2} - 1 \right]$$

$$= \frac{1}{\pi} + \frac{1}{\pi} \int \frac{\alpha}{1 - \alpha^2}$$

$$= \frac{1}{\pi} - \frac{1}{2\pi} \log(1 - \alpha^2).$$

In conclusion, the Bergman kernel for the order 1 Sobolev space using only the basis elements with even index is

$$K'(z, \zeta) = \frac{1}{\pi} - \frac{1}{2\pi} \log(1 - z \cdot \overline{\zeta}) - \frac{1}{2\pi} \log(1 + z \cdot \overline{\zeta}).$$

In short, there are singularities as z and ζ tend to the *same* disc boundary point and also as z and ζ tend to antipodal disc boundary points.

1.12   Ramadanov's Theorem

In the noted paper [RAM1], Ramadanov proved a very useful result about the limit of the Bergman kernels in an increasing sequence of domains. A version of Ramadanov's classical result is this:

Theorem 1.12.1. *Let $\Omega_1 \subseteq \Omega_2 \subseteq \cdots \subseteq$ be an increasing sequence of bounded domains in $\mathbb{C}^n$ and let $\Omega = \cup_{j=1}^{\infty} \Omega_j$. Assume also that Ω is bounded. Then*

$$K_\Omega(z, \zeta) = \lim_{j \to \infty} K_{\Omega_j}(z, \zeta),$$

with the limit being uniform on compact subsets of $\Omega \times \Omega$.

In the paper [KRA8], Krantz generalizes Ramadanov's result to a sequence of domains that is not necessarily increasing. Here we present the statement of his theorem and the proof.

Theorem 1.12.2. *Let Ω_j be a sequence of domains that converges to a limit domain Ω in the Hausdorff metric of domains (see [KRPA1]). Then $K_{\Omega_j} \to K_\Omega$ uniformly on compact subsets of $\Omega \times \Omega$.*

Proof. Let $\Omega_1, \Omega_2, \ldots$ be domains in $\mathbb{C}^n$ and let $\Omega_j \to \Omega$ in the topology of the Hausdorff metric. For convenience, we let $\Phi_j : \Omega \to \Omega_j$ be diffeomorphisms such that the Φ_j converge to the identity in a suitable topology.

Now fix a point z that lies in all the Ω_j and in Ω as well. Then $K_\Omega(z, \cdot)$ is the Hilbert space representative (according to the Riesz's theorem) of the point evaluation linear functional

$$A^2(\Omega) \ni f \mapsto f(z).$$

Of course it is also the case that $K_{\Omega_j}(z, \cdot)$ is the Hilbert space representative (according to the Riesz's theorem) of the point evaluation linear functional

$$A^2(\Omega_j) \ni f \mapsto f(z).$$

Of course the standard lemma for the Bergman theory (using the mean-value property) tells us that the point evaluation functional at z is bounded with a bound that is independent of j (in fact it only depends on the $(-n)^{\text{th}}$ power of the distance of z to the boundary and that may be taken to be uniform in j). Thus if we set $\psi_j(z) = K_{\Omega_j}(z, \Phi_j(\cdot))$, then $\|\psi_j\|_{L^2(\Omega)}$ is bounded, independent of j.

By the Banach–Alaoglu theorem, we may conclude that there is a weak-$*$ convergent subsequence ψ_{j_k}. Call the weak-$*$ limit ψ_0. But then, for $g \in A^2(\Omega_{j_k})$, we see that

$$g(z) = \int_{\Omega_{j_k}} K_{\Omega_{j_k}}(z, \zeta) g(\zeta) \, dV(\zeta)$$

$$= \int_\Omega \psi_{j_k}(\xi) g(\Phi_{j_k}(\xi)) \Phi_{j_k}'(\xi) \overline{\Phi_{j_k}'(\xi)} \, dV(\xi)$$

$$\equiv \int_\Omega \psi_{j_k}(\xi) h_{j_k}(\xi) \, dV(\xi)$$

Notice that the h_{j_k} are all defined on Ω and they converge in the strong topology of $L^2(\Omega)$ to some limit function $\tilde{h}$. In fact, by applying the $\overline{\partial}$ operator, one can see that $\tilde{h}$ is a *holomorphic* function on Ω.

Thus our expression converges to

$$\int_{\Omega} \psi_0(\xi)\tilde{h}(\xi)\,dV(\xi). \tag{1.12.2.1}$$

Now in fact we may apply this preceding argument to see that every subsequence of the index j has a subsequence so that we get the indicated convergence. The conclusion is that the ψ_j converge weak-$*$ to ψ_0 and the h_j converge strongly to $\tilde{h}$ so that

$$g(z) = \int_{\Omega} \psi_0(\xi)\tilde{h}(\xi)\,dV(\xi).$$

Next we examine the definition of the h_j and the Φ_j to see that in fact $\tilde{h}(z) = g(z)$. Thus we may write the last line as

$$g(z) = \int_{\Omega} \psi_0(\xi)\tilde{h}(\xi)\,dV(\xi) = \tilde{h}(z).$$

Finally, one can reason backwards to see that any L^2 holomorphic function $\tilde{h}$ on Ω can arise in this way. The only possible conclusion is that ψ_0 is the representing function for point evaluation at z. So ψ_0 is the Bergman kernel for Ω. In conclusion $K_{\Omega_j}(z, \cdot)$ converges weak-$*$ to $K_\Omega(z, \cdot)$.

Certainly we may now apply the weak-$*$ convergence to a testing function consisting of a C_c^∞ radial function to conclude that ψ_{j_k} in fact converges uniformly on compact sets to ψ_0. So we see that ψ_0 is conjugate holomorphic. We may therefore conclude that ψ_0 is the Bergman kernel of Ω.

Thus we see that the $K_{\Omega_j}(z, \cdot)$ converge uniformly on compact sets to $K_\Omega(z, \cdot)$. This is the desired conclusion. $\square$

1.13   Coda on the Szegő Kernel

It is not difficult to see that, suitably formulated, there is a version of Theorem 1.12.2 for the Szegő kernel (see [KRA8]). To wit,

Theorem 1.13.1. *Let Ω_j be a sequence of domains with C^2 boundary that converges to a limit domain Ω in the C^2 topology of domains. This means that each $\Omega_j = \{z \in \mathbb{C}^n : \rho_j(z) < 0\}$, $\nabla\rho_j \neq 0$ on $\partial\Omega_j$, and the defining functions*

ρ_j converge in the C^2 topology. Then the Szegő kernels $S_{\Omega_j} \to S_\Omega$ uniformly on compact subsets of $\Omega \times \Omega$.

Without much effort, it can also be seen that there is a version of our theorem in the rather general setting of Hilbert space with reproducing kernel. See [ARO] for a thorough treatment of this abstract concept.

1.14 Boundary Localization

We have already seen in Sect. 1.9 a suggestion that the Bergman kernel of a multiply connected domain ought to be expressible in terms of the Bergman kernels of component domains. In this section we briefly explore that idea.

We begin by examining a slightly different avenue for getting one's hands on the Bergman kernel of a domain. The general approach is perhaps best illustrated with an example. Let

$$\Omega = \{\zeta \in \mathbb{C} : 1 < |\zeta| < 2\}.$$

This is the annulus, and any explicit representation of its Bergman kernel will involve elliptic functions (see [BER1] and [BER2]). One might hope, however, to relate the Bergman kernel K_Ω of Ω to the Bergman kernels K_{Ω_1} and K_{Ω_2} of

$$\Omega_1 = \{\zeta \in \mathbb{C} : |\zeta| < 2\}$$

and

$$\Omega_2 = \{\zeta \in \mathbb{C} : 1 < |\zeta|\}.$$

The first of these has an explicitly known Bergman kernel (see [KRA1]) and the second domain is the inversion of a disc, so its kernel is known explicitly as well.

One could pose a similar question for domains of higher connectivity. The question also makes sense, with a suitable formulation, in several complex variables. Our purpose here is to come up with precise formulations of results such as these and to prove them. In one complex variable, we can make decisive use of classical results relating the Bergman kernel to the Green's function (see [KRA2]). In several complex variables there are analogous results of Garabedian (see [GARA]) that will serve in good stead.

In Sect. 1.14.1 we introduce appropriate definitions and notation. In Sect. 1.14.2 we prove a basic, representative result in the plane. Section 1.14.3 proves a more general result in the plane. Section 1.14.3 treats the multidimensional result.

We thank Richard Rochberg for bringing these questions to our attention.

1.14.1   Definitions and Notation

If $\Omega \subseteq \mathbb{C}^n$ is a bounded domain, then we let $K_\Omega(z, \zeta)$ denote its Bergman kernel. This is the reproducing kernel for

$$A^2(\Omega) \equiv \{ f \in L^2(\Omega) : f \text{ is holomorphic on } \Omega \}.$$

It is known, for planar domains, that $K_\Omega(z, \zeta)$ is related to the Green's function $G_\Omega(z, \zeta)$ for Ω by this formula:

$$K_\Omega(z, \zeta) = 4 \cdot \overline{\frac{\partial^2}{\partial \zeta \partial \overline{z}} G_\Omega(\zeta, z)}.$$

Of course it is essential for our analysis to realize that the Green's function is known quite explicitly on any given domain. If

$$\Gamma(\zeta, z) = \frac{1}{2\pi} \log |\zeta - z|$$

is the fundamental solution for the Laplacian (on all of $\mathbb{C}$), then we construct the Green's function as follows:

Given a domain $\Omega \subseteq \mathbb{C}$ with smooth boundary, the *Green's function* is posited to be a function $G_\Omega(\zeta, z)$ that satisfies

$$G_\Omega(\zeta, z) = \Gamma(\zeta, z) - F_z^\Omega(\zeta),$$

where $F_z^\Omega(\zeta) = F^\Omega(\zeta, z)$ is a particular harmonic function in the ζ variable. It is mandated that F^Ω be chosen (and is in fact uniquely determined by the condition) so that $G(\,\cdot\,, z)$ vanishes on the boundary of Ω. One constructs the function $F^\Omega(\,\cdot\,, z)$, for each fixed z, by solving a suitable Dirichlet problem. Again, the reference [KRA1, p. 40] has all the particulars. It is worth noting that the Green's function is a symmetric function of its arguments.

In our proof, we shall be able to exploit known properties of the Poisson kernel (see especially [KRA3]) and of the solution to the Dirichlet problem (see [KRA4]) to get the estimates that we need.

1.14.2   A Representative Result

We first prove our main result for the domain

$$\Omega = \{ \zeta \in \mathbb{C} : 1 < |\zeta| < 2 \}.$$

This argument will exhibit all the key ideas—at least in one complex variable. The later exposition will be clearer because we took the time to treat this case carefully.

Let

$$\Omega_1 = \{\zeta \in \mathbb{C} : |\zeta| < 2\}$$

and

$$\Omega_2 = \{\zeta \in \mathbb{C} : 1 < |\zeta|\}.$$

For convenience in what follows, we let S_1 be the boundary curve of Ω_1 and S_2 be the boundary curve of Ω_2. Of course it then follows that $\partial\Omega = S_1 \cup S_2$.

We claim that

$$K_\Omega(z, \zeta) = \frac{1}{2}[K_{\Omega_1}(z, \zeta) + K_{\Omega_2}(z, \zeta)] + \mathcal{E}(z, \zeta),$$

where $\mathcal{E}$ is an error term that is smooth on $\overline{\Omega} \times \overline{\Omega}$. In particular, $\mathcal{E}$ is bounded with all derivatives bounded on that domain.

For the proof, we write

$$\frac{1}{8}\left[\overline{K_{\Omega_1}(z, \zeta) + K_{\Omega_2}(z, \zeta)}\right]$$

$$= \frac{1}{2}\frac{\partial^2}{\partial\zeta\partial\overline{z}}\left[(\Gamma(\zeta, z) - F^{\Omega_1}(\zeta, z)) + (\Gamma(\zeta, z) - F^{\Omega_2}(\zeta, z))\right]$$

$$= \frac{\partial^2}{\partial\zeta\partial\overline{z}}\left(\Gamma(\zeta, z) - \frac{1}{2}[F^{\Omega_1}(\zeta, z) + F^{\Omega_2}(\zeta, z)]\right).$$

Now we claim that

$$F^{\Omega_1}(\zeta, z) + F^{\Omega_2}(\zeta, z) = 2F^\Omega(\zeta, z) + \mathcal{E}(z, \zeta)$$

for a suitable error term $\mathcal{E}$. We must analyze

$$G(\zeta, z) \equiv [F^{\Omega_1}(\zeta, z) + F^{\Omega_2}(\zeta, z)] - 2F^\Omega(\zeta, z).$$

We think of G as the solution of a Dirichlet problem on Ω, and we must analyze the boundary data. What we see is this:

- For z near S_1, F^Ω and F^{Ω_1} agree on S_1 (in the variable ζ) and equal 0. And F^{Ω_2} is smooth and bounded by $C \cdot |\log(1/2)|$, just by the form of the Green's function. All three functions are plainly smooth and bounded on S_2 (for z still near S_1) by similar reasoning. In conclusion, G is smooth and bounded on $\overline{\Omega}$ for z near S_2.

- For z near S_2, F^Ω and F^{Ω_2} agree on S_2 (in the variable ζ) and equal 0. And F^{Ω_1} is smooth and bounded by $C \cdot |\log(1/2)|$, just by the form of the Green's function. All three functions are plainly smooth and bounded on S_1 (for z still near S_1) by similar reasoning. In conclusion, G is smooth and bounded on $\overline{\Omega}$ for z near S_2.
- For z away from both S_1 and S_2—in the interior of Ω—it is clear that all the terms are bounded and smooth on $\partial\Omega$. So the solution G of the Dirichlet problem will also be smooth as desired.

As a result of these considerations, G is smooth on $\overline{\Omega}$.

That completes our argument and gives, altogether, the error term $\mathcal{E}$. Thus

$$F^{\Omega_1} + F^{\Omega_2} - 2F^\Omega = \mathcal{E}.$$

It follows that

$$\frac{1}{2}[K_{\Omega_1}(z,\zeta) + K_{\Omega_2}(z,\zeta)] = 4\overline{\frac{\partial^2}{\partial\zeta\partial\overline{z}}\left(\Gamma(\zeta,z) - F^\Omega(\zeta,z)\right)} + \mathcal{E}'$$

$$= K_\Omega(z,\zeta) + \mathcal{E}''$$

1.14.3   The More General Result in the Plane

Now consider a smoothly bounded domain $\Omega \subseteq \mathbb{C}$ with k connected components in its boundary, $k \geq 2$. We denote the boundary components by $S_1, \ldots, S_k$; for specificity, we let S_1 be the component of the boundary that bounds the unbounded component of the complement of Ω. Let Ω_1 be the *bounded* region in the plane bounded by the single Jordan curve S_1. Let $\Omega_2, \ldots, \Omega_k$ be the unbounded regions bounded by $S_2, S_3, \ldots, S_k$, respectively.

Then we may analyze, just as in the last subsection, the expression

$$K_\Omega - \frac{1}{k}\left[K_{\Omega_1} + K_{\Omega_2} + \cdots + K_{\Omega_k}\right]$$

to obtain a smooth error term

$$\mathcal{E} = \mathcal{E}_1 + \mathcal{E}_2 + \cdots + \mathcal{E}_k.$$

That completes our analysis of a smooth, finitely connected domain in the plane.

1.14.4   Domains in Higher-Dimensional Complex Space

The elegant paper [GARA] contains the necessary information about the relationship of the Bergman kernel and a certain Green's function in several complex variables so that we may carry out our program in that more general context.

Fix a smoothly bounded domain Ω in $\mathbb{C}^k$. Let $t = (t_1, \ldots, t_k)$ be a fixed point in Ω. Following Garabedian's notation, we set

$$r = \sqrt{\sum_{j=1}^{k} |z_j - t_j|^2}.$$

Let σ_k be constants chosen so that

$$\lim_{\epsilon \to 0} \sigma_k \int_{\Gamma_\epsilon} B \cdot \sum_{j=1}^{k} \frac{\partial r^{-2k+2}}{\partial z_j} \alpha_j \, d\sigma + B(t) = 0,$$

where Γ_ϵ is the sphere of radius ϵ about t, B is some continuous function, and $(\alpha_1, \ldots, \alpha_k)$ is a collection of complex-valued direction cosines.

Now set $\theta(z, t)$ to be that function

$$\theta = \sigma_k r^{-2k+2} + \text{regular terms} \tag{1.14.1}$$

on Ω so that

$$\sum_{j=1}^{k} \frac{\partial \theta}{\partial \bar{z}_j} \cdot \bar{\alpha}_j = 0$$

on $\partial \Omega$,

$$\frac{\partial}{\partial \bar{z}_j} \Delta \theta = 0$$

on Ω (for $j = 1, \ldots, k$), and such that

$$\int_\Omega \theta \bar{f} \, dV = 0,$$

for all functions f analytic in Ω. It follows from standard elliptic theory that such a θ exists.

In fact, according to [GAR], this function θ that we have constructed is a Green's function for the boundary value problem

$$\frac{\partial}{\partial \bar{z}_j} \Delta \beta = 0 \qquad \text{on } \Omega, \quad j = 1, \ldots, k$$

$$\sum_{j=1}^{k} \frac{\partial \beta}{\partial \bar{z}_j} \cdot \bar{\alpha}_j = 0 \qquad \text{on } \partial \Omega.$$

Garabedian goes on to prove that the Bergman kernel for Ω is related to the Green's function θ in this way:

$$K_\Omega(z,t) = \Delta_z \theta(z,t) .$$

This is just the information that we need to apply the machinery that has been developed here.

In order to flesh out the argument in the context of several complex variables, our primary task is to argue that our new Green's function has a form similar to the classical Green's function from one complex variable. But in fact this is immediate from (1.14.1). It follows from this that the argument in Sect. 1.14.3 using the maximum principle will go through as before, and we may establish a version of the result in Sects. 1.14.2 and 1.14.3 in the context of several complex variables. The theorem is this:

Theorem 1.14.2. *Let Ω be a smoothly bounded domain in $\mathbb{C}^n$ with boundary having connected components S_1, S_2, ..., S_k. For specificity, say that S_1 is the boundary component that bounds the unbounded portion of the complement of $\overline{\Omega}$. Let K_Ω be the Bergman kernel for Ω, let K_1 be the Bergman kernel for the bounded domain having S_1 as its single boundary element, and let K_j, for $j \geq 2$, be the Bergman kernel for the unbounded domain having S_j as its single boundary component. Then*

$$K_\Omega = K_1 + K_2 + \cdots + K_k + \mathcal{E},$$

where $\mathcal{E}$ is an error term that is bounded with bounded derivatives.

The reader can see that this new theorem is completely analogous to the results presented earlier in the one-variable setting. But it must be confessed that this theorem is something of a *canard*. For, when $j \geq 2$, any function holomorphic on the unbounded domain with boundary S_j will (by the Hartogs extension phenomenon) extend analytically to all of $\mathbb{C}^n$. And of course there are no L^2 holomorphic functions on all of $\mathbb{C}^n$. So it follows that $K_j \equiv 0$. So the theorem really says that

$$K_\Omega = K_1 + \mathcal{E}.$$

This is an interesting fact, but not nearly as important or provocative as the one-variable result. The one other point worth noting is that the statement of the result is now a bit different from that in one complex variable, just because we are dealing with a different Green's function for a different boundary value problem. Basically what we are seeing is that $K_2, \ldots, K_k$ do not count at all, and K_1 is the principal and only term.

Exercises

1. *The automorphism group of the ball.* Let $B \subseteq \mathbb{C}^2$ be the unit ball. Complete the following outline to calculate the set of biholomorphic self-maps of B.

    (a) Let $a \in \mathbb{C}, |a| < 1$. Then

    $$\phi_a(z_1, z_2) = \left(\frac{z_1 - a}{1 - \bar{a}z_1}, \frac{\sqrt{1 - |a|^2}z_2}{1 - \bar{a}z_1} \right)$$

    is a biholomorphic mapping of B to itself.

    (b) If ρ is a unitary mapping of $\mathbb{C}^2$ (i.e., the inverse of ρ is its conjugate transpose; equivalently, ρ preserves the Hermitian inner product on $\mathbb{C}^2$), then ρ is a biholomorphic mapping of the ball.

    (c) Prove that if $\psi : B \to B$ is holomorphic and $\psi(0) = 0$, then all eigenvalues of Jac $\psi(0)$ have modulus not exceeding 1. (*Hint:* Apply the one-variable Schwarz lemma in a clever way.)

    (d) Apply part (c) to any biholomorphic mapping of B that preserves the origin, and to its inverse, to see that such a mapping has Jacobian matrix at the origin with all eigenvalues of modulus 1.

    (e) Use the result of part (d) to see that any biholomorphic mapping of the ball that preserves the origin must be linear, indeed unitary.

    (f) If now α is any biholomorphic mapping of the ball, choose a unitary mapping ρ and a complex constant a such that $\phi_a \circ \rho \circ \alpha$ is a biholomorphic mapping that preserves the origin. Thus by part (e), $\phi_a \circ \rho \circ \alpha$ is unitary.

    (g) Conclude that the mappings ϕ_a and the unitary mappings generate the automorphism group of the ball in $\mathbb{C}^2$.

2. Generalize the result of Exercise 1 to n complex dimensions.

3. Let $\Omega \subseteq \mathbb{R}^N$ be a domain. Suppose that $\partial\Omega$ is a regularly imbedded C^j manifold, $j = 1, 2, \ldots$. This means that, for each $P \in \partial\Omega$, there is a neighborhood $U_P \subseteq \mathbb{R}^N$ and a C^j function $f_P : U_P \to \mathbb{R}$ with $\nabla f_P \neq 0$ and $\{x \in U_P : f_P(x) = 0\} = U_P \cap \partial\Omega$. Prove that there is a function $\rho : \mathbb{R}^N \to \mathbb{R}$ satisfying

    (a) $\nabla\rho \neq 0$ on $\partial\Omega$.
    (b) $\{x \in \mathbb{R}^N : \rho(x) < 0\} = \Omega$.
    (c) ρ is C^j.

    Prove that if Ω has a C^j defining function, then $\partial\Omega$ is a regularly imbedded C^j submanifold of $\mathbb{R}^N$.
    Prove that both of the preceding concepts are equivalent to the following: For each $P \in \partial\Omega$ there is a neighborhood U_P, a coordinate system $t_1, \ldots, t_N$ on U_P, and a C^j function $\phi(t_1, \ldots, t_N)$ such that $\{(t_1, \ldots, t_N) \in U_P : t_N = \phi(t_1, \ldots, t_{N-1})\} = \partial\Omega \cap U_P$. This means that $\partial\Omega$ is locally the graph of a C^j function.

4. Let $\tilde{\mathrm{h}}^2(\partial D)$ be the space of those continuous functions that are the boundary functions of harmonic functions on the disc. Mimic the construction of the Szegő kernel to obtain a reproducing kernel. What reproducing kernel do you obtain? Why is the space $\tilde{\mathrm{h}}^2(\partial D)$ defined incorrectly (see [EPS])?
5. Derive a formula for the Green's function and the Poisson kernel for the ball $B(x_0, r) \subseteq \mathbb{R}^N$ by using invariance properties of the Laplacian.
6. Let $\Omega \subseteq \mathbb{R}^N$ be a bounded domain with C^2 boundary. Let $G(x, y)$ be the Green's function for the Laplacian on Ω. Prove that $G(P, Q) = G(Q, P)$, all $P \neq Q$. In particular, G extends to a smooth ($C^{2-\epsilon}$) function on $\overline{\Omega} \times \overline{\Omega} \setminus \{(x, x) : x \in \overline{\Omega}\}$ and is harmonic in each variable. *Hint:* Fix $P, Q \in \Omega, P \neq Q$. Apply Green's theorem on

$$\Omega_\epsilon \equiv \Omega \setminus \left(\overline{B}(P, \epsilon) \cup \overline{B}(Q, \epsilon) \right),$$

ϵ small, to the functions $G(P, \cdot)$ and $G(Q, \cdot)$.
7. Hopf's lemma, originally proved (see [COH]) for the sake of establishing the maximum principle for solutions of second order elliptic equations, has proved to be a powerful tool in the study of functions of several complex variables. Here we state and outline a proof of this result. While simpler proofs are available (see [KRA4]), this one has the advantage of applying in rather general circumstances.

Theorem: Let $\Omega \subseteq \mathbb{R}^N$ be a bounded domain with C^2 boundary. Let $f : \overline{\Omega} \to \mathbb{R}$ be harmonic and nonconstant on Ω, C^1 on $\overline{\Omega}$. Suppose that f assumes a (not necessarily strict) maximum at $P \in \partial\Omega$. If $v = v_P$ is the unit outward normal to $\partial\Omega$ at P, then $(\partial f / \partial v)(P) > 0$.

Outline of Proof

(a) Let B_1 be a ball internally tangent to $\partial\Omega$ at P with $\partial B_1 \cap \partial\Omega = \{P\}$. Let $r > 0$ be the radius of B_1 (see Fig. 1.1). Assume without loss of generality that the center of B_1 is at the origin. Let B_2 be a ball centered at P of radius $r_1 < r$. Let $B' = B_1 \cap B_2$. Notice that $\partial B' = S_1' \cup S_2'$.
(b) Let $\alpha > 0$ and set $h(x) = e^{-\alpha|x|^2} - e^{-\alpha r^2}$. Then

$$\Delta h = e^{-\alpha|x|^2} \{4\alpha^2 |x|^2 - 2\alpha N\}.$$

(c) If $\alpha > 0$ is sufficiently large, then $\Delta h > 0$ on B'.
(d) Set $v(x) \equiv f(x) + \epsilon h(x)$. If $\epsilon > 0$ is sufficiently small, then $v(x) < f(P)$ for $x \in S_1'$. Also $v(x) = f(x) < f(P)$ for $x \in S_2' \setminus \{P\}$. Use the maximum principle.
(e) $\max_{x \in \overline{B'}} v(x) = f(P)$.
(f) $\frac{\partial v}{\partial v}(P) = \frac{\partial f}{\partial v}(P) + \epsilon \frac{\partial h}{\partial v}(P) \geq 0$.
(g) $\frac{\partial f}{\partial v}(P) > 0$.

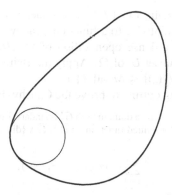

Fig. 1.1 An internally tangent ball

Now suppose only that $f \in C(\overline{\Omega})$ and harmonic on Ω and that the point $P \in \partial\Omega$ is a local (not necessarily strict) maximum of f. Modify the preceding argument to prove that

$$\liminf_{\epsilon \to 0^+} \left(\frac{f(P) - f(P - \epsilon v)}{\epsilon} \right) > 0.$$

8. Holomorphic mappings in $\mathbb{C}^2$ are not necessarily conformal (i.e., infinitesimally angle preserving). Show that the example $F(z_1, z_2) = (z_1^2, z_1^2 + z_2)$ confirms this statement.

9. Use the classical Hurwitz theorem of one complex variable to give another proof that a holomorphic function of several variables cannot have an isolated zero.

10. Are holomorphic functions (of several variables) open? Are holomorphic mappings (of several variables) open?

11. Let $z \in \Omega_1 \subseteq \Omega_2 \subseteq \mathbb{C}^n$. Let K_j be the Bergman kernel for Ω_j. Then show that $K_2(z, z) \le K_1(z, z)$ for all $z \in \Omega_1$. Further show that $\|K_2(z, \cdot)\|_{L^2(\Omega_2)} \le \|K_1(z, \cdot)\|_{L^2(\Omega_1)}$.

12. Let $\{a_{ij}\}_{i,j=1}^n$ be complex constants such that $\sum a_{ij} w_i \overline{w}_j \ge C|w|^2$ for some $C > 0$ and all $w \in \mathbb{C}^n$. Let

$$\Omega = \left\{ z \in \mathbb{C}^n : -2\operatorname{Re} z_1 + \sum_{i,j=1}^n a_{ij} z_i \overline{z}_j < 0 \right\}.$$

Prove that the Bergman kernel for Ω is given on the diagonal by

$$K(z, z) = \frac{n! \det (a_{ij})_{i,j=1}^n}{\pi^n (2\operatorname{Re} z_1 - \sum_{i,j=1}^n a_{ij} z_i \overline{z}_j)^{n+1}}.$$

(*Hint:* The domain Ω is biholomorphic to the ball. See [GRA] for details.)

13. Let $\Omega \subseteq \mathbb{C}^n$ be a domain. Let $\{f_j\}$ be a sequence of holomorphic functions on Ω that converges pointwise to a function f at every point of Ω. Prove that f is holomorphic on a dense open subset of Ω. (*Hint:* Let $\overline{U} \subseteq \Omega$ be the closure of any open subset U of Ω. Apply the Baire category theorem to the sets $S_M = \{z \in \overline{U} : |f_j(z)| \le M, \text{all } j\}$.)

14. Complete the following outline to prove the Cauchy–Fantappiè formula:

> **Theorem:** Let $\Omega \subset\subset \mathbb{C}^n$ be a domain with C^1 boundary. Let $w(z, \zeta) = (w_1(z, \zeta), \dots, w_n(z, \zeta))$ be a C^1, vector-valued function on $\overline{\Omega} \times \overline{\Omega} \setminus \{\text{diagonal}\}$ that satisfies
>
> $$\sum_{j=1}^{n} w_j(z, \zeta)(\zeta_j - z_j) \equiv 1.$$
>
> Then we have for any $f \in C^1(\overline{\Omega}) \cap \{\text{holomorphic functions on } \Omega\}$ and any $z \in \Omega$ the formula
>
> $$f(z) = \frac{1}{nW(n)} \int_{\partial\Omega} f(\zeta)\eta(w) \wedge \omega(\zeta).$$
>
> Here $\omega(\zeta) \equiv d\zeta_1 \wedge d\zeta_2 \wedge \cdots \wedge d\zeta_n$ and $\eta(w) \equiv \sum_{j=1}^{n}(-1)^{j+1}w_j dw_1 \wedge \cdots \wedge dw_{j-1} \wedge dw_{j+1} \wedge \cdots \wedge dw_n$.

Proof. We may assume that $z = 0 \in \Omega$.

(a) If $\alpha^1 = (a_1^1, \dots, a_n^1), \dots, \alpha^n = (a_1^n, \dots, a_n^n)$ are n-tuples of C^1 functions on $\overline{\Omega}$ that satisfy $\sum_{j=1}^{n} a_i^j(\zeta) \cdot (\zeta_j - z_j) = 1$, let

$$B(\alpha^1, \dots, \alpha^n) = \sum_{\sigma \in S_n} \epsilon(\sigma)a_{\sigma(1)}^1 \wedge \overline{\partial}(a_{\sigma(2)}^2) \wedge \cdots \wedge \overline{\partial}(a_{\sigma(n)}^n),$$

where S_n is the symmetric group on n letters and $\epsilon(\sigma)$ is the signature of the permutation σ. Prove that B is independent of α^1.

(b) It follows that $\overline{\partial}B = 0$ on $\overline{\Omega} \setminus \{0\}$ (indeed $\overline{\partial}B$ is an expression like B with the expression $a_{\sigma(1)}^1$ replaced by $\overline{\partial}a_{\sigma(1)}^1$).

(c) Use (b), especially the parenthetical remark, to prove inductively that if $\beta^1 = (b_2^1, \dots, b_n^1), \dots, \beta^n = (b_2^n, \dots, b_n^n)$, then there is a form γ on $\Omega \setminus \{0\}$ such that

$$\left[B(\alpha^1, \dots, \alpha^n) - B(\beta^1, \dots, \beta^n) \right] \wedge \omega(\zeta) = \overline{\partial}\gamma = d\gamma.$$

(d) Prove that if $\alpha^1 = \cdots = \alpha^n = (w_1, \dots, w_n)$, then $B(\alpha^1, \dots, \alpha^n)$ simplifies

$$B(\alpha^1, \dots, \alpha^n) \wedge \omega(\zeta) = (n-1)!\eta(w) \wedge \omega(\zeta).$$

(e) Let $\mathcal{S}$ be a small sphere of radius $\epsilon > 0$ centered at 0 such that $\mathcal{S} \subseteq \Omega$. Use part (c) to see that

$$\int_{\partial\Omega} f(\zeta)\eta(w) \wedge \omega(\zeta) = \int_{S} f(\zeta)\eta(w) \wedge \omega(\zeta).$$

(f) Now use (c) and (d) to see that

$$\int_{S} f(\zeta)\eta(w) \wedge \omega(\zeta) = \int_{S} f(\zeta)\eta(v) \wedge \omega(\zeta),$$

where

$$v(z, \zeta) = \frac{\overline{\zeta}_j - \overline{z}_j}{|\zeta - z|^2}.$$

(Warning: Be careful if you decide to apply Stokes's theorem.) We know that the last line is $n \cdot W(n) \cdot f(0)$.

15. Use a limiting argument to show that the hypotheses of the Cauchy–Fantappiè formula (the preceding exercise) may be weakened to $f \in C(\overline{\Omega})$, $w \in C(\overline{\Omega} \times \partial\Omega)$.

   Prove, using only linear algebra, that if w is as in the statement of the Cauchy–Fantappiè formula, then there are functions $\psi_1, \ldots, \psi_n, \Psi$ such that $w_j = \psi_j/\Psi, j = 1, \ldots, n$.

16. Show that $K(0, 0) = 1/V(B)$. Here K is the Bergman kernel for the ball B. Use the automorphism group of B, together with the invariance of the kernel, to calculate $K(z, z)$ for every $z \in B$. The values of K on the diagonal then completely determine $K(z, \zeta)$. Why?

17. Let Ω be the Hartogs triangle $\{(z_1, z_2) : |z_1| < |z_2| < 1\}$. Every analytic function on a neighborhood of $\overline{\Omega}$ continues to the bidisc. Find an orthonormal basis for $A^2(\Omega)$.

Chapter 2
The Bergman Metric

2.1 Smoothness to the Boundary of Biholomorphic Mappings

Poincaré's theorem (see [KRA1, KRA9] for discussion) that the ball and polydisc are biholomorphically inequivalent shows that there is no Riemann mapping theorem (at least in the traditional sense) in several complex variables. More recent results of Burns, Shnider, and Wells [BSW] and of Greene and Krantz [GRK1, GRK2] confirm how truly dismal the situation is. First, we need a definition.

Definition 2.1.1. Let ρ_0 be a C^k defining function for a bounded domain $\Omega_0 \subseteq \mathbb{R}^N, k \geq 2$. We define a neighborhood basis for Ω_0 in the C^k topology as follows: Let $\epsilon > 0$ be so small that if $\|\rho - \rho_0\|_{C^k} < \epsilon$, then ρ has nonvanishing gradient on $\{x : \rho(x) = 0\}$. For any such ρ, let $\Omega_\rho = \{x \in \mathbb{R}^N : \rho(x) < 0\}$. Define

$$\mathcal{U}_\epsilon^k(\Omega_0) = \{\Omega_\rho \subseteq \mathbb{R}^N : \|\rho - \rho_0\|_{C^k} < \epsilon\}.$$

Observe that $\mathcal{U}_\epsilon^k(\Omega_0)$ is a *collection of domains*. Then the sets $\mathcal{U}_\epsilon \equiv \mathcal{U}_\epsilon^k(\Omega_0)$ are called *neighborhoods of Ω_0 in the C^k topology*. Of course neighborhoods in the C^∞ topology are defined similarly.

If Ω_1, Ω_2 are domains in $\mathbb{C}^n$, then we say that $\Omega_1 \sim \Omega_2$ if Ω_1 is biholomorphic to Ω_2.

Theorem 2.1.2 (Burns–Shnider–Wells). *Let $k \in \mathbb{N}$. Let $\epsilon > 0$ be small. Let $\mathcal{U}_\epsilon^k \equiv \mathcal{U}_\epsilon^k(B)$ be any neighborhood of the ball $B \subseteq \mathbb{C}^n$ in the C^k topology as defined above. If $n \geq 2$, then $\mathcal{U}_\epsilon^k / \sim$ is uncountable, no matter how small $\epsilon > 0$ is or how large k is (even $k = \infty$ or $k = \omega$). By contrast, if $n = 1$ and $\epsilon < 1/5$, then $\mathcal{U}_\epsilon^k / \sim$ has just one element.*

The last statement of the theorem perhaps merits some explanation. If $n = 1$ and $\epsilon < 1/5$, then perforce any equivalence class in $\mathcal{U}_\epsilon^k / \sim$ will contain only bounded

S.G. Krantz, *Geometric Analysis of the Bergman Kernel and Metric*,
Graduate Texts in Mathematics 268, DOI 10.1007/978-1-4614-7924-6_2,
© Springer Science+Business Media New York 2013

domains that are simply connected. Thus any such domain will, by the Riemann mapping theorem, be conformally equivalent to the disc.

Greene and Krantz [GRK1, GRK2] have refined the theorem to show that, when $n \geq 2$, in fact each of the equivalence classes is closed and nowhere dense.

We now give a brief accounting of some of the differences between $n = 1$ and $n > 1$. A more detailed discussion appears in Greene and Krantz [GRK9], in [GKK] and also in the original work of Poincaré. This subject begins with the following breakthrough of Fefferman [FEF1, Part I]:

Theorem 2.1.3. *Let $\Omega_1, \Omega_2 \subseteq \mathbb{C}^n$ be strictly pseudoconvex domains with C^∞ boundary. If $\phi : \Omega_1 \to \Omega_2$ is biholomorphic, then ϕ extends to a C^∞ diffeomorphism of $\overline{\Omega}_1$ onto $\overline{\Omega}_2$.*

The Fefferman's theorem enables one to see that, if Ω_1 and Ω_2 are biholomorphic under ϕ, then there are certain differential invariants of $\partial\Omega_1, \partial\Omega_2$ that must be preserved under ϕ. More precisely, if k is large, then the kth-order Taylor expansion of the defining function ρ_1 for Ω_1 (resp. of the defining function ρ_2 for Ω_2) has more coefficients than the kth-order Taylor expansion for ϕ (the disparity in the number grows rapidly with k). Since ρ_1 is mapped to ρ_2 under composition with ϕ^{-1}, it follows that some of these coefficients, or combinations thereof, must be invariant under biholomorphic mappings. Tanaka [TAN] and Chern–Moser [CHM] have made these remarks precise and have shown how to calculate these invariants. A more leisurely discussion of these matters appears in Greene and Krantz [GRK9].

Now it is easy to see intuitively that two domains Ω and Ω_0 can be close in the C^k topology, any k, and have entirely different Chern–Moser–Tanaka invariants. This notion is made precise, for instance, in Burns, Shnider, and Wells [BSW], by using a transversality argument. [Note that everything we are saying is vacuous in $\mathbb{C}^1$ because the invariants must live in the complex tangent space to the boundary— which is empty in dimension one. See [KRA1].] It is essentially a foregone conclusion that things will go badly in higher dimensions.

If one seeks positive results in the spirit of the Riemann mapping theorem in dimension $n \geq 2$, then one must find statements of a different nature. Fridman [FRI] has constructed a "universal domain" Ω^* which can be used to exhaust any other. He has obtained a number of variants of this idea, using elementary but clever arguments. Semmes [SEM] has yet another approach to the Riemann mapping theorem that is more in the spirit of the work of Lempert [LEM1]. We next present, mainly for background, a substitute for the Riemann mapping theorem whose statement and proof is more in the spirit of the Fefferman's theorem. We continue to use the notation $\Omega_1 \sim \Omega_2$ to mean that Ω_1 is biholomorphic to Ω_2.

In what follows, we let $\text{Aut}(\Omega)$ denote the group (under composition of mappings) of biholomorphic self-maps of the domain Ω.

Theorem 2.1.4 (Greene–Krantz [GRK2]). *Let $B \subseteq \mathbb{C}^n$ be the unit ball. Let $\rho_0(z) = |z|^2 - 1$ be the usual defining function for B. If $\epsilon > 0$ is sufficiently small, $k = k(n)$ is sufficiently large, and $\Omega \in \mathcal{U}_\epsilon^k(B)$ then either*

$$\Omega \sim B \tag{2.1.4.1}$$

or

$$\Omega \text{ is not biholomorphic to the ball and} \qquad (2.1.4.2)$$

(a) *Aut(Ω) is compact.*
(b) *Aut(Ω) has a fixed point. Moreover,*

> *If $K \subset\subset B, \epsilon > 0$ is sufficiently small (depending on K), and $\Omega \in \mathcal{U}_\epsilon^k(B)$ has the property that its fixed point set lies in K, then there is a biholomorphic mapping $\Phi : \Omega \rightarrow \Phi(\Omega) \equiv \Omega' \subseteq \mathbb{C}^n$ such that $Aut(\Omega')$ is the restriction to Ω' of a subgroup of the group of unitary matrices.*

The collection of domains to which (2.1.4.2) applies is both dense and open.

Theorem 2.1.4 shows, in a weak sense, that domains near the ball that have any automorphisms other than the identity are (biholomorphic to) domains with only Euclidean automorphisms. It should be noted that (2.1.4.2a) is already contained in the theorem of Bun Wong and Rosay [WON, ROS] and that the denseness of the domains to which (2.1.4.2) applies is contained in the work of Burns–Shnider–Wells. The proof of Theorem 2.1.4 involves a detailed analysis of Fefferman's asymptotic expansion for the Bergman kernel and of the $\bar{\partial}$-Neumann problem and would double the length of this book if we were to treat it in any detail.

The purpose of this lengthy introduction has been to establish the importance of Theorem 2.1.4 and to set the stage for what follows. It may be noted that the proof of the result analogous to Fefferman's in $\mathbb{C}^1$, that a biholomorphic mappings of smooth domains extends smoothly to the boundary, was proved in the nineteenth century by Painlevé [PAI]. The result in one complex dimension has been highly refined, beginning with work of Kellogg [KEL] and more recently by Warschawski [WAR1, WAR2, WAR3], Rodin and Warschawski [ROW], and others. This classical work uses harmonic estimation, potential theory, and the Jordan curve theorem, devices which have no direct analogue in higher dimensions. A short, self-contained, proof of the one-variable result—using ideas closely related to those presented here—appears in [BEK].

We conclude this section by presenting a short and elegant proof of the Fefferman's Theorem 2.1.3. The techniques are due to Bell [BEL1] and Bell and Ligocka [BELL]. The proof uses an important and nontrivial fact (known as "Condition R" of Bell and Ligocka) about the $\bar{\partial}$-Neumann problem. We will actually prove Condition R for a strictly pseudoconvex domain in Theorem 4.4.5. (Condition R, and more generally the solution of the $\bar{\partial}$-Neumann problem, *is* considered in detail in the book Krantz [KRA4].)

Let $\Omega \subset\subset \mathbb{C}^n$ be a domain with C^∞ boundary. We define

Condition R (Bell [BEL1]) Define an operator on $L^2(\Omega)$ by

$$Pf(z) = \int_\Omega K(z, \zeta) f(\zeta) dV(\zeta) ,$$

where $K(z, \zeta)$ is the Bergman kernel for Ω. This is the Bergman projection. Then, for each $j > 0$, there is an $m = m(j) > 0$ such that P satisfies the estimates

$$\| Pf \|_{W^j(\Omega)} \le C_j \| f \|_{W^m(\Omega)}$$

for all testing functions f.

Using a little Sobolev theory (see [KRA4]), one can easily see that this formulation of Condition R is equivalent to the condition that the Bergman kernel map $C^\infty(\overline{\Omega})$ to $C^\infty(\overline{\Omega})$.

The deep fact, which we shall prove in Sect. 4.4, is that *Condition R holds on any strictly pseudoconvex domain.*

In fact we can and should say what is the key idea in establishing this last assertion. Let $P : L^2(\Omega) \to L^2(\Omega)$. The operator

$$\overline{\partial} : \bigwedge\nolimits^{0,j} \to \bigwedge\nolimits^{0,j+1} \tag{2.1.5}$$

is the usual exterior differential operator of complex analysis. One may show that the second-order, elliptic partial differential operator $\Box = \overline{\partial} \, \overline{\partial}^* + \overline{\partial}^* \, \overline{\partial}$ has a canonical right inverse called N. This is the $\overline{\partial}$-Neumann operator. These operators are treated in detail in [FOK] and [KRA4]. Then it is a straightforward exercise in Hilbert space theory to verify that

$$P = I - \overline{\partial}^* N \overline{\partial},$$

where P is the Bergman projection. Now the references [FOK] and [KRA4] prove in detail that N maps W^s (the Sobolev space of order s) to W^{s+1} for every s. It follows from this and formula (2.1.5) that P maps W^s to W^{s-1}. That is enough to verify Condition R.

We remark in passing that, in general, it does not matter whether $m(j)$ is much larger than j or whether the $m(j)$ in Condition R depends polynomially on j or exponentially on j. It so happens that, for a strictly pseudoconvex domain, we may take $m(j) = j$. This assertion is proved in [KRA4] in detail. On the other hand, Barrett [BAR1] has shown that, on the Diederich–Fornaess worm domain [DIF1], we must take $m(j) > j$. Later on, Christ [CHR1] showed that Condition R fails altogether on the worm.

Now we build a sequence of lemmas leading to the Fefferman's theorem. First we record some notation.

We let $W^j(\Omega)$ be the usual Sobolev space. See [KRA4] for this idea.

If $\Omega \subset\subset \mathbb{C}^n$ is any smoothly bounded domain and if $j \in \mathbb{N}$, we let

$$WH^j(\Omega) = W^j(\Omega) \cap \{\text{holomorphic functions on } \Omega\},$$

$$WH^\infty(\Omega) = \bigcap_{j=1}^{\infty} WH^j(\Omega) = C^\infty(\overline{\Omega}) \cap \{\text{holomorphic functions on } \Omega\}.$$

Here W^j is the standard Sobolev space on a domain (for which see [KRA4, ADA]). Let $W_0^j(\Omega)$ be the W^j closure of $C_c^\infty(\Omega)$. [Exercise: if j is sufficiently large, then the Sobolev embedding theorem implies trivially that $W_0^j(\Omega)$ is a proper subset of $W^j(\Omega)$.[1]]

Let us say that $u, v \in C^\infty(\overline{\Omega})$ agree up to order k on $\partial\Omega$ if

$$\left(\frac{\partial}{\partial z}\right)^\alpha \left(\frac{\partial}{\partial \overline{z}}\right)^\beta (u - v)\Bigg|_{\partial\Omega} = 0 \quad \forall \alpha, \beta \quad \text{with} \quad |\alpha| + |\beta| \leq k.$$

Lemma 2.1.6. *Let* $\Omega \subset\subset \mathbb{C}^n$ *be smoothly bounded and strictly pseudoconvex. Let* $w \in \Omega$ *be fixed. Let* K *denote the Bergman kernel. There is a constant* $C_w > 0$ *such that*

$$\|K(w, \cdot)\|_{\sup} \leq C_w.$$

Proof. The function $K(z, \cdot)$ is harmonic. Let $\phi : \Omega \to \mathbb{R}$ be a radial, C_c^∞ function centered at w. Assume that $\phi \geq 0$ and $\int \phi(\zeta) dV(\zeta) = 1$. Then the mean value property implies that

$$K(z, w) = \int_\Omega K(z, \zeta)\phi(\zeta) dV(\zeta).$$

But the last expression equals $P\phi(z)$. Therefore

$$\|K(w, \cdot)\|_{\sup} = \sup_{z \in \Omega} |K(w, z)|$$

$$= \sup_{z \in \Omega} |K(z, w)|$$

$$= \sup_{z \in \Omega} |P\phi(z)|.$$

By Sobolev's Theorem, this is

$$\leq C(\Omega) \cdot \|P\phi\|_{WH^{2n+1}}.$$

By Condition R, this is

$$\leq C(\Omega) \cdot \|\phi\|_{W^{m(2n+1)}} \equiv C_w. \qquad \square$$

---

[1] For the readers's convenience, we recall here that the Sobolev embedding theorem says that if a function on $\mathbb{R}^N$ has more than $N/2$ derivatives in L^2, then in fact it has a continuous derivative. See [STE1], for instance, for the details.

Lemma 2.1.7. *Let $u \in C^\infty(\overline{\Omega})$ be arbitrary. Let $s \in \{0, 1, 2, \dots\}$. Then there is a $v \in C^\infty(\overline{\Omega})$ such that $Pv = 0$ and the functions u and v agree to order s on $\partial\Omega$.*

Proof. After a partition of unity, it suffices to prove the assertion in a small neighborhood U of $z_0 \in \partial\Omega$. After a rotation, we may suppose that $\partial\rho/\partial z_1 \neq 0$ on $U \cap \overline{\Omega}$, where ρ is a defining function for Ω. Define the differential operator

$$\nu = \frac{\mathrm{Re}\left\{\sum_{j=1}^{n} \frac{\partial\rho}{\partial z_j} \frac{\partial}{\partial \overline{z}_j}\right\}}{\sum_{j=1}^{n} \left|\frac{\partial\rho}{\partial z_j}\right|^2}.$$

Notice that $\nu\rho = 1$. Now we define v by induction on s.

For the case $s = 0$, let

$$w_1 = \frac{\rho u}{\partial\rho/\partial\zeta_1}.$$

Define

$$v_1 = \frac{\partial}{\partial\zeta_1} w_1$$
$$= u + O(\rho).$$

Then u and v_1 agree to order 0 on $\partial\Omega$. Also

$$Pv_1(z) = \int K(z, \zeta) \frac{\partial}{\partial\zeta_1} w_1(\zeta) dV(\zeta).$$

This equals, by integration by parts,

$$-\int \frac{\partial}{\partial\zeta_1} K(z, \zeta) w_1(\zeta) dV(\zeta).$$

Notice that the integration by parts is valid by Lemma 2.1.6 and because $w_1|_{\partial\Omega} = 0$. Also, the integrand in this last line is zero because $K(z, \cdot)$ is conjugate holomorphic.

Suppose inductively that $w_{s-1} = w_{s-2} + \theta_{s-1}\rho^s$ and $v_{s-1} = (\partial/\partial z_1)(w_{s-1})$ have been constructed. We show that there is a w_s of the form

$$w_s = w_{s-1} + \theta_s \cdot \rho^{s+1}$$

such that $v_s = (\partial/\partial z_1)(w_s)$ agrees to order s with u on $\partial\Omega$. By the inductive hypothesis,

$$v_s = \frac{\partial}{\partial z_1} w_s$$

$$= \frac{\partial w_{s-1}}{\partial z_1} + \frac{\partial}{\partial z_1}[\theta_s \cdot \rho^{s+1}]$$

$$= v_{s-1} + \rho^s\left[(s+1)\theta_s\frac{\partial \rho}{\partial z_1} + \rho \cdot \frac{\partial \theta_s}{\partial z_1}\right]$$

agrees to order $s-1$ with u on $\partial\Omega$ so long as θ_s is smooth. So we need to examine $D(u-v_s)$, where D is an s-order differential operator. But if D involves a tangential derivative D_0, then write $D = D_0 \cdot D_1$. It follows that $D(u-v_s) = D_0(\alpha)$, where α vanishes on $\partial\Omega$ so that $D_0\alpha = 0$ on $\partial\Omega$. So we need only check $D = v^s$.

We have seen that θ_s must be chosen so that

$$v^s(u-v_s) = 0 \qquad \text{on} \quad \partial\Omega.$$

Equivalently,

$$v^s(u-v_{s-1}) - v^s\left(\frac{\partial}{\partial z_1}\right)(\theta_s\rho^{s+1}) = 0 \quad \text{on} \qquad \partial\Omega$$

or

$$v^s(u-v_{s-1}) - \theta_s\left(v^s\frac{\partial}{\partial z_1}\rho^{s+1}\right) = 0 \quad \text{on} \qquad \partial\Omega$$

or

$$v^s(u-v_{s-1}) - \theta_s \cdot (s+1)!\frac{\partial \rho}{\partial z_1} = 0 \quad \text{on} \qquad \partial\Omega.$$

It follows that we must choose

$$\theta_s = \frac{v^s(u-v_{s-1})}{(s+1)!\frac{\partial \rho}{\partial z_1}},$$

which is indeed smooth on U. As in the case $s = 0$, it holds that $Pv_s = 0$. This completes the induction and the proof. $\qquad\square$

Remark 2.1.8. A retrospection of the proof reveals that we have constructed v by subtracting from u a Taylor-type expansion in powers of ρ. $\qquad\square$

Lemma 2.1.9. *For each $s \in \mathbb{N}$ we have $WH^\infty(\Omega) \subseteq P(W_0^s(\Omega))$.*

Proof. Let $u \in C^\infty(\overline{\Omega})$. Choose v according to Lemma 2.1.7. Then $u-v \in W_0^s$ and $Pu = P(u-v)$. Therefore

$$P(W_0^s) \supseteq P(C^\infty(\overline{\Omega})) \supseteq P(WH^\infty(\Omega)) = WH^\infty(\Omega). \qquad\square$$

Henceforth, let Ω_1, Ω_2 be fixed C^∞ strictly pseudoconvex domains in $\mathbb{C}^n$, with K_1, K_2 their Bergman kernels and P_1, P_2 the corresponding Bergman projections. Let $\phi : \Omega_1 \to \Omega_2$ be a biholomorphic mapping, and let $u = \det \mathrm{Jac}_{\mathbb{C}}\phi$. For $j = 1, 2$, let $\delta_j(z) = \delta_{\Omega_j}(z) = \mathrm{dist}(z, {}^c\Omega_j)$.

Lemma 2.1.10. *For any $g \in L^2(\Omega_2)$, we have*

$$P_1(u \cdot (g \circ \phi)) = u \cdot ((P_2(g)) \circ \phi).$$

Proof. Notice that $u \cdot (g \circ \phi) \in L^2(\Omega_1)$ by change of variables. Therefore

$$P_1(u \cdot (g \circ \phi))(z) = \int_{\Omega_1} K_1(z, \zeta) u(\zeta) g(\phi(\zeta)) dV(\zeta)$$

$$= \int_{\Omega_1} u(z) K_2(\phi(z), \phi(\zeta)) \overline{u(\zeta)} u(\zeta) g(\phi(\zeta)) dV(\zeta)$$

by Proposition 1.1.14. Change of variable now yields

$$P_1(u \cdot (g \circ \phi))(z) = u(z) \int_{\Omega_2} K_2(\phi(z), \xi) g(\xi) dV(\xi)$$

$$= u(z) \cdot [(P_2(g)) \circ \phi](z). \qquad \square$$

Lemma 2.1.11. *Let $\psi : \Omega_1 \to \Omega_2$ be a C^j diffeomorphism that satisfies*

$$\left| \frac{\partial^\alpha \psi}{\partial z^\alpha}(z) \right| \leq C \cdot (\delta_1(z))^{-|\alpha|}, \qquad (2.1.11.1)$$

for all multi-indices α with $|\alpha| \leq j \in \mathbb{N}$ and

$$|\nabla \psi^{-1}(w)| \leq C(\delta_2(w))^{-1}. \qquad (2.1.11.2)$$

Suppose also that

$$\delta_2(\psi(z)) \leq C\delta_1(z). \qquad (2.1.11.3)$$

Then there is a number $J = J(j)$ such that whenever $g \in W_0^{j+J}(\Omega_2)$, then $g \circ \psi \in W_0^j(\Omega_1)$.

Proof. The subscript 0 causes no trouble by the definition of W_0^j. Therefore it suffices to prove an estimate of the form

$$\|g \circ \psi\|_{W_0^j} \leq C \|g\|_{W_0^{j+J}}, \quad \text{all } g \in C_c^\infty(\Omega).$$

By the chain rule and Leibniz's rule, if α is a multi-index of modulus not exceeding j, then

$$\left(\frac{\partial}{\partial z}\right)^\alpha (g \circ \psi) = \sum \left[(D^\beta g) \circ \psi\right] \cdot D^{\gamma_1}\psi \cdots D^{\gamma_\ell}\psi,$$

where $|\beta| \le |\alpha|, \sum |\gamma_i| \le |\alpha|$, and the number of terms in the sum depends only on α (a classical formula of Faà de Bruno—see [ROM]—actually gives this sum quite explicitly, but we do not require such detail). Note here that $D^{\gamma_i}\psi$ is used to denote a derivative of *some component* of ψ. By hypothesis, it follows that

$$\left|\left(\frac{\partial}{\partial z}\right)^\alpha (g \circ \psi)\right| \le C \sum |(D^\beta g) \circ \psi| \cdot (\delta_1(z))^{-j}.$$

Therefore

$$\int_{\Omega_1} \left|\left(\frac{\partial}{\partial z}\right)^\alpha (g \circ \psi)\right|^2 dV \le$$

$$\le C \sum \int_{\Omega_1} |(D^\beta g) \circ \psi|^2 (\delta_1(z))^{-2j} dV(z)$$

$$= C \sum \int_{\Omega_2} |D^\beta g(w)|^2 \delta_1 \left(\psi^{-1}(w)\right)^{-2j}$$

$$\times |\det J_{\mathbb{C}}\psi^{-1}|^2 dV(w).$$

But (2.1.11.2) and (2.1.11.3) imply that the last line is majorized by

$$C \sum \int_{\Omega_2} |D^\beta g(w)|^2 \delta_2(w)^{-2j} \delta_2(w)^{-2n} dV(w). \tag{2.1.11.4}$$

Now if J is large enough, depending on the Sobolev embedding Theorem, then

$$|D^\beta g(w)| \le C \|g\|_{W_0^{j+J}} \cdot \delta_2(w)^{2n+2j}.$$

(Remember that g is compactly supported in Ω_2.) Hence, (2.1.11.4) is majorized by $C \|g\|_{W_0^{j+J}}$. $\qquad\square$

Lemma 2.1.12. *For each $j \in \mathbb{N}$, there is an integer J so large that if $g \in W_0^{j+J}(\Omega_2)$, then $g \circ \phi \in W_0^j(\Omega_1)$.*

Proof. The Cauchy estimates give (since ϕ is bounded) that

$$\left|\frac{\partial^\alpha \phi_\ell}{\partial z^\alpha}(z)\right| \le C \cdot (\delta_1(z))^{-|\alpha|}, \qquad \ell = 1, \dots, n \tag{2.1.12.1}$$

and

$$|\nabla \phi^{-1}(w)| \leq C(\delta_2(w))^{-1}, \tag{2.1.12.2}$$

where $\phi = (\phi_1, \ldots, \phi_n)$. We will prove that

$$C \cdot \delta_1(z) \geq \delta_2(\phi(z)). \tag{2.1.12.3}$$

Then Lemma 2.1.9 gives the result.

To prove (2.1.12.3), let ρ be a smooth strictly plurisubharmonic defining function for Ω_1. Then $\rho \circ \phi^{-1}$ is a smooth plurisubharmonic function on Ω_2. Since ρ vanishes on $\partial \Omega_1$ and since ϕ^{-1} is proper, we conclude that $\rho \circ \phi^{-1}$ extends continuously to $\overline{\Omega}_2$. If $P \in \partial \Omega_2$ and ν_P is the unit outward normal to $\partial \Omega_2$ at P, then Hopf's lemma implies that the (lower) one-sided derivative $(\partial / \partial \nu_P)(\rho \circ \phi^{-1})$ satisfies

$$\frac{\partial}{\partial \nu_P}(\rho \circ \phi^{-1}(P)) \geq C.$$

So, for $w = P - \epsilon \nu_P, \epsilon$ small, it holds that

$$-\rho \circ \phi^{-1}(w) \geq C \cdot \delta_2(w).$$

These estimates are uniform in $P \in \partial \Omega_2$. Using the comparability of $|\rho|$ and δ_1 yields

$$C\delta_1(\phi^{-1}(w)) \geq \delta_2(w).$$

Setting $z = \phi^{-1}(w)$ now gives

$$C'\delta_1(z) \geq \delta_2(\phi(z)),$$

which is (2.1.12.3). $\square$

Exercise for the Reader: Let $\Omega \subset\subset \mathbb{C}^n$ be a smoothly bounded domain. Let $j \in \mathbb{N}$. There is an $N = N(j)$ so large that $g \in W_0^N$ implies that g vanishes to order j on $\partial \Omega$.

Lemma 2.1.13. *The function u is in $C^\infty(\overline{\Omega}_1)$.*

Proof. It suffices to show that $u \in W^j(\Omega_1)$, every j. So fix j. Let $m = m(j)$ as in Condition R. According to (2.1.12.1), $|u(z)| \leq C\delta_1(z)^{-2n}$. Then, by Lemma 2.1.12 and the Exercise for the Reader following it, there is a J so large that $g \in W_0^{m+J}(\Omega_2)$ implies $u \cdot (g \circ \phi) \in W^m(\Omega_1)$. Choose, by Lemma 2.1.7, a $g \in W_0^{m+J}(\Omega_2)$ such that $P_2 g \equiv 1$. Then Lemma 2.1.10 yields

$$P_1(u \cdot (g \circ \phi)) = u.$$

By Condition R, it follows that $u \in W^j(\Omega_1)$. $\qquad\qquad\square$

Lemma 2.1.14. *The function u is bounded from 0 on $\overline{\Omega}_1$.*

Proof. By symmetry, we may apply Lemma 2.1.13 to ϕ^{-1} and $\det J_\mathbb{C}(\phi^{-1}) = 1/u$. We conclude that $1/u \in C^\infty(\overline{\Omega}_2)$. Thus u is nonvanishing on $\overline{\Omega}$. $\qquad\square$

Proof of the Fefferman's Theorem (Theorem 2.1.3): Use the notation of the proof of Lemma 2.1.12. Choose $g_1, \ldots, g_n \in W_0^{m+J}(\Omega_2)$ such that $P_2 g_i(w) = w_i$ (here w_i is the ith coordinate function). Then Lemma 2.1.10 yields that $u \cdot \phi_i \in W^j(\Omega_1), i = 1, \ldots, n$. By Lemma 2.1.12, $\phi_i \in W^j(\Omega_1), i = 1, \ldots, n$. By symmetry, $\phi^{-1} \in W^j(\Omega_2)$. Since j is arbitrary, the Sobolev embedding theorem finishes the proof. $\qquad\qquad\square$

It is important to understand the central role of Condition R in this proof. With some emendations, the proof we have presented shows that, if $\Omega_1, \Omega_2 \subset\subset \mathbb{C}^n$ are smoothly bounded, pseudoconvex, and both satisfy Condition R, then a biholomorphic mapping from Ω_1 to Ω_2 extends smoothly to the boundary (in fact Bell [BEL1] has shown that it suffices for just one of the domains to satisfy Condition R). Condition R is known to hold on domains that have real analytic boundaries (see Diederich and Fornæss [DIF2]), and more generally on domains of finite type (see [CAT1, CAT2]). There are a number of interesting examples of non-pseudoconvex domains on which Condition R fails (see Barrett [BAR1] and Kiselman [KIS]). It had been conjectured that Condition R holds on all smoothly bounded pseudoconvex domains. But Christ [CHR1] showed that in fact Condition R fails on the Diederich–Fornæss worm domain (which is smoothly bounded and pseudoconvex).

Lempert [LEM1] has derived a sharp boundary regularity result for biholomorphic mappings of strictly pseudoconvex domains with C^k boundary. Pinchuk and Tsyganov [PIT] have an analogous result. The correct conclusion turns out to be that there is a loss of smoothness in some directions. So the sharp regularity result is formulated in terms of nonisotropic spaces. It is too technical to describe here.

2.2 Boundary Behavior of the Bergman Metric

The Bergman metric on the disc D is given by

$$g_{jk} = \frac{\partial}{\partial z} \frac{\partial}{\partial \overline{z}} \log K(z, z) = \frac{\partial}{\partial z} \frac{\partial}{\partial \overline{z}} \left[-\log \pi - 2\log(1 - |z|^2) \right] = \frac{2}{(1 - |z|^2)^2}.$$

Thus the length of a curve $\gamma : [0, 1] \to D$ is given by

$$\ell_B(\gamma) = \int_0^1 \frac{2\|\gamma'(t)\|}{(1 - |\gamma(t)|^2)} \, dt \, .$$

It is natural to wonder what one can say about the Bergman metric on a more general class of domains. Let $\Omega \subseteq \mathbb{C}$ be a bounded domain with C^4 boundary, and suppose for the moment that Ω is simply connected. Then the Riemann mapping theorem tells us that there is a conformal mapping $\Phi : \Omega \to D$. Of course then we know that

$$K_\Omega(z, z) = |\Phi'(z)|^2 K_D(\Phi(z), \Phi(z)) = \frac{|\Phi'(z)|^2}{\pi \cdot (1 - |\Phi(z)|^2)^2} \, .$$

Then the Bergman metric on Ω is given by

$$g_{jk} = \frac{\partial}{\partial z} \frac{\partial}{\partial \overline{z}} \log \left[\frac{|\Phi'(z)|^2}{\pi \cdot (1 - |\Phi(z)|^2)^2} \right]$$

$$= \frac{\partial}{\partial z} \frac{\partial}{\partial \overline{z}} \left[\log |\Phi'(z)|^2 - \log \pi - 2 \log(1 - |\Phi(z)|^2) \right] \, .$$

Of course $|\Phi'(z)|$ is bounded and bounded from 0 (see [BEK]). Also the derivatives of Φ, up to order 3, are bounded. So the second derivative of $\log |\Phi'(z)|^2$ is a bounded term. The second derivative of $\log \pi$ is of course 0.

The second derivative of the remaining (and most interesting) term may be calculated to be

$$\frac{\partial^2}{\partial z \partial \overline{z}} \log K_\Omega(z, z) = \frac{\partial^2}{\partial \overline{z} \partial z} \left[-2 \log(1 - |\Phi(z)|^2) \right]$$

$$= \frac{\partial}{\partial \overline{z}} \left(\frac{-2}{1 - |\Phi(z)|^2} \cdot \left[-\Phi'(z) \overline{\Phi}(z) \right] \right)$$

$$= \frac{2}{(1 - |\Phi(z)|^2)^2} \cdot \left[-\overline{\Phi}'(z) \Phi(z) \right] \cdot \left[-\Phi'(z) \overline{\Phi}(z) \right]$$

$$- \frac{2}{1 - |\Phi(z)|^2} \cdot \left[-\Phi'(z) \overline{\Phi}'(z) \right] \, .$$

This in turn, after some simplification, equals

$$\frac{2|\Phi'(z)|^2}{(1 - |\Phi(z)|^2)^2} \, . \tag{2.2.1}$$

As previously noted, the numerator is bounded and bounded from 0. Hopf's lemma (see [KRA1]) tells us that $(1 - |\Phi(z)|^2) \approx \mathrm{dist}_{\partial\Omega}(z)$. So that the Bergman metric on Ω blows up like the reciprocal of the square of the distance to the boundary—just as on the disc.

In the case that Ω has C^2 boundary and is finitely connected—not necessarily simply connected—then one may use the Ahlfors map (see [KRA5]) instead of the Riemann mapping and obtain a result similar to that in (2.2.1). We omit the details.

Bremermann [BRE] showed that any domain in $\mathbb{C}^n$ with complete Bergman metric is a domain of holomorphy. This is in fact not difficult, as one can use the hypothesis of the completeness of the metric to confirm the Kontinuitätssatz (see [KRA1]), hence derive pseudoconvexity. Bremerman also gave an example to demonstrate that the converse is not true. However, Ohsawa [OHS] has shown that any pseudoconvex domain with C^1 boundary has complete Bergman metric. This result is important both conceptually and practically. There is no analogous result for either the Carathéodory or Kobayashi metrics.

In the paper [KOB1, Theorem 9.2], Kobayashi shows that any bounded analytic polyhedron has complete Bergman metric.

2.3   The Biholomorphic Inequivalence of the Ball and the Polydisc

In this section we give a Bergman-geometric proof of the following classical result of Poincaré (Poincaré's original proof was more group theoretic).

Theorem 2.3.1. *There is no biholomorphic map of the bidisc $D^2(0, 1)$ to the ball $B(0, 1) \subseteq \mathbb{C}^2$.*

Proof. Suppose, seeking a contradiction, that there is such a map. Since Möbius transformations act transitively on the disc, pairs of them act transitively on the bidisc. Therefore we may compose ϕ with a self-map of the bidisc and assume that ϕ maps 0 to 0.

If $Y \in \partial B$, then the disc $d_Y = \{z \in B : z = \zeta Y, \zeta \in \mathbb{C}, |\zeta| < 1\}$ is a totally geodesic submanifold of B (informally, this means that if P, Q are points of d_Y, then the geodesic connecting them in the Riemannian manifold d_Y is the same as the geodesic connecting them in the Riemannian manifold B—see Kobayashi and Nomizu [KON]).

By our discussion in the calculation of the Poincaré metric, we may conclude that the geodesics, or paths of least length, emanating from the origin in the ball are the rays $\tau_Y : t \mapsto tY$. (This assertion may also be derived from symmetry considerations.)

Likewise, if $\alpha, \beta \in \mathbb{C}, |\alpha| = 1, |\beta| = 1$, then the disc $e_0 = \{(\zeta\alpha, \zeta\beta) : \zeta \in D\} \subseteq D^2(0, 1)$ is a totally geodesic submanifold of $D^2(0, 1)$. Again we may apply our discussion of the Poincaré metric on the disc to conclude that the geodesic curve emanating from the origin in the bidisc in the direction $X = (\alpha, \beta)$ is $\psi_{\alpha\beta} : t \mapsto tX$. A similar argument shows that the curve $t \mapsto (t, 0)$ is a geodesic in the bidisc.

Now if $t \mapsto tX$ is one of the above-mentioned geodesics on the bidisc, then it will be mapped under ϕ to a geodesic $t \mapsto tY$ in the ball. If $0 < t_1 < t_2 < 1$, then the points $t_1 X, t_2 X \in D^2$ will be mapped to points $t_1' Y, t_2' Y \in B$ and it must be that

$0 < t_1' < t_2' < 1$ since ϕ is an isometry hence must map the point $t_2 X$ to a point
further from the origin than it maps $t_1 X$ (because $t_2 X$ is further from the origin than
$t_1 X$). It follows that the limit

$$\lim_{t \to 1^-} \phi(t X)$$

exists for every choice of X and the limit lies in ∂B. After composing ϕ with a
rotation, we may suppose that $\{\phi(t(1,0))\}$ terminates at $(1,0)$.

Now consider the function $f(z_1, z_2) = (z_1 + 1)/2$ on B. This function has the
property that $f(1,0) = 1$, f is holomorphic on a neighborhood of $\overline{B}$, and $|f(z)| <$
1 for $z \in \overline{B} \sum \{(1,0)\}$. For $0 < r < 1$ we invoke the mean-value property for a
harmonic function to write

$$\frac{1}{2\pi} \int_0^{2\pi} f \circ \phi(r, re^{i\theta}) d\theta = f \circ \phi(r, 0). \qquad (2.3.1.1)$$

As $r \to 1^-$ the right-hand side tends to $\lim_{t \to 1^-} f(t, 0) = 1$. However, each of the
paths $r \to (r, re^{i\theta})$ is a geodesic in the bidisc, as discussed above, and for different
$\theta \in [0, 2\pi)$ they are distinct. Thus the curves $r \to \phi(r, re^{i\theta})$ have distinct limits
in ∂B, and these limits will be different from the point $(1,0) \in \partial B$. In particular,
$\lim_{r \to 1^-} f \circ \phi(r, re^{i\theta})$ exists for each $\theta \in [0, 2\pi)$ and assumes a value of modulus
strictly less than 1.

By the Lebesgue dominated convergence theorem, we may pass to the limit as
$r \to 1^-$ in the left side of (2.3.1.1) to obtain a limit that must be strictly less than
one in absolute value. That is the required contradiction. $\qquad\qquad\qquad\qquad\square$

Exercises

1. Let us construct an invariant metric on the disc D in the complex plane by hand.
   Take it that the length of the vector $\langle 1, 0 \rangle$ at the base point 0 is 1. Now use the
   conformal invariance of the metric to calculate the length of any other vector at
   any other base point. Of course the metric that you obtain in this way should be
   (a constant multiple of) the Poincaré metric.
2. Imitate Exercise 1 for the unit ball in $\mathbb{C}^n$.
3. Consider the domain $\Omega = \{\zeta \in \mathbb{C} : \text{Re}\,\zeta < 0 \text{ or } \text{Im}\,\zeta < 0\}$. This domain is
   conformally equivalent to the upper half plane, which is in turn conformally
   equivalent to the unit disc. Therefore one can calculate the Bergman kernel for
   Ω using its mapping invariance property. What can you say about the boundary
   behavior of this kernel at the origin? What can you say about the boundary
   behavior of the Bergman metric at the origin?
4. Calculate the Bergman kernel and metric for the punctured disc $D' \equiv \{\zeta \in \mathbb{C} :$
   $0 < |\zeta| < 1\}$. How do they differ from the Bergman kernel and metric for the
   usual disc D?

Calculate the Bergman kernel and metric for the punctured ball $B' \equiv \{z \in \mathbb{C}^n : 0 < |z| < 1\}$. How do they differ from the Bergman kernel and metric for the usual ball B?

5. Use the results from the text about the Bergman kernel for the annulus to calculate the Bergman metric on the annulus.

6. Let $\Omega = B(0,1) \setminus B(1,1/2)$, where $B(P,r)$ denotes the open ball with center P and radius r. Estimate the distance from the origin to the point $(1/2,0)$ in the Bergman metric of Ω.

7. The infinitesimal Kobayashi–Royden metric on a domain $\Omega \subseteq \mathbb{C}^n$ is defined by $F_K : \Omega \times \mathbb{C}^n \to \mathbb{R}$, where

$$F_K(z,\xi) \equiv \inf\{\alpha : \alpha > 0 \text{ and } \exists f \in \Omega(B)$$

$$\text{with } f(0) = z, \left(f'(0)\right)(e_1) = \xi/\alpha\}$$

$$= \inf\left\{ \frac{|\xi|}{|(f'(0))(e_1)|} : f \in \Omega(B), (f'(0))(e_1) \text{ is a}\right.$$

$$\text{constant multiple of } \xi\}$$

$$= \frac{|\xi|}{\sup\{|(f'(0))(e_1)| : f \in \Omega(B), (f'(0))(e_1) \text{ is a constant multiple of } \xi\}} \, .$$

Use the Schwarz lemma to calculate the Kobayashi–Royden metric on the disc. How does it compare with the Poincaré metric? Now calculate the Kobayashi–Royden metric on the ball.

8. The infinitesimal Carathéodory metric on a domain $\Omega \subseteq \mathbb{C}^n$ is defined by $F_C : \Omega \times \mathbb{C}^n \to \mathbb{R}$, where

$$F_C(z,\xi) = \sup_{\substack{f \in B(\Omega) \\ f(z)=0}} |f_*(z)\xi| \equiv \sup_{\substack{f \in B(\Omega) \\ f(z)=0}} \left| \sum_{j=1}^{n} \frac{\partial f}{\partial z_j}(z) \cdot \xi_j \right| \, .$$

Use the Schwarz lemma to calculate the Carathéodory metric on the disc. How does it compare with the Poincaré metric? Now calculate the Carathéodory metric on the ball.

9. Refer to Exercises 7 and 8 for terminology. Prove that the Carathéodory metric is always less than or equal to the Kobayashi metric.

10. Refer to Exercises 7 and 8 for terminology. Prove that, if $\Phi : \Omega_1 \to \Omega_2$ is a holomorphic mapping of complex domains, then its distance is non-increasing in the Carathéodory metric. Prove an analogous statement for the Kobayashi–Royden metric.

Chapter 3
Further Geometric and Analytic Theory

3.1 Bergman Representative Coordinates

The theory of the Bergman kernel gives rise to many important geometric invariants. Among these are the not-very-well-known Bergman representative coordinates. This is a local coordinate system in which a biholomorphic mapping is realized as a linear mapping. Such a result, while initially quite startling, is in fact completely analogous to the result in the Riemannian geometry regarding geodesic normal coordinates. But geodesic normal coordinates are almost never holomorphic—unless the Kähler metric is flat. By contrast, the Bergman representative coordinates are always holomorphic. Our presentation in this section is inspired by [GKK].

The Bergman representative coordinates are of considerable intrinsic and theoretical interest. But they are also useful in understanding the boundary behavior of biholomorphic mappings. And Greene–Krantz made good use of them in proving their noted semicontinuity theorem (see Sect. 5.2).

The Bergman representative coordinates are also essential in the proof of Lu Qi-Keng's theorem on bounded domains with the Bergman metrics of constant holomorphic sectional curvature. This result will be stated and proved in the present section.

Now let Ω be a bounded domain in $\mathbb{C}^n$ and let q be a point of Ω. The Bergman kernel $K_\Omega(q,q)$ on the diagonal is of course real and positive so that there is a neighborhood U of q such that, for all z, w in U, $K_\Omega(z,w) \neq 0$. Then for all z, w in U, we define

$$b_j(z) = b_{j,q}(z) = \frac{\partial}{\partial \overline{w}_j} \log \frac{K(z,w)}{K(w,w)}\bigg|_{w=q}.$$

Note that these coordinates are well defined, independent of the choice of logarithmic branch. Each representative coordinate $b_j(z)$ is clearly a holomorphic function of z.

S.G. Krantz, *Geometric Analysis of the Bergman Kernel and Metric*,
Graduate Texts in Mathematics 268, DOI 10.1007/978-1-4614-7924-6_3,
© Springer Science+Business Media New York 2013

The mapping

$$z \longmapsto (b_1(z), \ldots, b_n(z)) \in \mathbb{C}^n$$

is defined and holomorphic in a neighborhood of the point q (a neighborhood on which the kernel does not vanish). Note also that $(b_1(q), \ldots, b_n(q)) = (0, \ldots, 0)$.

We are hoping to use these functions as holomorphic local coordinates in a neighborhood of q. By the holomorphic inverse function theorem, these functions give local coordinates if the holomorphic Jacobian

$$\det \left(\frac{\partial b_j}{\partial z_k} \right)_{j,k=1,\ldots,n}$$

is nonzero at q.

But in fact the nonvanishing of this determinant at q is an immediate consequence of the fact that the Bergman metric is positive definite. To see this relationship, observe that

$$\left. \frac{\partial b_j}{\partial z_k} \right|_{z=q} = \left. \frac{\partial}{\partial z_k} \left(\frac{\partial}{\partial \overline{w}_j} \log K(z, w) \right) \right|_{z=w=q}$$

$$= \left. \frac{\partial^2}{\partial z_k \partial \overline{z}_j} \log K(z, z) \right|_{z=q}.$$

This last term is of course the Hermitian inner product $\left\langle \frac{\partial}{\partial z_k}, \frac{\partial}{\partial z_j} \right\rangle \Big|_q$ with respect to the Bergman metric. Thus the expression

$$\det \left(\frac{\partial b_j}{\partial z_k} \right) \Big|_q$$

is the determinant of the inner product matrix of a positive definite Hermitian inner product. Hence this determinant is positive.

The utility of the new coordinates in studying biholomorphic mappings comes from the following:

Lemma 3.1.1. *Let Ω_1 and Ω_2 be two bounded domains in $\mathbb{C}^n$ with $q^1 \in \Omega_1$ and $q^2 \in \Omega_2$ fixed points. Denote by $b_1^1, \ldots, b_n^1$ the Bergman coordinates as defined near q^1 in Ω_1 and $b_1^2, \ldots, b_n^2$ the Bergman coordinates defined in the same way near q^2 in Ω_2. Suppose that there is a biholomorphic mapping $F : \Omega_1 \to \Omega_2$ with $F(q^1) = q^2$. Then the function defined near $0 \in \mathbb{C}^n$ by*

$$b^1 \text{ coordinate } \alpha \longmapsto b^2 \text{ coordinate of the image of the } \alpha\text{-point under } F$$

is a $\mathbb{C}$-linear transformation.

In short, we say that the biholomorphic mapping F is linear when expressed in the Bergman representative coordinates b^j. In point of fact, the linear mapping induced by the introduction of the Bergman representative coordinates is nothing other than the complex Jacobian of the mapping F at the point q^1.

Proof of the Lemma. To avoid confusion, we write $(z_1, \ldots, z_n)$ and $(w_1, \ldots, w_n)$ for the $\mathbb{C}^n$-coordinates in Ω_1 and $(Z_1, \ldots, Z_n)$ and $(W_1, \ldots, W_n)$ for the $\mathbb{C}^n$-coordinates in Ω_2. Now observe that, for each $j = 1, \ldots, n$,

$$\frac{\partial}{\partial \overline{w}_j} \log \frac{K_{\Omega_2}(F(z), F(w))}{K_{\Omega_2}(F(w), F(w))} = \frac{\partial}{\partial \overline{w}_j} \log \frac{K_{\Omega_1}(z, w)}{K_{\Omega_1}(w, w)}.$$

The reason for this identity is

$$\frac{K_{\Omega_2}(F(z), F(w))}{K_{\Omega_2}(F(w), F(w))} = \frac{K_{\Omega_1}(z, w)}{K_{\Omega_1}(w, w)} \times \text{(a holomorphic function of } z)$$

$$\times \text{(a holomorphic function of } w).$$

This last follows from the transformation law. Thus we obtain (from the complex chain rule) that

$$b_j^1(z) \overset{\text{def}}{=} \frac{\partial}{\partial \overline{w}_j} \log \frac{K_{\Omega_1}(z, w)}{K_{\Omega_1}(w, w)} \bigg|_{w=q_1}$$

$$= \frac{\partial}{\partial \overline{w}_j} \left[\log K_{\Omega_2}(F(z), F(w)) - \log K_{\Omega_2}(F(w), F(w)) \right] \bigg|_{w=q_1}$$

$$= \sum_k \left[\frac{\partial \overline{F}^k}{\partial \overline{w}_j} \cdot \frac{\partial}{\partial \overline{W}_k} \log \frac{K_{\Omega_2}(F(z), W)}{K_{\Omega_2}(W, W)} \right] \bigg|_{W=F(q_1)},$$

where F^k is the kth coordinate of $F(w_1, \ldots, w_k)$. But this last expression is exactly

$$\sum_k \frac{\partial \overline{F}^k}{\partial \overline{w}_j} \bigg|_{w=q_1} \cdot b_k^2(F(z)).$$

Hence

$$b_j^1(z) = \sum_k \frac{\partial \overline{F}^k}{\partial \overline{w}_j} \bigg|_{w=q_1} \cdot b_k^2(F(z)).$$

Since the Jacobian matrix $(\partial F^k / \partial w_j)$ of F is invertible at q, it follows that the $b_k^2(F)$ are linear functions of the b_j^1 coordinates. $\qquad \square$

The reader may wish to try calculating the Bergman representative coordinates on the unit ball. You may then use these to rediscover the Möbius transformations on the ball.

Note that the whole concept of representative coordinates extends essentially automatically to complex manifolds for which the Bergman metric construction for $(n, 0)$ forms already discussed above yields a positive definite metric. The construction can still be done locally, using general local holomorphic coordinates, and it remains true that the Bergman coordinates linearize holomorphic mappings.

3.2   The Berezin Transform

3.2.1   Preliminary Remarks

Let $\Omega \subseteq \mathbb{C}^n$ be a bounded domain (i.e., a connected open set) with C^2 boundary. Following the general rubric of "Hilbert space with reproducing kernel" laid down by Nachman Aronszajn [ARO], both the Bergman space $A^2(\Omega)$ and the Hardy space $H^2(\Omega)$ have reproducing kernels.

The Bergman kernel (for A^2) and the Szegő kernel (for H^2) both have the advantage of being canonical. But neither is positive, and this makes them tricky to handle. The Bergman kernel can be treated with the theory of the Hilbert integral (see [PHS]), and the Szegő kernel can often be handled with a suitable theory of singular integrals (see [KRA2]).

It is a classical construction of Hua (see [HUA]) that one can use the Szegő kernel to produce another reproducing kernel $\mathcal{P}(z, \zeta)$ which also reproduces H^2 but which is positive. In this sense it is more like the Poisson kernel of harmonic function theory. In point of fact, this so-called Poisson–Szegő kernel coincides with the Poisson kernel when the domain is the disc D in the complex plane $\mathbb{C}$. Furthermore, the Poisson–Szegő kernel solves the Dirichlet problem for the invariant Laplacian (i.e., the Laplace–Beltrami operator for the Bergman metric) on the ball in $\mathbb{C}^n$. Unfortunately a similar statement about the Poisson–Szegő kernel cannot be made on any other domain (although we shall explore substitute results on strictly pseudoconvex domains later in this book). See [GRA1, GRA2] for the full story of these matters.

We want to develop these ideas with the Szegő kernel replaced by the Bergman kernel. This notion was developed independently by Berezin [BERE] in the context of quantization of Kähler manifolds. Indeed, one assigns to a bounded function on the manifold the corresponding Toeplitz operator. This process of assigning a linear operator to a function is called *quantization*. A nice exposition of the ideas appears in [PEE]. Further basic properties may be found in [ZHU].

Approaches to the Berezin transform are often operator-theoretic (see [ENG1, ENG2]), or sometimes geometric [PEE]. Our point of view here will be more function-theoretic. We shall repeat (in perhaps new language) some results that are known in other contexts. And we shall also enunciate and prove new results. We hope that the mix serves to be both informative and useful.

3.2.2 Introduction to the Poisson–Bergman Kernel

In the seminal work [HUA], Hua proposed a program for producing a positive kernel from a canonical kernel. He defined

$$\mathcal{P}(z, \zeta) = \frac{|S(z, \zeta)|^2}{S(z, z)},$$

where S is the standard Szegő kernel on a given bounded domain Ω. Now we have

Proposition 3.2.1. *Let Ω be a bounded domain with C^2 boundary and S its Szegő kernel. With $\mathcal{P}(z, \zeta)$ as defined above, and with $f \in C(\overline{\Omega})$ holomorphic on Ω, we have*

$$f(z) = \int_{\partial\Omega} \mathcal{P}(z, \zeta) f(\zeta) \, d\sigma(\zeta)$$

for all $z \in \Omega$.

Proof: See Proposition 1.2.9. $\square$

The integral with kernel $\mathcal{P}(z, \zeta)$ is called the *Berezin transform*.

It is natural to ask whether the result of the proposition extends to all functions $f \in H^2(\Omega)$. For this, it would suffice to show that $C(\overline{\Omega}) \cap \mathcal{O}(\Omega)$ is dense in $H^2(\Omega)$. In fact this density result is known to be true, because of the regularity theory for the $\overline{\partial}_b$ operator, when Ω is either strictly pseudoconvex or of finite type in the sense of Catlin–D'Angelo–Kohn. One can reason as follows (and we thank Harold Boas for this argument): Let $f \in H^2(\Omega)$. Then certainly $f \in L^2(\partial\Omega)$ and, just by measure theory, one can approximate f in L^2 norm by a function $\varphi \in C^\infty(\partial\Omega)$. Let $\Phi = P_S\varphi$, the Szegő projection of φ. Then, since P_S is a continuous operator on $L^2(\partial\Omega)$, the function Φ is an $L^2(\partial\Omega)$ approximant of f. But it is also the case, by regularity theory of the $\overline{\partial}_b$ operator, that $\Phi = P_S\varphi$ is in $C^\infty(\overline{\Omega})$. That proves the needed approximation result. Of course a similar argument would apply on any domain on which the Szegő projection maps smooth functions to smooth functions. See [STE2] for some observations about this matter.

Now Hua did not consider his construction for the Bergman kernel, but in fact it is just as valid in that context. We may define

$$\mathcal{B}(z, \zeta) = \frac{|K(z, \zeta)|^2}{K(z, z)}.$$

We call this the *Poisson–Bergman kernel*. It is also sometimes called the *Berezin kernel*. Then we have

Proposition 3.2.2. *Let Ω be a bounded domain and K its Bergman kernel. With $\mathcal{B}(z, \zeta)$ as defined above, and with $f \in C(\overline{\Omega})$ holomorphic on Ω, we have*

$$f(z) = \int_{\partial\Omega} \mathcal{B}(z, \zeta) f(\zeta) \, dV(\zeta)$$

for all $z \in \Omega$.

The proof is just the same as that for Proposition 1.2.9, and we omit the details. One of the purposes of the present discussion is to study properties of the Poisson–Bergman kernel $\mathcal{B}$.

Of course the Poisson–Bergman kernel is real, so it will also reproduce the real parts of holomorphic functions. Thus in one complex variable, the integral reproduces harmonic functions. In several complex variables, it reproduces pluriharmonic functions.

Again, it is natural to ask under what circumstances Proposition 3.2.2 holds for all functions in the Bergman space $A^2(\Omega)$. The question is virtually equivalent to asking when the elements that are continuous on $\overline{\Omega}$ are dense in A^2. Catlin [CAT3] has given an affirmative answer to this query on any smoothly bounded pseudoconvex domain.

One of the features that makes the Bergman kernel both important and useful is its invariance under biholomorphic mappings. This fact is useful in conformal mapping theory, and it also gives rise to the Bergman metric. The fundamental result is this:

Proposition 3.2.3. *Let Ω_1, Ω_2 be domains in $\mathbb{C}^n$. Let $f : \Omega_1 \to \Omega_2$ be biholomorphic. Then*

$$\det J_{\mathbb{C}} f(z) K_{\Omega_2}(f(z), f(\zeta)) \det \overline{J_{\mathbb{C}} f(\zeta)} = K_{\Omega_1}(z, \zeta).$$

Here $J_{\mathbb{C}} f$ is the complex Jacobian matrix of the mapping f. Refer to [KRA1, KRA4] for more on this topic.

It is useful to know that the Poisson–Bergman kernel satisfies a similar transformation law:

Proposition 3.2.4. *Let Ω_1, Ω_2 be domains in $\mathbb{C}^n$. Let $f : \Omega_1 \to \Omega_2$ be biholomorphic. Then*

$$\mathcal{B}_{\Omega_2}(f(z), f(\zeta)) |\det J_{\mathbb{C}} f(\zeta)|^2 = \mathcal{B}_{\Omega_1}(z, \zeta).$$

Proof: Of course we use the result of Proposition 3.2.3. Now

$$\mathcal{B}_{\Omega_1}(z, \zeta) = \frac{|K_{\Omega_1}(z, \zeta)|^2}{K_{\Omega_1}(z, z)}$$

$$= \frac{|\det J_{\mathbb{C}} f(z) \cdot K_{\Omega_2}(f(z), f(\zeta)) \cdot \overline{\det J_{\mathbb{C}} f(\zeta)}|^2}{\det J_{\mathbb{C}} f(z) \cdot K_{\Omega_2}(f(z), f(z)) \cdot \overline{\det J_{\mathbb{C}} f(z)}}$$

$$= \frac{|\det J_{\mathbb{C}} f(\zeta)|^2 \cdot |K_{\Omega_2}(f(z), f(\zeta))|^2}{K_{\Omega_2}(f(z), f(z))}$$

$$= |\det J_{\mathbb{C}} f(\zeta)|^2 \cdot \mathcal{B}_{\Omega_2}(f(z), f(\zeta)). \qquad \Box$$

We conclude this section with an interesting observation about the Berezin transform—see [ZHU].

Proposition 3.2.5. *The operator*

$$\mathcal{B} f(z) = \int_B \mathcal{B}(z, \zeta) f(\zeta) \, dV(\zeta),$$

acting on $L^1(B)$*, is univalent.*

Proof: In fact it is useful to take advantage of the symmetry of the ball. We can rewrite the Poisson–Bergman integral as

$$\int_B f \circ \Phi_z(\zeta) \, dV(\zeta),$$

where Φ_z is a suitable automorphism of the ball. Then it is clear that this integral can be identically zero in z only if $f \equiv 0$. That completes the proof. $\qquad \Box$

Another, slightly more abstract, way to look at this matter is as follows (we thank Richard Rochberg for this idea, and see also [ENG1]). Let f be any L^1 function on B. For $w \in B$ define

$$g_w(\zeta) = \frac{1}{(1 - \overline{w} \cdot \zeta)^{n+1}}.$$

If f is bounded on the ball, let

$$T_f : g \mapsto P_B(fg).$$

We may write the Berezin transform now as

$$\Lambda f(w, z) = \frac{\langle T_f g_z, g_w \rangle}{\langle g_w, g_w \rangle}.$$

This function is holomorphic in z and conjugate holomorphic in w. The statement that the Berezin transform $\mathcal{B}f(z) \equiv 0$ is the same as $\Lambda f(z, z) = 0$. But it is a standard fact (see [KRA1]) that we may then conclude that $\Lambda f(w, z) \equiv 0$. Then $T_f g_z \equiv 0$ and so $f \equiv 0$. So the Berezin transform is univalent. $\square$

3.2.3 Boundary Behavior

It is natural to want information about the boundary limits of potentials of the form $\mathcal{B}f$ for $f \in L^2(\Omega)$. We begin with a simple lemma:

Lemma 3.2.6. *Let Ω be a bounded domain and $\mathcal{B}$ its Poisson–Bergman kernel. If $z \in \Omega$ is fixed, then*

$$\int_\Omega \mathcal{B}(z, \zeta) \, dV(\zeta) = 1 \, .$$

Proof: Certainly the function $f(\zeta) \equiv 1$ is an element of the Bergman space on Ω. As a result,

$$1 = f(z) = \int_\Omega \mathcal{B}(z, \zeta) f(\zeta) \, dV(\zeta) = \int_\Omega \mathcal{B}(z, \zeta) \, dV(\zeta)$$

for any $z \in \Omega$. $\square$

Our first result is as follows:

Proposition 3.2.7. *Let Ω be the ball B in $\mathbb{C}^n$. Then the mapping*

$$f \mapsto \int_\Omega \mathcal{B}(z, \zeta) f(\zeta) \, dV(\zeta)$$

sends $L^p(\Omega)$ to $L^p(\Omega)$, $1 \leq p \leq \infty$.

Proof: We know from the lemma that

$$\|\mathcal{B}(z, \cdot)\|_{L^1(\Omega)} = 1$$

for each fixed z. An even easier estimate shows that

$$\|\mathcal{B}(\cdot, \zeta)\|_{L^1(\Omega)} \leq 1$$

for each fixed ζ. Now Schur's lemma, or the generalized Minkowski inequality, gives the desired conclusion. [Note here that we made decisive use of the fact that $\mathcal{B}(z, \zeta) \geq 0$.] $\square$

Proposition 3.2.8. *Let $\Omega \subseteq \mathbb{C}^n$ be the unit ball B. Let $f \in C(\overline{\Omega})$. Let $F = \mathcal{B}f$. Then F extends to a function that is continuous on $\overline{\Omega}$. Moreover, if $P \in \partial\Omega$, then*

$$\lim_{\Omega \ni z \to P} F(z) = f(P).$$

Proof: Let $\epsilon > 0$. Choose $\delta > 0$ such that if $z, w \in \overline{\Omega}$ and $|z - w| < \delta$, then $|f(z) - f(w)| < \epsilon$. Let $M = \sup_{\zeta \in \overline{\Omega}} |f(\zeta)|$. Now, for $z \in \Omega$, $P \in \partial\Omega$, and $|z - P| < \epsilon$, we have that

$$|F(z) - f(P)| = \left| \int_\Omega \mathcal{B}(z, \zeta) f(\zeta) \, dV(\zeta) - f(P) \right|$$

$$= \left| \int_\Omega \mathcal{B}(z, \zeta) f(\zeta) \, dV(\zeta) - \int_\Omega \mathcal{B}(z, \zeta) f(P) \, dV(\zeta) \right|$$

$$\leq \int_{\substack{\zeta \in \Omega \\ |\zeta - P| < \delta}} \mathcal{B}(z, \zeta) |f(\zeta) - f(P)| \, dV(\zeta)$$

$$+ \int_{\substack{\zeta \in \Omega \\ |\zeta - P| \geq \delta}} \mathcal{B}(z, \zeta) |f(\zeta) - f(P)| \, dV(\zeta)$$

$$\leq \int_{\substack{\zeta \in \Omega \\ |\zeta - P| < \delta}} \mathcal{B}(z, \zeta) \cdot \epsilon \, dV(\zeta) + \int_{\substack{\zeta \in \Omega \\ |\zeta - P| \geq \delta}} \mathcal{B}(z, \zeta) \cdot 2M \, dV(\zeta)$$

$$\equiv I + II.$$

Now the lemma tells us that $I = \epsilon$. Also we know that the Poisson–Bergman kernel for the ball is

$$\mathcal{B}(z, \zeta) = c_n \frac{(1 - |z|^2)^{n+1}}{|1 - z \cdot \overline{\zeta}|^{2n+2}}.$$

Thus by inspection, $\mathcal{B}(z, \zeta) \to 0$ as $z \to P$ for $|\zeta - P| \geq \delta$. Thus II is smaller than ϵ as soon as z is close enough to P.

In summary, for z sufficiently close to P, $|F(z) - f(P)| < 2\epsilon$. That is what we wished to prove. $\qquad\square$

Arazy and Engliš have in fact shown that the last result is true on any pseudoconvex domain for which each boundary point is a peak point (for the algebra $A(\Omega)$ of functions continuous on the closure and holomorphic inside). Thus the result is true in particular on strictly pseudoconvex domains (see [KRA1]) and finite type domains in $\mathbb{C}^2$ (see [BEF1]).

Here is another way to look at the matter on strictly pseudoconvex domains. In fact our observation, at the end of the proof of the last proposition, about the vanishing of $\mathcal{B}(z, \zeta)$ for $z \to P$ and $|\zeta - P| \geq \delta$ is a tricky point and not generally known. On a strictly pseudoconvex domain Ω, we have Fefferman's asymptotic expansion [FEF1, Part I]. This says that, in suitable local holomorphic coordinates near a boundary point P, we have

$$K_\Omega(z, \zeta) = \frac{c_n}{(1 - z \cdot \overline{\zeta})^{n+1}} + k(z, \zeta) \cdot \log |1 - z \cdot \overline{\zeta}| + \mathcal{E}(z, \zeta). \qquad (3.2.9)$$

Thus using an argument quite similar to the one that we carry out in detail in Sect. 1.2 for the Poisson–Szegő kernel, one can obtain an asymptotic expansion for the Poisson–Bergman kernel. One sees that, in local coordinates near the boundary.

$$\mathcal{B}_\Omega(z, \zeta) = c_n \cdot \frac{(1 - |z|^2)^{n+1}}{|1 - z \cdot \overline{\zeta}|^{2n+2}} + \mathcal{E}(z, \zeta),$$

where $\mathcal{E}$ is a kernel that induces a smoothing operator. In particular, the singularity of $\mathcal{E}$ will be measurably less than the singularity of the lead term. So it will still be the case that $\mathcal{B}(z, \zeta) \to 0$ as $z \to P \in \partial\Omega$ and $|\zeta - P| \geq \delta$. So we have:

Proposition 3.2.10. *Let $\Omega \subseteq \mathbb{C}^n$ be a smoothly bounded, strictly pseudoconvex domain in $\mathbb{C}^n$. Let $f \in C(\overline{\Omega})$. Then the function $\mathcal{B}f$ extends to be continuous on $\overline{\Omega}$. Moreover, if $P \in \partial\Omega$, then*

$$\lim_{\Omega \ni z \to P} \mathcal{B}f(z) = f(P).$$

It is natural, from the point of view of measure theory and harmonic analysis, to want to extend the result of Proposition 3.2.10 to a broader class of functions. To this end we introduce a maximal function to use as a tool.

Definition 3.2.11. Let Ω be a smoothly bounded, strictly pseudoconvex domain in $\mathbb{C}^n$. If $z, \zeta \in \overline{\Omega}$, then we set

$$\rho(z, \zeta) = |1 - z \cdot \overline{\zeta}|^{1/2}.$$

Proposition 3.2.12. *When $\Omega = B$, the unit ball, then the function ρ is a metric on ∂B. For a more general smoothly bounded, strictly pseudoconvex domain, the function ρ is a pseudometric. That is to say, there is constant $C \geq 1$ such that*

$$\rho(z, \zeta) \leq C \left(\rho(z, \xi) + \rho(\xi, \zeta) \right).$$

Proof: The first assertion is Proposition 6.5.1 in [KRA4]. The second assertion is proved in [KRA1, pp. 357–358]. We shall provide the details of this argument in Proposition 3.5.2. $\square$

Proposition 3.2.13. *The balls*

$$\beta_2(z, r) = \{\zeta \in \Omega : \rho(z, \zeta) < r\},$$

together with ordinary Euclidean volume measure dV, form a space of homogeneous type in the sense of Coifman and Weiss [COW].

Proof: This is almost immediate from the preceding proposition, but details may be found in [KRA1, Sect. 8.6]. □

Definition 3.2.14. For $z \in \Omega$ and $f \in L^1_{loc}(\Omega)$ we define

$$\mathcal{M}f(z) = \sup_{r>0} \frac{1}{V(\beta_2(z,r))} \int_{\beta_2(z,r)} |f(\zeta)| \, dV(\zeta).$$

Theorem 3.2.15. *The operator $\mathcal{M}$ is of weak type $(1, 1)$ and of strong type (p, p), $1 < p \leq \infty$.*

Proof: Again this is a standard consequence of the previous proposition in the context of spaces of homogeneous type. See [COW]. □

Theorem 3.2.16. *Let Ω be the unit ball B in $\mathbb{C}^n$. Let f be a locally integrable function on Ω. Then there is a constant $C > 0$ such that, for $z \in \Omega$,*

$$|\mathcal{B}f(z)| \leq C \cdot \mathcal{M}f(z).$$

Proof: It is easy to see that $|1 - z \cdot \bar{\zeta}| \geq (1/2)(1 - |z|^2)$. Therefore we may perform these standard estimates:

$$|\mathcal{B}f(z)| = \left| \int_\Omega \mathcal{B}(z, \zeta) f(\zeta) \, dV(\zeta) \right|$$

$$\leq \sum_{j=-1}^\infty \int_{2^j(1-|z|^2) \leq |1-z\cdot\bar\zeta| \leq 2^{j+1}(1-|z|^2)} \mathcal{B}(z, \zeta) |f(\zeta)| \, dV(\zeta)$$

$$\leq \sum_{j=-1}^\infty \int_{|1-z\cdot\bar\zeta| \leq 2^{j+1}(1-|z|^2)} \frac{(1 - |z|^2)^{n+1}}{[2^j(1 - |z|^2)]^{2n+2}} \, dV(\zeta)$$

$$\leq C \cdot \sum_{j=-1}^\infty 2^{-j(n+1)} \cdot \left[\frac{1}{(1 - |z|^2)^{n+1} 2^{(j+1)(n+1)}} \right] \int_{|1-z\cdot\bar\zeta| \leq 2^{j+1}(1-|z|^2)} |f(\zeta)| \, dV(\zeta)$$

$$\leq C \cdot \sum_{j=-1}^\infty 2^{-j(n+1)} \cdot \left[\frac{1}{V(\beta_2(z, \sqrt{2^{j+1}(1 - |z|^2)}))} \right] \int_{\beta_2(z, \sqrt{2^{j+1}(1-|z|^2)})} |f(\zeta)| \, dV(\zeta)$$

$$\tag{3.2.16.1}$$

The last line is majorized by

$$\leq C' \cdot \sum_{j=-1}^\infty 2^{-j(n+1)} \mathcal{M}f(z)$$

$$\leq C \cdot \mathcal{M}f(z). \qquad \square$$

Theorem 3.2.17. *Let Ω be the unit ball B in $\mathbb{C}^n$. Let f be an $L^p(\Omega, dV)$ function, $1 \leq p \leq \infty$. Then $\mathcal{B}f$ has radial boundary limits almost everywhere on $\partial\Omega$.*

Proof: The proof follows standard lines, using Theorems 3.2.15 and 3.2.16. See the detailed argument in [KRA1, Theorem 8.6.11].   □

In fact a slight emendation of the arguments just presented allows a more refined result.

Definition 3.2.18. Let $P \in \partial B$ and $\alpha > 1$. Define the *admissible approach region of aperture α* by

$$\mathcal{A}_\alpha(P) = \{z \in B : |1 - z \cdot \overline{\zeta}| < \alpha(1 - |z|^2)\}.$$

Admissible approach regions are a new type of region for Fatou-type theorems. These were first introduced in [KOR1, KOR2] and generalized and developed in [STE2] and later in [KRA6]. Now we have

Theorem 3.2.19. *Let f be an $L^p(B)$ function, $1 \leq p \leq \infty$. Then, for almost every $P \in \partial B$,*

$$\lim_{\mathcal{A}_\alpha(P) \ni z \to P} \mathcal{B}f(z)$$

exists.

In fact, using the Fefferman asymptotic expansion (as discussed in detail in the next section), we may imitate the development of Theorems 3.2.15 and 3.2.16 and prove a result analogous to Theorem 3.2.17 on any smoothly bounded, strictly pseudoconvex domain. We omit the details, as they would repeat ideas that we present elsewhere in the present book for slightly different purposes.

3.3 Ideas of Fefferman

In the seminal paper [FEF1, Part I], Charles Fefferman produced an asymptotic expansion for the Bergman kernel of a strictly pseudoconvex domain. He used this expansion to get detailed information about the boundary behavior of certain"pseudotransversal" geodesics in the metric; this data in turn was used to show that biholomorphic mappings of strictly pseudoconvex domains continue smoothly to the boundaries.

Fefferman's ideas have been quite influential. For instance, Paul Klembeck [KLE] used the Fefferman expansion to calculate the boundary asymptotics of Bergman metric curvature on a strictly pseudoconvex domain. See Chap. 7 below. This in turn led him to a new and very natural proof of the Bun Wong–Rosay theorem:

Theorem: Let $\Omega \subseteq \mathbb{C}^n$ be a bounded domain with C^2 boundary. Let $P \in \partial\Omega$ be a point of strong pseudoconvexity. Suppose that φ_j are biholomorphic self-maps of Ω with the property that there is a point $X \in \Omega$ such that $\lim_{j \to \infty} \varphi_j(X) = P$. Then Ω is biholomorphic to the unit ball B_n in $\mathbb{C}^n$.

Greene and Krantz [GRK1, GRK2, GRK3] showed that the Fefferman asymptotic expansion deforms stably under smooth deformation of the boundary of a strictly pseudoconvex domain $\Omega \subseteq \mathbb{C}^n$. They used that information to prove a variety of results about the Bergman geometry and also about automorphism groups of domains. We state just two of their results here:

Theorem: Let $\Omega_0 \subseteq \mathbb{C}^n$ be a fixed strictly pseudoconvex domain with smooth boundary. If Ω is another smoothly bounded strictly pseudoconvex domain with boundary sufficiently close to $\partial\Omega$ in the smooth domain topology, then the automorphism group (group of biholomorphic self-mappings of Ω) of Ω is a subgroup of the automorphism group of Ω_0. Indeed, there is a smooth mapping $\Phi : \Omega \to \Omega_0$ such that

$$\mathrm{Aut}(\Omega) \ni \varphi \longmapsto \Phi \circ \varphi \circ \Phi^{-1} \in \mathrm{Aut}(\Omega_0)$$

conjugates the automorphism group of Ω into the automorphism group of Ω_0.

Theorem: Let $\Omega \subseteq \mathbb{C}^n$ be a smoothly bounded domain which is sufficiently close to the unit ball $B_n \subseteq \mathbb{C}^n$ in the smooth domain topology and is biholomorphically inequivalent to the ball (such domains are generic—see [GRK1, GRK2]). Then there is a holomorphic embedding $\Psi : \Omega \to \mathbb{C}^n$ so that the automorphism group of $\Omega' \equiv \Psi(\Omega)$ is the restriction to Ω' of a subgroup of the unitary group on n letters.

We cannot provide all the details of the Fefferman's construction here. It is a long and tedious argument. But we can discuss and describe the asymptotic expansion and indicate some of its uses.

So fix a strictly pseudoconvex domain Ω with smooth boundary and fix a point $P \in \partial\Omega$. For z and ζ in Ω and sufficiently near P, Fefferman tells us that (in suitable local coordinates)

$$K_\Omega(z, \zeta) = \frac{c_n}{(1 - z \cdot \bar{\zeta})^{n+1}} + k(z, \zeta) \cdot \log|1 - z \cdot \bar{\zeta}| + \mathcal{E}(z, \zeta). \qquad (3.3.1)$$

Here $\mathcal{E}$ is an error term that is smaller, in a measurable sense, than the lead terms.

One can see from formula (3.3.1) that calculations with the Bergman kernel of a strictly pseudoconvex domain are tantamount to calculations with the Bergman kernel for the ball (up to a calculable and estimable error).[1]

---

[1]The logarithmic term was one of the big surprises of the Fefferman's work. It was quite unexpected. And it does not conform to the paradigm that "the Bergman kernel of a strictly pseudoconvex domain is just like that for the ball." But the logarithmic term has only a weak singularity and is easily estimated. Fefferman provided, in his paper [FEF1, Part I], a concrete example of a domain in which the logarithmic term actually arises. See Sect. 6.7 below. And Burns later proved that the logarithmic term is generic. That is to say, if Ω is a strictly pseudoconvex domain with smooth boundary and none of the Fefferman asymptotic expansions near any of the boundary points have logarithmic terms, then Ω is biholomorphic to the unit ball B. Burns

Thus one can prove, in the vein of Klembeck, that the curvature tensor of the Bergman metric is asymptotically, as the base point p approaches the boundary of the domain, equal to the curvature tensor for the Bergman metric of the ball. Because the automorphism group of the ball acts transitively on the unit sphere bundle in the tangent bundle to the ball, it follows that the latter curvature tensor must be constant.

3.4   Results on the Invariant Laplacian

If $g = (g_{jk})$ is a Riemannian metric on a domain Ω in complex Euclidean space, then there is a second-order partial differential operator, known as the *Laplace–Beltrami operator*, that is invariant under isometries of the metric. In fact, if g denotes the determinant of the metric matrix g, and if (g^{jk}) denotes the inverse matrix, then this partial differential operator is defined to be

$$\mathcal{L} = \frac{2}{g} \sum_{j,k} \left\{ \frac{\partial}{\partial \bar{z}_j} \left(g g^{jk} \frac{\partial}{\partial z_k} \right) + \frac{\partial}{\partial z_k} \left(g g^{jk} \frac{\partial}{\partial \bar{z}_k} \right) \right\} .$$

Now of course we are interested in artifacts of the Bergman theory. If $\Omega \subseteq \mathbb{C}^n$ is a bounded domain and $K = K_\Omega$ its Bergman kernel, then it is well known (see [KRA1]) that $K(z, z) > 0$ for all $z \in \Omega$. Then it makes sense to define

$$g_{jk}(z) = \frac{\partial^2}{\partial z_j \, \partial \bar{z}_k} \log K(z, z)$$

for $j, k = 1, \ldots, n$. Then Proposition 1.1.14 can be used to demonstrate that this metric—which is in fact a Kähler metric on Ω—is invariant under biholomorphic mappings of Ω. In other words, any biholomorphic $\Phi : \Omega \to \Omega$ is an isometry in the metric g. This is the celebrated *Bergman metric*.

If $\Omega \subseteq \mathbb{C}^n$ is the unit ball B, then the Bergman kernel is given by

$$K_B(z, \zeta) = \frac{1}{V(B)} \cdot \frac{1}{(1 - z \cdot \bar{\zeta})^{n+1}} ,$$

where $V(B)$ denotes the Euclidean volume of the domain B. Then

$$\log K(z, z) = -\log V(B) - (n + 1) \log(1 - |z|^2).$$

---

never published this result. Boutet de Monvel [BOU] in dimension 2 and Robin Graham [GRA3] in general gave rigorous proofs of the result. See also the work of Hirachi [HIR1]. There are unbounded domains and also roughly bounded domains on which the analogue of this result for the Szegő is known to fail—see [HIR2].

Further,

$$\frac{\partial}{\partial z_j}(-(n+1)\log(1-|z|^2)) = (n+1)\frac{\bar{z}_j}{1-|z|^2}$$

and

$$\frac{\partial^2}{\partial z_j \partial \bar{z}_k}(-(n+1)\log(1-|z|^2)) = (n+1)\left[\frac{\delta_{jk}}{1-|z|^2} + \frac{\bar{z}_j z_k}{(1-|z|^2)^2}\right]$$

$$= \frac{(n+1)}{(1-|z|^2)^2}\left[\delta_{jk}(1-|z|^2) + \bar{z}_j z_k\right]$$

$$\equiv g_{jk}(z).$$

When $n = 2$ we have

$$g_{jk}(z) = \frac{3}{(1-|z|^2)^2}\left[\delta_{jk}(1-|z|^2) + \bar{z}_j z_k\right].$$

Thus

$$\left(g_{jk}(z)\right) = \frac{3}{(1-|z|^2)^2}\begin{pmatrix} 1-|z_2|^2 & \bar{z}_1 z_2 \\ \bar{z}_2 z_1 & 1-|z_1|^2 \end{pmatrix}.$$

Let

$$\left(g^{jk}(z)\right)_{j,k=1}^2$$

represent the inverse of the matrix

$$\left(g_{jk}(z)\right)_{j,k=1}^2 .$$

Then an elementary computation shows that

$$\left(g^{jk}(z)\right)_{j,k=1}^2 = \frac{1-|z|^2}{3}\begin{pmatrix} 1-|z_1|^2 & -z_2\bar{z}_1 \\ -z_1\bar{z}_2 & 1-|z_2|^2 \end{pmatrix} = \frac{1-|z|^2}{3}(\delta_{jk} - \bar{z}_j z_k)_{j,k}.$$

Let

$$g \equiv \det\left(g_{jk}(z)\right).$$

Then

$$g = \frac{9}{(1 - |z|^2)^3}.$$

Now let us calculate. If $(g_{jk})_{j,k=1}^2$ is the Bergman metric on the ball in $\mathbb{C}^2$, then we have

$$\sum_{j,k} \frac{\partial}{\partial \bar{z}_j} (g g^{jk}) = 0$$

and

$$\sum_{j,k} \frac{\partial}{\partial z_j} (g g^{jk}) = 0.$$

We verify these assertions in detail in dimension 2: Now

$$g g^{jk} = \frac{9}{(1 - |z|^2)^3} \cdot \frac{1 - |z|^2}{3} (\delta_{jk} - \bar{z}_j z_k)$$

$$= \frac{3}{(1 - |z|^2)^2} (\delta_{jk} - \bar{z}_j z_k).$$

It follows that

$$\frac{\partial}{\partial \bar{z}_j} \left[g g^{jk} \right] = \frac{6 z_j}{(1 - |z|^2)^3} (\delta_{jk} - \bar{z}_j z_k) - \frac{3 z_k}{(1 - |z|^2)^2}.$$

Therefore

$$\sum_{j,k=1}^2 \frac{\partial}{\partial \bar{z}_j} \left[g g^{jk} \right] = \sum_{j,k=1}^2 \left[\frac{6 z_j (\delta_{jk} - \bar{z}_j z_k)}{(1 - |z|^2)^3} - \frac{3 z_j}{(1 - |z|^2)^2} \right]$$

$$= 6 \sum_k \frac{z_k}{(1 - |z|^2)^3} - 6 \sum_{j,k} \frac{|z_j|^2 z_k}{(1 - |z|^2)^3} - 6 \sum_k \frac{z_k}{(1 - |z|^2)^2}$$

$$= 6 \sum_j \frac{z_j}{(1 - |z|^2)^2} - 6 \sum_k \frac{z_k}{(1 - |z|^2)^2}$$

$$= 0.$$

The other derivative is calculated similarly.

Our calculations show that, on the ball in $\mathbb{C}^2$,

$$\mathcal{L} \equiv \frac{2}{g} \sum_{j,k} \left\{ \frac{\partial}{\partial \bar{z}_j} \left(g g^{jk} \frac{\partial}{\partial z_k} \right) + \frac{\partial}{\partial z_k} \left(g g^{jk} \frac{\partial}{\partial \bar{z}_j} \right) \right\}$$

$$= 4 \sum_{j,k} g^{jk} \frac{\partial}{\partial \bar{z}_j} \frac{\partial}{\partial z_k}$$

$$= 4 \sum_{j,k} \frac{1 - |z|^2}{3} (\delta_{jk} - \bar{z}_j z_k) \frac{\partial^2}{\partial z_k \partial \bar{z}_j}.$$

Now the interesting fact for us is encapsulated in the following proposition:

Proposition 3.4.1. *The Poisson–Szegő kernel on the ball B solves the Dirichlet problem for the invariant Laplacian $\mathcal{L}$. That is to say, if f is a continuous function on ∂B, then the function*

$$u(z) = \begin{cases} \int_{\partial B} \mathcal{P}(z, \zeta) \cdot f(\zeta) \, d\sigma(\zeta) & \text{if } z \in B \\ f(z) & \text{if } z \in \partial B \end{cases}$$

is continuous on $\overline{B}$ and is annihilated by $\mathcal{L}$ on B.

This fact is of more than passing interest. In one complex variable, the study of holomorphic functions on the disc and the study of harmonic functions on the disc are inextricably linked because the real part of a holomorphic function is harmonic and conversely. Such is not the case in several complex variables. Certainly the real part of a holomorphic function is harmonic. But in fact it is more: Such a function is *pluriharmonic*. For the converse direction, any real-valued pluriharmonic function is locally the real part of a holomorphic function. This assertion is false if "pluriharmonic" is replaced by "harmonic."

And the result of Proposition 3.4.1 should not really be considered to be surprising. For the invariant Laplacian is invariant under isometries of the Bergman metric, hence invariant under automorphisms of the ball. And the Poisson–Szegő kernels behave nicely under automorphisms. E. M. Stein was able to take advantage of these invariance properties to give a proof of Proposition 3.4.1 using Godement's theorem—that any function that satisfies a suitable mean-value property must be harmonic (i.e., annihilated by the relevant Laplace operator). See [STE2] for the details.

Sketch of the Proof of Proposition 3.4.1   Now

$$\mathcal{L}u = \mathcal{L} \int_{\partial B} \mathcal{P}(z, \zeta) \cdot f(\zeta) \, d\sigma(\zeta) = \int_{\partial B} \left[\mathcal{L}_z \mathcal{P}(z, \zeta) \right] \cdot f(\zeta) \, d\sigma(\zeta) .$$

Thus it behooves us to calculate $\mathcal{L}_z \mathcal{P}(z, \zeta)$. Now we shall calculate this quantity for each fixed ζ. Thus without loss of generality, we may compose with a unitary rotation and suppose that $\zeta = (1 + i0, 0 + i0)$ so that (in complex dimension 2)

$$\mathcal{P} = c_2 \cdot \frac{(1 - |z|^2)^2}{|1 - z_1|^4} .$$

This will make our calculations considerably easier.

By brute force, we find that

$$\frac{\partial \mathcal{P}}{\partial \bar{z}_1} = -2(1 - z_1)(1 - |z|^2) \cdot \left[\frac{-1 + z_1 + |z_2|^2}{|1 - z_1|^6} \right]$$

$$\frac{\partial^2 \mathcal{P}}{\partial \bar{z}_1 \partial z_1} = \frac{-2}{|1 - z_1|^6} \cdot \left[-|z_1|^2 - |z_1|^2 |z_2|^2 + 3|z_2|^2 - z_1 |z_2|^2 \right.$$
$$\left. -2|z_2|^4 - 1 + z_1 + \bar{z}_1 - \bar{z}_1 |z_2|^2 \right]$$

$$\frac{\partial^2 \mathcal{P}}{\partial \bar{z}_1 \partial z_2} = \frac{-2(1 - z_1)}{|1 - z_1|^6} \cdot \left[2\bar{z}_2 - \bar{z}_2 z_1 - 2\bar{z}_2 |z_2|^2 - \bar{z}_2 |z_1|^2 \right]$$

$$\frac{\partial^2 \mathcal{P}}{\partial z_1 \partial \bar{z}_2} = \frac{-2(1 - \bar{z}_1)}{|1 - z_1|^6} \cdot \left[2z_2 - z_2 \bar{z}_1 - 2z_2 |z_2|^2 - z_2 |z_1|^2 \right]$$

$$\frac{\partial \mathcal{P}}{\partial z_2} = \frac{-2z_2 + 2|z_1|^2 z_2 + 2|z_2|^2 z_2}{|1 - z_1|^4}$$

$$\frac{\partial^2 \mathcal{P}}{\partial z_2 \partial \bar{z}_2} = \frac{-2 + 2|z_1|^2 + 4|z_2|^2}{|1 - z_1|^4} \tag{3.4.1.1}$$

Now we know that, in complex dimension two,

$$\mathcal{L}_z \mathcal{P}(z, \zeta) = \frac{4}{3}(1 - |z|^2) \cdot (1 - |z_1|^2) \cdot \frac{\partial^2 \mathcal{P}_z}{\partial z_1 \partial \bar{z}_1} + \frac{4}{3}(1 - |z|^2) \cdot (-\bar{z}_1 z_2) \cdot \frac{\partial^2 \mathcal{P}_z}{\partial z_2 \partial \bar{z}_1}$$
$$+ \frac{4}{3}(1 - |z|^2) \cdot (-\bar{z}_2 z_1) \cdot \frac{\partial^2 \mathcal{P}_z}{\partial z_1 \partial \bar{z}_2} + \frac{4}{3}(1 - |z|^2) \cdot (1 - |z_2|^2) \cdot \frac{\partial^2 \mathcal{P}_z}{\partial z_2 \partial \bar{z}_2} .$$

Plugging the values from (3.4.1.1) into this last equation gives

$$\mathcal{L}_z \mathcal{P}(z, \zeta) = \frac{4}{3}(1 - |z|^2) \cdot (1 - |z_1|^2) \cdot \frac{-2}{|1 - z_1|^6} \cdot \left[-|z_1|^2 - |z_1|^2 |z_2|^2 \right.$$
$$\left. + 3|z_2|^2 - z_1 |z_2|^2 - 2|z_2|^4 - 1 + z_1 + \bar{z}_1 - \bar{z}_1 |z_2|^2 \right]$$

$$+ \frac{4}{3}(1 - |z|^2) \cdot (-\bar{z}_1 z_2)$$

$$\times \frac{-2(1 - z_1)}{|1 - z_1|^6} \cdot \left[2\bar{z}_2 - \bar{z}_2 z_1 - 2\bar{z}_2 |z_2|^2 - \bar{z}_2 |z_1|^2 \right]$$

$$+ \frac{4}{3}(1 - |z|^2) \cdot (-\bar{z}_2 z_1)$$

$$\times \frac{-2(1-\bar{z}_1)}{|1-z_1|^6} \cdot \left[2z_2 \quad z_2\bar{z}_1 - 2z_2|z_2|^2 - z_2|z_1|^2 \right]$$

$$+ \frac{4}{3}(1-|z|^2) \cdot (1-|z_2|^2) \cdot |1-z_1|^2 \cdot \frac{-2+2|z_1|^2+4|z_2|^2}{|1-z_1|^6}.$$

Multiplying out the terms, we find that

$$\mathcal{L}_z \mathcal{P}(z,\zeta) = \frac{-2}{|1-z_1|^6} \cdot \left[-|z_1|^2 - 4|z_1|^2|z_2|^2 + 3|z_2|^2 - z_1|z_2|^2 - 2|z_2|^4 - 1 \right.$$

$$+ z_1 + \bar{z}_1 - \bar{z}_1|z_2|^2 + |z_1|^4 + |z_1|^4|z_2|^2 + z_1|z_1|^2|z_2|^2$$

$$\left. + 2|z_1|^2|z_2|^4 + |z_1|^2 - z_1|z_1|^2 - \bar{z}_1|z_1|^2 + \bar{z}_1|z_1|^2|z_2|^2 \right]$$

$$- \frac{2}{|1-z_1|^6} \cdot \left[-2\bar{z}_1|z_2|^2 + 3|z_1|^2|z_2|^2 + 2|z_2|^4\bar{z}_1 + \bar{z}_1|z_2|^2|z_1|^2 \right.$$

$$\left. - z_1|z_1|^2|z_2|^2 - 2|z_1|^2|z_2|^4 - |z_2|^2|z_1|^4 \right]$$

$$- \frac{2}{|1-z_1|^6} \cdot \left[-2z_1|z_2|^2 + 3|z_1|^2|z_2|^2 + 2|z_2|^4z_1 + z_1|z_2|^2|z_1|^2 \right.$$

$$\left. - \bar{z}_1|z_1|^2|z_2|^2 - 2|z_1|^2|z_2|^4 - |z_2|^2|z_1|^4 \right]$$

$$- \frac{2}{|1-z_1|^6} \cdot \left[1 - |z_1|^2 - 3|z_2|^2 + |z_1|^2|z_2|^2 + 2|z_2|^4 - z_1 + z_1|z_1|^2 \right.$$

$$+ 3z_1|z_2|^2 - z_1|z_1|^2|z_2|^2 - 2z_1|z_2|^4 - \bar{z}_1 + \bar{z}_1|z_1|^2 + 3\bar{z}_1|z_2|^2 - \bar{z}_1|z_1|^2|z_2|^2$$

$$\left. - 2\bar{z}_1|z_2|^4 + |z_1|^2 - |z_1|^4 - 3|z_1|^2|z_2|^2 + |z_1|^4|z_2|^2 + 2|z_1|^2|z_2|^4 \right].$$

And now, if we combine all the terms in brackets, a small miracle happens: Everything cancels. The result is

$$\mathcal{L}_z \mathcal{P}(z,\zeta) \equiv 0. \qquad \square$$

Thus in some respects, it is inappropriate to study holomorphic functions on the ball in $\mathbb{C}^n$ using the Poisson kernel. The classical Poisson integral does *not* create pluriharmonic functions, and it does not create functions that are annihilated by the invariant Laplacian. In view of Proposition 3.4.1, the Poisson–Szegő kernel is much more apposite. As an instance, Adam Koranyi [KOR1, KOR2] made decisive use of this observation in his study (proving boundary limits of H^2 functions through admissible approach regions $\mathcal{A}_\alpha$) of the boundary behavior of $H^2(B)$ functions.

It is known that the property described in Proposition 3.4.1 is special to the ball—it is simply untrue on any other domain (see [GRA1, GRA2] for more detail on this matter). Now one of the points that we want to make in this section is that the result of the proposition can be extended—in an approximate sense—to a broader class of domains.

Proposition 3.4.2. *Let* $\Omega \subseteq \mathbb{C}^n$ *be a smoothly bounded, strictly pseudoconvex domain and* $\mathcal{P}$ *its Poisson–Szegő kernel. Then, if* $f \in C(\partial\Omega)$, *we may write*

$$\mathcal{P}f(z) = \mathcal{P}_1 f(z) + \mathcal{E} f(z),$$

where

(i) *The term* $\mathcal{P}_1 f$ *is "approximately annihilated" by the invariant Laplacian on* Ω.
(ii) *The operator* $\mathcal{E}$ *is smoothing in the sense of pseudodifferential operators.*

We shall explain the meaning of (i) and (ii) in the course of the proofs of these statements.

Proof of Proposition 3.4.2: We utilize of course the asymptotic expansion for the Szegő kernel on a smoothly bounded, strictly pseudoconvex domain (see [FEF1, Part I], [BOS]). It says that, for z, ζ near a boundary point P, we have (in suitable biholomorphic local coordinates)

$$S_\Omega(z,\zeta) = \frac{c_n}{(1 - z \cdot \bar{\zeta})^n} + h(z,\zeta) \cdot \log|1 - z \cdot \bar{\zeta}| + \mathcal{E}(z,\zeta). \tag{3.4.2.1}$$

Here h is a smooth function on $\overline{\Omega} \times \overline{\Omega}$.
Now we calculate $\mathcal{P}(z,\zeta)$ in the usual fashion:

$$\mathcal{P}_\Omega(z,\zeta) = \frac{|S(z,\zeta)|^2}{S(z,z)} = \frac{\left| \dfrac{c_n}{(1 - z \cdot \overline{\bar{\zeta}})^n} + h(z,\zeta) \cdot \log|1 - z \cdot \bar{\zeta}| \right|^2}{\dfrac{c_n}{(1 - |z|^2)^n} + h(z,z) \cdot \log(1 - |z|^2)} + \mathcal{F}(z,\zeta).$$

$$\tag{3.4.2.2}$$

One can use just elementary algebra to simplify this last expression and obtain that, in suitable local coordinates near the boundary,

$$\mathcal{P}_\Omega(z,\zeta) = c_n \cdot \frac{(1 - |z|^2)^n}{|1 - z \cdot \bar{\zeta}|^{2n}}$$

$$+ \frac{2(1 - |z|^2)^n}{|1 - z \cdot \bar{\zeta}|^n} \log|1 - z \cdot \bar{\zeta}| + \mathcal{O}\left[(1 - |z|^2)^n \cdot \log|1 - z \cdot \bar{\zeta}| \right]$$

$$\equiv c_n \cdot \frac{(1 - |z|^2)^n}{|1 - z \cdot \bar{\zeta}|^{2n}} + \mathcal{G}(z,\zeta). \tag{3.4.2.3}$$

Now the first expression on the right-hand side of (3.4.2.3) is (in the local coordinates in which we are working) the usual Poisson–Szegő kernel for the unit ball in $\mathbb{C}^n$. The second is an error term which we now analyze.

In fact we claim that the error term is integrable in ζ, uniformly in z, and the same can be said for the gradient (in the z variable) of the error term. The first of these statements is obvious, as both parts of the error term are clearly majorized by the Poisson–Szegő kernel itself. As for the second part, we note that the gradient of the error gives rise to three types of terms:

$$\nabla \mathcal{E} \approx \frac{(1-|z|^2)^{n-1}}{|1-z \cdot \bar{\zeta}|^n} \cdot \log|1-z \cdot \bar{\zeta}|$$

$$+ \frac{(1-|z|^2)^n}{|1-z \cdot \bar{\zeta}|^{n+1}} \cdot \log|1-z \cdot \bar{\zeta}|$$

$$+ \frac{(1-|z|^2)^n}{|1-z \cdot \bar{\zeta}|^{n+1}}$$

$$\equiv I + II + III. \tag{3.4.2.4}$$

Now it is clear by inspection that I and II are majorized by the ordinary Poisson–Szegő kernel, so they are both integrable in ζ as claimed. As for III, we must calculate:

$$\int_{\zeta \in \partial \Omega} \frac{(1-|z|^2)^{n-1}}{|1-z \cdot \bar{\zeta}|^{n+1}} \, d\sigma(\zeta) \leq \sum_{j=-1}^{\infty} \int_{2^j(1-|z|^2) \leq |1-z \cdot \bar{\zeta}| \leq 2^{j+1}(1-|z|^2)}$$

$$\frac{(1-|z|^2)^{n-1}}{[2^j(1-|z|^2)]^{n+1}} \, d\sigma(\zeta)$$

$$\leq \sum_{j=-1}^{\infty} \frac{1}{(1-|z|^2)^2} \int_{|1-z \cdot \bar{\zeta}| \leq 2^{j+1}(1-|z|^2)} 2^{-j(n+1)} \, d\sigma(\zeta)$$

$$\leq \sum_{j=-1}^{\infty} C \cdot \frac{2^{-j(n+1)}}{(1-|z|^2)^2} \cdot \left[\sqrt{2^{j+1}(1-|z|^2)} \right]^{2n-2}$$

$$\times \left[2^{j+1} \cdot (1-|z|^2) \right]$$

$$\leq \sum_{j=-1}^{\infty} \frac{1}{(1-|z|^2)^2} \cdot (1-|z|^2)^{n-1} \cdot (1-|z|^2)$$

$$\times 2^{-j(n+1)} \cdot 2^{(j+1)(n-1)} \cdot 2^{j+1}$$

$$\leq C \cdot 2^n (1-|z|^2)^{n-2} \cdot \sum_{j=-1}^{\infty} 2^{-j}$$

$$< \infty.$$

Thus we see that the Poisson–Szegő kernel for our strictly pseudoconvex domain Ω can be expressed, in suitable local coordinates, as the Poisson–Szegő kernel for the ball plus an error term whose gradient induces a bounded operator on L^p. This means that the error term itself maps L^p to a Sobolev space. In other words, it is a smoothing operator (hence negligible from our point of view).

In fact there are several fairly well-known results about the interaction of the Poisson–Bergman kernel and the invariant Laplacian. We summarize some of the basic ones here.

Proposition 3.4.3. *Let f be a C^2 function on the unit ball that is annihilated by the invariant Laplacian $\mathcal{L}$. Then, for any $0 < r < 1$ and S the unit sphere,*

$$\int_S f(r\zeta)\,d\sigma(\zeta) = c(r) \cdot f(0)\,.$$

Here $d\sigma$ is a rotationally invariant measure on the sphere S.

Proof: Replacing f with the average of f over the orthogonal group, this just becomes a calculation to determine the exact value of the constant $c(r)$—see [RUD2, p. 51]. $\qquad\square$

Proposition 3.4.4. *Suppose that f is a C^2 function on the unit ball B that is annihilated by the invariant Laplacian $\mathcal{L}$. Then f satisfies the identity $\mathcal{B}f = f$. In other words, for any $z \in B$,*

$$f(z) = \int_B \mathcal{B}(z,\zeta)f(\zeta)\,dV(\zeta)\,.$$

Proof: We have checked the result when $z = 0$ in the last proposition. For a general z, compose with a Möbius transformation and use the biholomorphic invariance of the kernel and the differential operator $\mathcal{L}$. $\qquad\square$

Remark 3.4.5. It is a curious fact (see [AFR]) that the converse of this last proposition is only true (as stated) in complex dimensions $1, 2, \ldots, 11$. It is false in dimensions 12 and higher. $\qquad\square$

Finally we need to address the question of whether the invariant Laplacian *for the domain Ω* annihilates the principal term of the right-hand side of the formula (3.4.2.3). The point is this. The biholomorphic change of variable that makes (3.4.2.3) valid is *local*. It is valid on a small, smoothly bounded subdomain $\Omega' \subseteq \Omega$ which shares a piece of boundary with $\partial\Omega$. According to Fefferman [FEF1, Part I] (see also the work in [GRK1,GRK2]), there is a smaller subdomain $\Omega'' \subseteq \Omega'$ (which also shares a piece of boundary with $\partial\Omega$ and $\partial\Omega'$) so that the Bergman metric of Ω' is close—in the C^2 topology—to the Bergman metric of Ω *on the smaller domain Ω''*. It follows then that the Laplace–Beltrami operator $Ł_{\Omega'}$ for the Bergman metric of Ω' will be close to the Laplace–Beltrami operator $Ł_\Omega$ of Ω on the smaller subdomain Ω''. Now, on Ω', the operator $Ł_{\Omega'}$ certainly annihilates the principal term of (3.4.2.3). It follows then that, on Ω'', the operator $\mathcal{L}_\Omega$ *nearly* annihilates the principal term of (3.4.2.3). We shall not calculate the exact sense in which this last statement is true, but leave details for the interested reader.

This discussion completes our consideration of (3.4.2.3). $\square$

It is natural to wonder whether the Poisson–Bergman kernel $\mathcal{B}$ has any favorable properties with respect to important partial differential operators. We have the following positive result:

Proposition 3.4.6. *Let $\Omega = B$, the unit ball in $\mathbb{C}^n$, and $\mathcal{B} = \mathcal{B}_B(z, \zeta)$ its Poisson–Bergman kernel. Then $\mathcal{B}$ is plurisubharmonic in the ζ variable.*

Proof: Fix a point $\zeta \in B$ and let Φ be an automorphism of B such that $\Phi(\zeta) = 0$. From Proposition 3.2.4, we then have

$$\mathcal{B}_B(z, \zeta) = \mathcal{B}_B(\Phi(z), \Phi(\zeta)) \cdot |\det J_{\mathbb{C}}\Phi(\zeta)|^2 = \mathcal{B}_B(\Phi(z), 0) \cdot |\det J_{\mathbb{C}}\Phi(\zeta)|^2 \,.$$

$$(3.4.6.1)$$

We see that the right-hand side is an expression that is independent of ζ multiplied times a plurisubharmonic function. A formula similar to (3.4.6.1) appears in [HUA].

The same argument shows that $\mathcal{B}(\zeta, \zeta)$ is plurisubharmonic. $\square$

3.5 The Dirichlet Problem for the Invariant Laplacian on the Ball

We will study the following Dirichlet problem on $B \subseteq \mathbb{C}^2$:

$$\begin{cases} \triangle_B u = 0 \text{ on } B \\ u\big|_{\partial B} = \phi, \end{cases}$$

$$(3.5.1)$$

where ϕ is a given continuous function on ∂B. Here $\triangle_B$ is the invariant Laplacian (i.e., the Laplace–Beltrami operator for the Bergman metric) on the unit ball in $\mathbb{C}^n$.

Exercise: Is this a well-posed boundary value problem (in the sense of Lopatinski)? Consult [KRA4] for more on this topic.

The remarkable fact about this relatively innocent-looking boundary value problem is that there exist data functions $\phi \in C^\infty(\partial B)$ with the property that the (unique) solution to the boundary value problem is not even C^2 on $\overline{B}$. This result appears in [GRA1,GRA2] and was also discovered independently by Garnett–Krantz (see [KRA4]). It is in striking contrast to the situation that obtains for the Dirichlet problem for a uniformly elliptic operator.

Observe that, for $n = 1$, our Dirichlet problem becomes

$$\begin{cases} (1 - |z|^2)^2 \, \triangle u = 0 \text{ on } D \subseteq \mathbb{C} \\ u|_{\partial D} = \phi, \end{cases}$$

which is just the same as

$$\begin{cases} \triangle u = 0 \quad \text{on } D \subseteq \mathbb{C} \\ u|_{\partial D} = \phi. \end{cases}$$

This is the standard Dirichlet problem for the Laplacian—a uniformly strongly elliptic operator. Thus there is a complete existence and regularity theory: the solution u will be as smooth on the closure as is the data ϕ (provided that we measure this smoothness in the correct norms). Our problem in dimensions $n > 1$ yields some surprises. We begin by developing some elementary geometric ideas.

Let $\zeta, \xi \in \partial B$. Define

$$\rho(\zeta, \xi) = |1 - \zeta \cdot \overline{\xi}|^{1/2},$$

where $\zeta \cdot \overline{\xi} \equiv \zeta_1 \overline{\xi}_1 + \zeta_2 \overline{\xi}_2$. Then we have

Proposition 3.5.2. *The binary operator ρ is a metric on ∂B.*

Proof: Let $z, w, \zeta \in \partial B$. We shall show that

$$\rho(z, \zeta) \leq \rho(z, w) + \rho(w, \zeta).$$

Assume for simplicity that the dimension $n = 2$. After applying a unitary rotation, we may suppose that $w = \mathbf{1} = (1, 0)$. Now $|1 - z \cdot \overline{\zeta}| = \frac{1}{2}|z - \zeta|^2$. Therefore it suffices for us to prove that

$$\frac{1}{\sqrt{2}}|z - \zeta| \leq \sqrt{|1 - z_1|} + \sqrt{|1 - \zeta_1|}.$$

But

$$\frac{1}{\sqrt{2}}|z - \zeta| \leq \frac{1}{\sqrt{2}}|z - \mathbf{1}| + \frac{1}{\sqrt{2}}|\mathbf{1} - \zeta|.$$

(Notice that for z, ζ symmetrically situated about $\mathbf{1}$ and very near to $\mathbf{1}$, this is nearly an equality.) Thus it is enough to prove that

$$|z - \mathbf{1}| \leq \sqrt{2}\sqrt{|1 - z_1|}.$$

Finally, we calculate that

$$|z - 1| = \sqrt{|1 - z_1|^2 + |z_2|^2}$$

$$= \sqrt{|1 - z_1|^2 + 1 - |z_1|^2}$$

$$= \sqrt{2 - 2\operatorname{Re} z_1}$$

$$= \sqrt{2}\sqrt{1 - \operatorname{Re} z_1}$$

$$\leq \sqrt{2}|1 - z_1|. \qquad \qquad \square$$

Now we define balls using ρ: For $P \in \partial B$ and $r > 0$ we define $\beta(P, r) = \{\zeta \in \partial B : \rho(P, \zeta) < r\}$. [These skew balls play a decisive role in the complex geometry of several variables. We shall get just a glimpse of their use here.] Let $0 \neq z \in B$ be fixed and let P be its orthogonal projection on the boundary: $\tilde{z} = z/|z|$. If we fix $r > 0$, then we may verify directly that

$$\mathcal{P}(z, \zeta) \to 0 \quad \text{uniformly in} \quad \zeta \in \partial B \setminus \beta(\tilde{z}, r) \text{ as } z \to \tilde{z}.$$

Proposition 3.5.2. *Let $B \subseteq \mathbb{C}^n$ be the unit ball and $g \in C(\partial B)$. Then the function*

$$G(z) = \begin{cases} \int_{\partial B} \mathcal{P}(z, \zeta) g(\zeta) \, d\sigma(\zeta) & \text{if } z \in B \\ g(z) & \text{if } z \in B \end{cases}$$

solves the Dirichlet problem (3.5.1) for the Laplace–Beltrami operator $\triangle_B$. Here $\mathcal{P}$ is the Poisson–Szegő kernel.

Proof: It is straightforward to calculate that

$$\triangle_B G(z) = \int_{\partial B} [\triangle_B \mathcal{P}(z, \zeta)] g(\zeta) \, d\sigma(\zeta)$$

$$= 0$$

because $\triangle_B \mathcal{P}(\cdot, \zeta) = 0$.

For simplicity let us now restrict attention once again to dimension $n = 2$. We wish to show that G is continuous on $\overline{B}$. First recall that

$$\mathcal{P}(z, \zeta) = \frac{1}{\sigma(\partial B)} \frac{(1 - |z|^2)^2}{|1 - z \cdot \overline{\zeta}|^4}.$$

Notice that

$$\int_{\partial B} |\mathcal{P}(z, \zeta)| \, d\sigma(\zeta) = \int_{\partial B} \mathcal{P}(z, \zeta) \, d\sigma(\zeta) = \int_{\partial B} \mathcal{P}(z, \zeta) \cdot 1 \, d\sigma(\zeta) = 1$$

since the identically 1 function is holomorphic on Ω and is therefore reproduced by integration against $\mathcal{P}$. We have used also the fact that $\mathcal{P} \geq 0$.

Now we enter the proof proper of the proposition. Fix $\epsilon > 0$. By the uniform continuity of g, we may select a $\delta > 0$ such that if $P \in \partial B$ and $\zeta \in \beta(P, \delta)$, then $|g(P) - g(\zeta)| < \epsilon$. Then, for any $0 \neq z \in B$ and P its projection to the boundary, we have

$$
|G(z) - g(P)| = \left| \int_{\partial B} \mathcal{P}(z, \zeta) g(\zeta) \, d\sigma(\zeta) - g(P) \right|
$$

$$
= \left| \int_{\partial B} \mathcal{P}(z, \zeta) g(\zeta) \, d\sigma(\zeta) - \int_{\partial B} \mathcal{P}(z, \zeta) g(P) \, d\sigma(\zeta) \right|
$$

$$
\leq \int_{\partial B} \mathcal{P}(z, \zeta) |g(\zeta) - g(P)| \, d\sigma
$$

$$
= \int_{\beta(P, \delta)} \mathcal{P}(z, \zeta) |g(\zeta) - g(P)| \, d\sigma(\zeta)
$$

$$
\int_{\partial B \setminus \beta(P, \delta)} \mathcal{P}(z, \zeta) |g(\zeta) - g(P)| \, d\sigma(\zeta)
$$

$$
\leq \epsilon + 2 \|g\|_{L^\infty} \int_{\partial B \setminus \beta(P, \delta)} \mathcal{P}(z, \zeta) \, d\sigma(\zeta).
$$

By the remarks preceding this argument, we may choose r sufficiently close to 1 such that $\mathcal{P}(z, \zeta) < \epsilon$ for $|z| > r$ and $\zeta \in \partial B \setminus \beta(P, \delta)$. Thus with these choices, the last line does not exceed $C \cdot \epsilon$.

We conclude the proof with an application of the triangle inequality: Fix $P \in \partial B$ and suppose that $0 \neq z \in B$ satisfies both $|P - z| < \delta$ and $|z| > r$. If $\widetilde{z} = z/|z|$ is the projection of z to ∂B, then we have

$$
|G(z) - g(P)| \leq |G(z) - g(\widetilde{z})| + |g(\widetilde{z}) - g(P)|.
$$

The first term is majorized by ϵ by the argument that we just presented. The second is less than ϵ by the uniform continuity of g on $\partial \Omega$.

That concludes the proof. $\square$

Now we know how to solve the Dirichlet problem for $\triangle_B$, and we want next to consider regularity for this operator. The striking fact, in contrast with the uniformly elliptic case, is that for g even in $C^\infty(\partial B)$, we may not conclude that the solution G of the Dirichlet problem is C^∞ on $\overline{B}$. In fact, in dimension n, the function G is not generally in $C^n(\overline{B})$. Consider the following example:

Example: Let $n = 2$. Define

$$
g(z_1, z_2) = |z_1|^2.
$$

Of course $g \in C^\infty(\partial B)$. We now calculate $\mathcal{P}g(z)$ rather explicitly. We have

$$\mathcal{P}g(z) = \frac{1}{\sigma(\partial B)} \int_{\partial B} \frac{(1-|z|^2)^2}{|1-z\cdot\bar{\zeta}|^4} |\zeta_1|^2 \, d\sigma(\zeta).$$

Let us restrict our attention to points z in the ball of the form $z = (r+i0,0)$. We set

$$\mathcal{P}g(r+i0) \equiv \phi(r).$$

We shall show that ϕ fails to be C^2 on the interval $[0,1]$ at the point 1. We have

$$
\begin{aligned}
\phi(r) &= \frac{1}{\sigma(\partial B)} \int_{\partial B} \frac{(1-r^2)^2}{|1-r\zeta_1|^4} |\zeta_1|^2 \, d\sigma(\zeta) \\
&= \frac{(1-r^2)^2}{\sigma(\partial B)} \int_{|\zeta_1|<1} \int_{|\zeta_2|=\sqrt{1-|\zeta_1|^2}} \frac{|\zeta_1|^2}{|1-r\zeta_1|^4} \cdot \frac{1}{\sqrt{1-|\zeta_1|^2}} \, ds(\zeta_2) \, dA(\zeta_1) \\
&= \frac{(1-r^2)^2}{\sigma(\partial B)} \int_{|\zeta_1|<1} \frac{|\zeta_1|^2}{|1-r\zeta_1|^4} \frac{2\pi\sqrt{1-|\zeta_1|^2}}{\sqrt{1-|\zeta_1|^2}} \, dA(\zeta_1) \\
&= \frac{2\pi}{\sigma(\partial B)} (1-r^2)^2 \int_{|\zeta_1|<1} \frac{|\zeta_1|^2}{|1-r\zeta_1|^4} \, dA(\zeta_1).
\end{aligned}
$$

Now we set $\zeta_1 = \tau e^{i\psi}, 0 \le \tau < 1, 0 \le \psi \le 2\pi$. The integral is then

$$\frac{2\pi}{\sigma(\partial B)} (1-r^2)^2 \int_0^{2\pi} \int_0^1 \frac{\tau^2}{|1-r\tau e^{i\psi}|^4} \tau \, d\tau \, d\psi.$$

We perform the change of variables $r\tau = s$ and set $C = 2\pi/\sigma(\partial B)$. The integral becomes

$$C\frac{(1-r^2)^2}{r^4} \int_0^{2\pi} \int_0^r \frac{s^3}{|1-se^{i\psi}|^4} \, ds\, d\psi$$

$$C\frac{(1-r^2)^2}{r^4} \int_0^r s^3 \int_0^{2\pi} \frac{1}{|1-se^{i\psi}|^4} \, d\psi\, ds.$$

Now let us examine the inner integral. It equals

$$
\begin{aligned}
&\int_0^{2\pi} \frac{1}{(1-se^{i\psi})^2(1-se^{-i\psi})^2} \, d\psi \\
&= \int_0^{2\pi} \frac{e^{2i\psi}}{(1-se^{i\psi})^2(e^{i\psi}-s)^2} \, d\psi \\
&= \int_0^{2\pi} \frac{\frac{e^{i\psi}}{(1-se^{i\psi})^2}}{(e^{i\psi}-s)^2} e^{i\psi} \, d\psi \\
&= 2\pi \cdot \frac{1}{2\pi i} \oint_{|\eta|=1} \frac{\frac{\eta}{(1-s\eta)^2}}{(\eta-s)^2} \, d\eta.
\end{aligned}
$$

Applying the theory of residues to this Cauchy integral, we find that the last line equals

$$2\pi \frac{d}{d\eta}\left(\frac{\eta}{(1-s\eta)^2}\right)\bigg|_{\eta=s} = 2\pi \frac{1+s^2}{(1-s^2)^3}.$$

Thus

$$\phi(r) \quad = \quad 2\pi C \frac{(1-r^2)^2}{r^4} \int_0^r \frac{s^3(1+s^2)}{(1-s^2)^3} \, ds$$

$$\overset{(s\mapsto\sqrt{s})}{=} \quad \pi C \frac{(1-r^2)^2}{r^4} \int_0^{r^2} \frac{s(1+s)}{(1-s)^3} \, ds$$

$$= \quad \pi C \frac{(1-r^2)^2}{r^4} \int_0^{r^2} \frac{2}{(1-s)^3} - \frac{3}{(1-s)^2} + \frac{1}{(1-s)} \, ds$$

$$= \quad \pi C \frac{(1-r^2)^2}{r^4} \left\{ \left[\frac{1}{(1-s)^2} - \frac{3}{(1-s)} - \log(1-s) \right]_0^{r^2} \right\}$$

$$= \quad \pi C \frac{(1-r^2)^2}{r^4} \left\{ \frac{1}{(1-r^2)^2} - \frac{3}{1-r^2} - \log(1-r^2) + 2 \right\}$$

$$= \quad \frac{\pi C}{r^4} \left\{ 1 - 3(1-r^2) - (1-r^2)^2 \log(1-r^2) + 2(1-r^2)^2 \right\}$$

$$= \quad \pi C \left\{ \frac{1 - 3(1-r^2) + 2(1-r^2)^2}{r^4} - \frac{(1-r^2)^2 \log(1-r^2)}{r^4} \right\}.$$

Thus we see that $\phi(r)$ is the sum of two terms. The first of these is manifestly smooth at the point 1. However, the second is not C^2 (from the left) at 1. Therefore the function ϕ is not C^2 at 1 and we conclude that $\mathcal{P}g$ is not smooth at the point $(1,0) \in \partial B$, even though g itself is.   □

The phenomenon described in this example was discovered by Garnett and Krantz in 1977 (unpublished) and independently by Graham [GRA1, GRA2]. Graham subsequently developed a regularity theory for $\triangle_B$ using weighted function spaces. He also used Fourier analysis to explain the failure of boundary regularity in the usual function space topologies.

It turns out that these matters were anticipated by Folland in 1975 (see [FOL]). Using spherical harmonics, one can see clearly that the Poisson–Szegő integral of a function $g \in C^\infty(\partial B)$ will be smooth on $\overline{B}$ if and only if g is the boundary function of a pluriharmonic function (these arise naturally as the real parts of holomorphic functions—see [KRA1]). We shall explicate these matters in Chap. 4 with a discussion of spherical harmonics.

3.6 Concluding Remarks

The idea of reproducing kernels in harmonic analysis is an old one. The Poisson and Cauchy kernels date back to the mid-nineteenth century.

Cauchy integral formula is special in that its kernel, which is

$$\frac{1}{2\pi i} \cdot \frac{1}{\zeta - z},$$

is the same on any domain. A similar statement is *not* true for the Poisson kernel, although see [KRA7] for a study of the asymptotics of this kernel.

The complex reproducing kernels that are indigenous to several complex variables are much more subtle. It was only in 1974 that C. Fefferman was able to calculate the Bergman kernel asymptotics on strictly pseudoconvex domains. Prior to that, the very specific calculations of Hua [HUA] on concrete domains with a great deal of symmetry were the standard in the subject. A variant of the Fefferman's construction also applies to the Szegő kernel (see also [BOS]). Carrying out an analogous program on a more general class of domains has proved to be challenging.

This portion of the book is an invitation to study yet another kernel—the Poisson–Bergman kernel. Inspired by the ideas of [HUA], this is a positive reproducing kernel for the Bergman space. There are many questions about the role of this new kernel that remain unanswered.

Exercises

1. Use results from the text to give an approximate formula for the Berezin kernel of the annulus.
2. Up to an error term, calculate the Bergman representative coordinates on the annulus.
3. Write down a biholomorphic mapping from the unit ball B in $\mathbb{C}^n$ to the domain

$$\mathcal{U} = \{z \in \mathbb{C}^n : \operatorname{Im} z_1 > |z_2|^2 + \cdots + |z_n|^2\}.$$

   [**Hint:** Imitate the Cayley transform from classical function theory.] Use the biholomorphic invariance of the Bergman kernel to determine the Bergman kernel for $\mathcal{U}$.
4. Calculate the Berezin kernel for the domain $\mathcal{U}$ in Exercise 3.
5. What can you say about the Berezin kernel for the domain

$$U = \{\zeta \in \mathbb{C} : 0 < \operatorname{Im} \zeta < 1\}?$$

   What is the asymptotic behavior of the kernel as the argument(s) tend to infinity?

6. Use the Fefferman asymptotic expansion for the Bergman kernel of a strictly pseudoconvex domain to give a new asymptotic expansion for the Berezin kernel of a strictly pseudoconvex domain. What sort of logarithmic term do you get?

7. It is certainly the case that the Bergman kernel of the unit disc is a derivative of the Szegő kernel for the disc. Something of this nature is also true on the ball. Explain.

8. Use the philosophy of Exercise 7 together with the Fefferman asymptotic expansion to guess what the asymptotic expansion of the Szegő kernel on a strictly pseudoconvex domain should look like. Refer to the paper [BOS] to verify your answer.

9. Let $\Omega \subseteq \mathbb{C}$ be a bounded domain with C^2 boundary. Let $S \subseteq \partial\Omega$ be an arc of the boundary. Then certainly there is a conformal map φ of the disc into Ω so that an arc of the boundary of the image U coincides with S. Thus the Bergman kernel of the disc can be transferred to U. And the Bergman kernel of U can be compared to the Bergman kernel of Ω (this is all a poor man's version of the Fefferman's procedure for strictly pseudoconvex domains). Fill in the details of this argument to obtain an asymptotic expansion for the Bergman kernel of Ω.

10. Let $\Omega \subseteq \mathbb{C}$ be a bounded domain with boundary point P at which the boundary has a corner with angle θ. What can you say about the asymptotic behavior of the Bergman kernel at P? How does your answer vary with θ? What happens as $\theta \to \pi$?

Chapter 4
Partial Differential Equations

4.1 The Idea of Spherical Harmonics

Spherical harmonics are for many purposes the natural generalization of the Fourier analysis of the circle to higher dimensions. Spherical harmonics are also intimately connected to the representation theory of the orthogonal group. As a result, analogues of the spherical harmonics play an important role in general representation theory.

Our presentation of spherical harmonics owes a debt to [STW]. In fact Chap. 4, Sect. 4.2 of [STW] contains all the most basic ideas about spherical harmonics, and we refer the reader to [STW] both for the standard notation and for details.

In the next section we begin our serious investigation of the more advanced properties of spherical harmonics.

4.2 Advanced Topics in the Theory of Spherical Harmonics: The Zonal Harmonics

Since the case $N \leq 2$ is very familiar and has been treated in some detail elsewhere, let us assume from now on that $N > 2$.

Fix a point $x' \in \Sigma_{N-1}$ and consider the linear functional on $\mathcal{H}_k$ given by

$$e_{x'} : Y \mapsto Y(x').$$

Of course $\mathcal{H}_k$ is a Hilbert space so there exists a unique spherical harmonic $Z_{x'}^{(k)}$ such that

$$Y(x') = e_{x'}(Y) = \int_{\Sigma_{N-1}} Y(t') Z_{x'}^{(k)}(t') \, dt'$$

S.G. Krantz, *Geometric Analysis of the Bergman Kernel and Metric*,
Graduate Texts in Mathematics 268, DOI 10.1007/978-1-4614-7924-6_4,
© Springer Science+Business Media New York 2013

for all $Y \in \mathcal{H}_k$. [The reader will note here some formal parallels between the zonal harmonic theory and the Bergman kernel theory covered earlier. In fact this parallel goes deeper. See, for instance, [ARO] for more on these matters.]

Definition 4.2.1. The function $Z_{x'}^{(k)}$ is called the *zonal harmonic* of degree k with pole at x'.

Lemma 4.2.2. *If $\{Y_1, \ldots, Y_{a_k}\}$ is an orthonormal basis for $\mathcal{H}_k$, then*

(a) $\sum_{m=1}^{a_k} \overline{Y_m(x')} Y_m(t') = Z_{x'}^{(k)}(t')$.
(b) $Z_{x'}^{(k)}$ *is real valued and* $Z_{x'}^{(k)}(t') = Z_{t'}^{(k)}(x')$.
(c) *If ρ is a rotation, then* $Z_{\rho x'}^{(k)}(\rho t') = Z_{x'}^{(k)}(t')$.

Proof. Let $Z_{x'}^{(k)} = \sum_{m=1}^{a_k} \langle Z_{x'}^{(k)}, Y_m \rangle Y_m$ be the standard representation of $Z_{x'}^{(k)}$ with respect to the orthonormal basis $\{Y_1, \ldots, Y_{a_k}\}$. Then

$$\langle Z_{x'}^{(k)}, Y_m \rangle = \int_{\Sigma_{N-1}} \overline{Y_m(t')} Z_{x'}^{(k)}(t') \, dt' = \overline{Y_m(x')};$$

we have used here the reproducing property of the zonal harmonic (note that since Y_m is harmonic then so is $\overline{Y_m}$). This proves (a), for we know that

$$Z_{x'}^{(k)}(t') = \sum_{m=1}^{a_k} \langle Z_{x'}^{(k)}, Y_m \rangle Y_m(t') = \overline{Y_m(x')} Y_m(t').$$

To prove (b), let $f \in \mathcal{H}_k$. Then

$$\overline{f}(x') = \int_{\Sigma_{N-1}} \overline{f}(t') Z_{x'}^{(k)}(t') \, dt'$$

$$= \overline{\int_{\Sigma_{N-1}} f(t') \overline{Z_{k'}^{(k)}(t')} \, dt'}.$$

That is,

$$f(x') = \int_{\Sigma_{N-1}} f(t') \overline{Z_{x'}^{(k)}(t')} \, dt'.$$

Thus we see that $\overline{Z_{x'}^{(k)}}$ reproduces $\mathcal{H}_k$ at the point x'. By the uniqueness of the zonal harmonic at x', we conclude that $Z_{x'}^{(k)} = \overline{Z_{x'}^{(k)}}$. Hence, $Z_{x'}^{(k)}$ is real valued. Now, using (a), we have

$$Z_{x'}^{(k)}(t') = \sum_{m=1}^{a_k} \overline{Y_m(x')}Y_m(t')$$

$$= \overline{\sum_{m=1}^{a_k} Y_m(x')\overline{Y_m(t')}}$$

$$= \overline{Z_{t'}^{(k)}(x')}$$

$$= Z_{t'}^{(k)}(x').$$

This establishes (b).

To check that (c) holds, it suffices by uniqueness to see that $Z_{\rho x'}^{(k)}(\rho t')$ reproduces $\mathcal{H}_k$ at x'. This is a formal exercise which we omit. □

Lemma 4.2.3. *Let* $\{Y_1, \ldots, Y_{a_k}\}$ *be any orthonormal basis for* $\mathcal{H}^k$. *The following properties hold for the zonal harmonics:*

(a) $Z_{x'}^{(k)}(x') = \frac{a_k}{\sigma(\Sigma_{N-1})}$, *where* $a_k = \dim \mathcal{A}_k = \dim \mathcal{H}_k$.

(b) $\sum_{m=1}^{a_k} |Y_m(x')|^2 = \frac{a_k}{\sigma(\Sigma_{N-1})}$.

(c) $|Z_{t'}^{(k)}(x')| \leq \frac{a_k}{\sigma(\Sigma_{N-1})}$.

Proof. Let $x_1', x_2' \in \Sigma_{N-1}$ and let ρ be a rotation such that $\rho x_1' = x_2'$. Then, by parts (a) and (c) of 4.2.2, we know that

$$\sum_{m=1}^{a_k} |Y_m(x_1')|^2 = Z_{x_1'}^{(k)}(x_1') = Z_{x_2'}^{(k)}(x_2') = \sum_{m=1}^{a_k} |Y_m(x_2')|^2 \equiv c.$$

Thus

$$a_k = \sum_{m=1}^{a_k} \int_{\Sigma_{N-1}} |Y_m(x')|^2 \, d\sigma(x')$$

$$= \int_{\Sigma_{N-1}} \sum_{m=1}^{a_k} |Y_m(x')|^2 \, dx'$$

$$= c\sigma(\Sigma_{N-1}).$$

This proves parts (a) and (b).

For part (c), notice that

$$\|Z_{x'}^{(k)}\|_{L^2}^2 = \int_{\Sigma_{N-1}} |Z_{x'}^{(k)}(t')|^2 \, dt'$$

$$= \int_{\Sigma_{N-1}} \left(\sum_m \overline{Y_m(x')} Y_m(t') \right) \left(\overline{\sum_\ell \overline{Y_\ell(x')} Y_\ell(t')} \right) \, dt'$$

$$= \sum_m |Y_m(x')|^2$$

$$= \frac{a_k}{\sigma(\Sigma_{N-1})}.$$

Finally, we use the reproducing property of the zonal harmonics to see that

$$|Z_{t'}^{(k)}(x')| = \left| \int_{\Sigma_{N-1}} Z_{t'}^{(k)}(w') Z_{x'}^{(k)}(w') \, dw' \right|$$

$$\leq \|Z_{t'}^{(k)}\|_{L^2} \cdot \|Z_{x'}^{(k)}\|_{L^2}$$

$$= \frac{a_k}{\sigma(\Sigma_{N-1})}. \qquad \square$$

Now we wish to present a version of the expansion of the Poisson kernel in terms of spherical harmonics in higher dimensions. Recall that the Poisson kernel for the ball in $\mathbb{R}^N$ is

$$P(x, t') = \frac{1}{\sigma(\Sigma_{N-1})} \frac{1 - |x|^2}{|x - t'|^N}$$

for $0 \leq |x| < 1$ and $|t'| = 1$. Now we have

Theorem 4.2.4. *If $x \in B$, then we write $x = rx'$ with $|x'| = 1$. It holds that*

$$P(x, t') = \sum_{k=0}^{\infty} r^k Z_{x'}^{(k)}(t') = \sum_{k=0}^{\infty} r^k Z_{t'}^{(k)}(x')$$

is the Poisson kernel for the ball. That is, if $f \in C(\partial B)$, then

$$\int_{\partial B} P(x, t') f(t') \, d\sigma(t') \equiv u(x)$$

solves the Dirichlet problem on the ball with data f.

Proof. Observe that

$$a_k = d_k - d_{k-2} = \binom{N+k-1}{k} - \binom{N+k-3}{k-2}$$

$$= \frac{(N+k-3)!}{(k-1)!(N-2)!} \left\{ \frac{(N+k-1)(N+k-2)}{k(N-1)} - \frac{k-1}{N-1} \right\}$$

$$= \binom{N+k-3}{k-1} \left\{ \frac{N+2k-2}{k} \right\}$$

$$\leq C \cdot \binom{N+k-3}{k-1}$$

$$\leq C \cdot k^{N-2}.$$

Here $C = C(N)$ depends on the dimension, but not on k. With this estimate, and the estimate on the size of the zonal harmonics from the preceding lemma, we see that the series

$$\sum_{k=0}^{\infty} r^k Z_{t'}^{(k)}(x')$$

converges uniformly on compact subsets of B. Indeed, for $|x| \leq s < 1, x = rx'$, we have that

$$\sum_{k=0}^{\infty} |r^k Z_{t'}^{(k)}(x')| \leq \sum_{k=0}^{\infty} s^k \frac{a_k}{\sigma(\Sigma_{N-1})} \leq \sum_{k=0}^{\infty} s^k \frac{C \cdot k^{N-2}}{\sigma(\Sigma_{N-1})} = C' \cdot \sum_{k=0}^{\infty} s^k k^{N-2} < \infty.$$

Now let $u(t') = \sum_{m=0}^{P} Y_m(t')$ be a finite linear combination of spherical harmonics with all $Y_m \in \mathcal{H}_k$. Then

$$\sum_{m=0}^{p} |x|^k Y_m \left(\frac{x}{|x|} \right) \equiv u(x) = \int_{\Sigma_{N-1}} u(t') P(x,t') \, dt'$$

is the solution to the classical Dirichlet problem with data Y. Here $P(x,t')$ is the classical Poisson kernel. On the other hand,

$$\int_{\Sigma_{N-1}} u(t') \sum_{k=0}^{\infty} |x|^k Z_{t'}^{(k)}(x')\, dt'$$

$$= \sum_{m=0}^{p} \int_{\Sigma_{N-1}} Y_m(t') \sum_{k=0}^{\infty} |x|^k Z_{t'}^{(k)}(x')\, dt'$$

$$= \sum_{m=0}^{p} \sum_{k=0}^{\infty} |x|^k \int_{\Sigma_{N-1}} Y_m(t') Z_{t'}^{(k)}(x')\, dt'$$

$$= \sum_{m=0}^{p} |x|^m Y_m(x')$$

$$= u(x).$$

Thus

$$\int_{\Sigma_{N-1}} \left[P(x,t') - \sum_k |x|^k Z_x'^{(k)}(t') \right] u(t')\, dt' = 0$$

for all finite linear cominations of spherical harmonics. Since the latter are dense in $L^2(\Sigma_{N-1})$, the desired assertion follows.   □

Our immediate goal now is to obtain an explicit formula for each zonal harmonic $Z_{x'}^{(k)}$. We begin this process with some generalities about polynomials.

Lemma 4.2.5. *Let P be a polynomial in $\mathbb{R}^N$ such that*

$$P(\rho x) = P(x)$$

for all $\rho \in O(N)$ and $x \in \mathbb{R}^N$. Then there exist constants $c_0, \ldots, c_p$ such that

$$P(x) = \sum_{m=0}^{p} c_m \left(x_1^2 + \cdots + x_N^2 \right)^m.$$

Proof. We write P as a sum of homogeneous terms:

$$P(x) = \sum_{\ell=0}^{q} P_\ell(x),$$

where P_ℓ is hologeneous of degree ℓ. Now for any $\epsilon > 0$ and $\rho \in O(N)$, we have

$$\sum_{\ell=0}^{q} \epsilon^{\ell} P_{\ell}(x) = \sum_{\ell=0}^{q} P_{\ell}(\epsilon x)$$

$$= P(\epsilon x)$$

$$= P(\epsilon \rho x)$$

$$= \sum_{\ell=0}^{q} P_{\ell}(\epsilon \rho x)$$

$$= \sum_{\ell=0}^{q} \epsilon^{\ell} P_{\ell}(\rho x).$$

For fixed x, we think of the far left and far right of this last sequence of equalities as identities *of polynomials in* ϵ. It follows that $P_{\ell}(x) = P_{\ell}(\rho x)$ for every ℓ. The result of these calculations is that we may concentrate our attentions on P_{ℓ}.

Consider the function $|x|^{-\ell} P_{\ell}(x)$. It is homogeneous of degree 0 and still invariant under the action of $O(N)$. Then

$$|x|^{-\ell} P_{\ell}(x) = c_{\ell},$$

for some constant c_{ℓ}. This forces ℓ to be even (since P_{ℓ} is a *polynomial* function); the result follows. □

Definition 4.2.6. Let $e \in \Sigma_{N-1}$. A *parallel of* Σ_{N-1} *orthogonal to* e is the intersection of Σ_{N-1} with a hyperplane (not necessarily through the origin) orthogonal to the line determined by e and the origin.

Notice that a parallel of Σ_{N-1} orthogonal to e is a set of the form

$$\{x' \in \Sigma : x' \cdot e = c\},$$

$-1 \leq c \leq 1$. Observe that a function F on Σ_{N-1} is constant on parallels orthogonal to $e \in \Sigma_{N-1}$ if and only if for all $\rho \in O(N)$ that fix e it holds that $F(\rho x') = F(x')$.

Lemma 4.2.7. *Let* $e \in \Sigma_{N-1}$. *An element* $Y \in \mathcal{H}_k$ *is constant on parallels of* Σ *orthogonal to* e *if and only if there exists a constant* c *such that*

$$Y = cZ_e^{(k)}.$$

Proof. Recall that we are assuming that $N \geq 3$. Let ρ be a rotation which fixes e. Then, for each $x' \in \Sigma$, we have

$$Z_e^{(k)}(x') = Z_{\rho e}^{(k)}(\rho x') = Z_e^{(k)}(\rho x').$$

Hence $Z_e^{(k)}$ is constant on the parallels of Σ orthogonal to e.

To prove the converse direction, assume that $Y \in \mathcal{H}_k$ is constant on the parallels of Σ orthogonal to e. Let $e_1 = (1, 0, \ldots, 0) \in \Sigma$ and let τ be a rotation such that $e = \tau e_1$. Define

$$W(x') = Y(\tau x').$$

Then $W \in \mathcal{H}_k$ is constant on the parallels of Σ orthogonal to e_1. Suppose we can show that $W = c Z_{e_1}^{(k)}(x')$ for some constant c. Then

$$Y(x') = W(\tau^{-1} x') = c Z_{e_1}^{(k)}(\tau^{-1} x')$$
$$= c Z_{\tau e_1}^{(k)}(x') = c Z_e^{(k)}(x').$$

So the lemma will follow. Thus we examine W and take $e = e_1$.

Define

$$P(x) = \begin{cases} |x|^k W(x/|x|) & \text{if } x \neq 0 \\ 0 & \text{if } x = 0. \end{cases}$$

Let ρ be a rotation that fixes e_1. We write

$$P(x) = \sum_{j=0}^{k} x_1^{k-j} P_j(x_2, \ldots, x_N).$$

Since ρ fixes the powers of x_1, it follows that ρ leaves each P_j invariant. Then each P_j is a polynomial in $(x_2, \ldots, x_N) \in \mathbb{R}^{N-1}$ that is invariant under the rotations of $\mathbb{R}^{N-1}$. We conclude that $P_j = 0$ for odd j and

$$P_j(x_2, \ldots, x_N) = c_j \left(x_2^2 + \cdots + x_N^2 \right)^{j/2} \equiv c_j R^j(x_2, \ldots, x_N)$$

for j even. Therefore

$$P(x) = c_0 x_1^k + c_2 x_1^{k-2} R^2 + \cdots c_{2\ell} x_1^{2\ell} R^{k-2\ell},$$

for some $\ell \leq k/2$. Of course P is harmonic, so $\triangle P \equiv 0$. A direct calculation then shows that

$$0 = \triangle P = \sum_p \left[c_{2p} \alpha_p + c_{2(p+1)} \beta_p \right] x_1^{k-2(p+1)} R^{2p},$$

where

$$\alpha_p \equiv (k - 2p)(k - 2p - 1)$$

and

$$\beta_p \equiv 2(p + 1)(N + 2p - 1).$$

Therefore we find the following recursion relation for the c's:

$$c_{2(p+1)} = -\frac{\alpha_p c_{2p}}{\beta_p}.$$

In particular, c_0 determines all the other c's.

From this it follows that all the elements of $\mathcal{H}_k$ which are constant on parallels of Σ orthogonal to e_1 are constant multiples of each other. Since $Z_{e_1}^{(k)}$ is one such element of $\mathcal{H}_k$, this proves our result.   □

Lemma 4.2.8. *Fix* k. *Let* $F_{y'}(x')$ *be defined for all* $x', y' \in \Sigma$. *Assume that*

(i) $F_{y'}(\cdot)$ *is a spherical harmonic of degree* k *for every* $y' \in \Sigma$.
(ii) *For every rotation* ρ *we have* $F_{\rho y'}(\rho x') = F_{y'}(x')$, *all* $x', y' \in \Sigma$.

Then there is a constant c *such that*

$$F_{y'}(x') = c Z_{y'}^{(k)}(x').$$

Exercise: Show that a function that is invariant under a Lie group action must be smooth (because the group is). Thus it follows immediately that the function F in the lemma is a priori smooth.

Proof of the Lemma: Fix $y' \in \Sigma$ and let $\rho \in O(N)$ be such that $\rho(y') = y'$. Then

$$F_{y'}(x') = F_{\rho y'}(\rho x') = F_{y'}(\rho x').$$

Therefore by the preceding lemma,

$$F_{y'}(x') = c_{y'} Z_{y'}^{(k)}(x').$$

(Here the constant $c_{y'}$ may in principle depend on y'.) We need to see that for $y_1', y_2' \in \Sigma$ arbitrary, it in fact holds that $c_{y_1'} = c_{y_2'}$. Let $\sigma \in O(N)$ be such that $\sigma(y_1') = y_2'$. By hypothesis (ii),

$$
\begin{aligned}
c_{y_2'} Z_{y_2'}^{(k)}(\sigma x') &= F_{y_2'}(\sigma x') \\
&= F_{\sigma y_1'}(\sigma x') \\
&= F_{y_1'}(x') \\
&= c_{y_1'} Z_{y_1'}^{(k)}(x') \\
&= c_{y_1'} Z_{\sigma y_1'}^{(k)}(\sigma x') \\
&= c_{y_1'} Z_{y_2'}^{(k)}(\sigma x').
\end{aligned}
$$

Since these equalities hold for all $x' \in \Sigma$, we conclude that

$$c_{y_2'} = c_{y_1'}.$$

That is,

$$F_{y'}(x') = c Z_{y'}^{(k)}(x').$$ $\qquad\square$

Definition 4.2.9. Let $0 \le |z| < 1, |t| \le 1$, and fix $\lambda > 0$. Consider the equation $z^2 - 2tz + 1 = 0$. Then $z = t \pm \sqrt{t^2 - 1}$ so that $|z| = 1$. Hence, $z^2 - 2tz + 1$ is zero-free in the disc $\{z : |z| < 1\}$ and the function $z \mapsto (1 - 2tz + z^2)^{-\lambda}$ is well defined and holomorphic in the disc. Set, for $0 \le r < 1$,

$$(1 - 2rt + r^2)^{-\lambda} = \sum_{k=0}^{\infty} P_k^\lambda(t) r^k.$$

Then $P_k^\lambda(t)$ is defined to be the *Gegenbauer polynomial of degree k* associated to the parameter λ.

Proposition 4.2.10. *The Gegenbauer polynomials satisfy the following properties:*

(1) $P_0^\lambda(t) \equiv 1$.
(2) $\frac{d}{dt} P_k^\lambda(t) = 2\lambda P_{k-1}^{\lambda+1}(t)$ *for* $k \ge 1$.
(3) $\frac{d}{dt} P_1^\lambda(t) = 2\lambda P_0^{\lambda+1}(t) = 2\lambda$.
(4) P_k^λ *is actually a polynomial of degree k in t.*
(5) *The monomials* $1, t, t^2, \ldots$ *can be obtained as finite linear combinations of* $P_0^\lambda, P_1^\lambda, P_2^\lambda, \ldots$.
(6) *The linear space spanned by the* P_k^λ*'s is uniformly dense in* $C[-1, 1]$.
(7) $P_k^\lambda(-t) = (-1)^k P_k^\lambda(t)$ *for all* $k \ge 0$.

Proof. We obtain (1) by simply setting $r = 0$ in the defining equation for the Gegenbauer polynomials.
   For (2), note that

$$2r\lambda \sum_{k=0}^{\infty} P_k^{\lambda+1}(t) r^k \equiv 2r\lambda (1 - 2rt + r^2)^{-(\lambda+1)}$$

$$= \frac{d}{dt}(1 - 2rt + r^2)^{-\lambda}$$

$$= \sum_{k=0}^{\infty} \frac{d}{dt} P_k^\lambda(t) r^k.$$

The result now follows by identifying coefficients of like powers of r.

For (3), observe that [using (1) and (2)]

$$\frac{d}{dt} P_1^\lambda(t) = 2\lambda P_0^{k+1}(t) = 2\lambda.$$

It follows from integration that P_1^λ is a polynomial of degree 1 in t. Applying (2) and iterating yields (4).

Now (5) follows from (4) (inductively) and (6) is immediate from (5) and the Weierstrass approximation theorem.

Finally,

$$\sum_{k=0}^\infty P_k^\lambda(-t)r^k \equiv \left(1 - 2r(-t) + r^2\right)^{-\lambda}$$

$$= \left(1 - 2t(-r) + (-r)^2\right)^{-\lambda}$$

$$= \sum_{k=0}^\infty P_k^\lambda(t)(-r)^k$$

$$= \sum_{k=0}^\infty (-1)^k P_k^\lambda(t)r^k.$$

Now (7) follows from comparing coefficients of like powers of r. $\square$

Theorem 4.2.11. *Let $N > 2, \lambda = (N-2)/2, k \in \{0, 1, 2, \dots\}$. Then there exists a constant $c_{k,N}$ such that*

$$Z_{y'}^{(k)}(x') = c_{k,N} P_k^\lambda(x' \cdot y').$$

Exercise: Compute by hand what the analogous statement is for $N = 2$. (Recall that the zonal harmonics in dimension 2 are just $\cos k\theta / \sqrt{\pi}$ and $\sin k\theta / \sqrt{\pi}$ for $k \geq 1$.)

Proof. Let $y' \in \Sigma$ be fixed. For $x \in \mathbb{R}^N$ define

$$F_{y'}(x) = |x|^k P_k^\lambda \left(\frac{x}{|x|} \cdot y' \right).$$

By part (7) of Proposition 4.2.10, if k is even, then

$$P_k^\lambda(t) = \sum_{j=0}^m d_{2j} t^{2j} \qquad \text{with } 2m = k;$$

also if k is odd, then

$$P_k^\lambda(t) = \sum_{j=0}^m d_{2j+1} t^{2j+1} \qquad \text{with } 2m+1 = k.$$

In both cases, $F_{y'}(x)$ is then a homogeneous polynomial of degree k. For instance, if k is even, then

$$F_{y'}(x) = |x|^k P_k^\lambda \left(\frac{x \cdot y'}{|x|} \right)$$

$$= |x|^{2m} \sum_{j=0}^m d_{2j} \left(\frac{x \cdot y'}{|x|} \right)^{2j}$$

$$= \sum_{j=0}^m d_{2j} (|x|^2)^{m-j} (x \cdot y')^{2j}.$$

We want to check that the hypotheses of Lemma 4.2.8 are satisfied when $F_{y'}(x')$ is so defined. Once this is done then the conclusion of our Proposition follows immediately. Thus we need to check that $F_{y'}(x')$ is rotationally invariant and that $F_{y'}(\cdot)$ is harmonic.

If $\rho \in O(N)$ and $x' \in \Sigma_{N-1}$ then

$$F_{\rho y'}(\rho x') = |x'|^k P_k^\lambda \left(\frac{\rho x' \cdot \rho y'}{|x|} \right)$$

$$= |x'|^k P_k^\lambda \left(\frac{x' \cdot y'}{|x|} \right)$$

$$= F_{y'}(x').$$

This establishes the rotational invariance.

To check harmonicity, recall that the map $x \mapsto |x - y'/s|^{2-N}$ is harmonic on $\mathbb{R}^N \setminus \{y'/s\}$ when $N \geq 3, s \neq 0$, and $y' \in \Sigma$. Then, with $\lambda = (N-2)/2$, we have

$$s^{2-N} \left| x - \frac{y'}{s} \right|^{2-N} = [(sx - y') \cdot (sx - y')]^{(2-N)/2}$$

$$= [|sx|^2 - 2(sx) \cdot y' + 1]^{(2-N)/2}$$

$$= \left[1 - 2(s|x|) \left(\frac{x}{|x|} \cdot y' \right) + (s|x|)^2 \right]^{-\lambda}$$

$$\equiv [1 - 2rt + r^2]^{-\lambda}$$

$$= \sum_{k=0}^\infty s^k |x|^k P_k^\lambda \left(\frac{x}{|x|} \cdot y' \right). \qquad (4.2.11.1)$$

Here we have taken $r = s|x|$ and $t = (x/|x|) \cdot y'$. Thus the sum at the end of this calculation is a harmonic function of x in $R_s = \{x \in \mathbb{R}^N : 0 < |x| < 1/s\}$ for $y' \in \Sigma$ fixed.

To see that each coefficient

$$|x|^k P_k^\lambda \left(\frac{x}{|x|} \cdot y' \right)$$

in the series is a harmonic function of $x \in \mathbb{R}^N$, we proceed as follows. Fix $0 \neq x_0 \in \mathbb{R}^N$. Then for every s such that $0 < s < 1/|x_0|$ formula (4.2.11.1) tells us that the function

$$x \mapsto \sum_{k=0}^\infty s^k |x|^k P_k^\lambda \left(\frac{x}{|x|} \cdot y' \right)$$

is harmonic. Therefore this function satisfies the mean-value property. By uniform convergence we can switch the order of summation and integration in the mean-value property to obtain

$$\sum_{k=0}^\infty s^k \frac{1}{\sigma(B(x_0,r))} \int_{\partial B(x_0,r)} |x|^k P_k^\lambda \left(\frac{x}{|x|} \cdot y' \right) d\sigma(x)$$

$$= \frac{1}{\sigma(B(x_0,r))} \int_{\partial B(x_0,r)} \sum_{k=0}^\infty s^k |x|^k P_k^\lambda \left(\frac{x}{|x|} \cdot y' \right) d\sigma(x)$$

$$= \sum_{k=0}^\infty s^k |x_0|^k P_k^\lambda \left(\frac{x_0}{|x_0|} \cdot y' \right)$$

for $0 < r < |x_0|$. Since this equality holds for $0 < s < 1/|x_0|$, the identity principle for power series tells us that

$$\frac{1}{\sigma(B(x_0,r))} \int_{\partial B(x_0,r)} |x|^k P_k^\lambda \left(\frac{x}{|x|} \cdot y' \right) d\sigma(x) = |x_0|^k P_k^\lambda \left(\frac{x_0}{|x_0|} \cdot y' \right)$$

for every $0 < r < |x_0|$. It is a standard fact (see [KRA1, Chap. 1]) that any function satisfying a mean-value property of this sort—for any x_0 and all small r—must be harmonic. We conclude that

$$F_{y'}(x) = |x|^k P_k^\lambda \left(\frac{x}{|x|} \cdot y' \right)$$

is harmonic. The theorem follows. $\qquad\square$

4.3   Spherical Harmonics in the Complex Domain and Applications

Now we give a rendition of "bigraded spherical harmonics" which is suitable for the study of functions of several complex variables. Our purpose is to return finally to the study of the regularity for the Laplace–Beltrami operator for the Bergman metric on the ball. Because of the detailed exposition that has gone on before, and because much of this new material is routine, we shall perform many calculations in $\mathbb{C}^2$ only and shall leave several others to the reader.

Definition 4.3.1. Let $\mathcal{H}^{p,q}$ be the space consisting of all restrictions to the unit sphere in $\mathbb{C}^n$ of harmonic (in the classical sense) polynomials that are homogeneous of degree p in z and homogeneous of degree q in $\bar{z}$.

Observe that

$$\mathcal{H}_k = \cup_{p+q=k} \mathcal{H}^{p,q}.$$

Proposition 4.3.2. *The spaces $\mathcal{H}^{p,q}$ enjoy the following properties:*

(1) $D(p,q;n) \equiv \dim_{\mathbb{C}} \mathcal{H}^{p,q} = \dfrac{(p+q+n-1)(p+n-2)!(q+n-2)!}{p!q!(n-1)!(n-2)!}.$

(2) *The space $\mathcal{H}^{p,q}$ is $U(n)$ irreducible. That is, $\mathcal{H}^{p,q}$ has no proper linear subspace L such that, for each $U \in U(n)$, U maps L into L.*

(3) *If $f_1, \ldots, f_D$ is an orthonormal basis for $\mathcal{H}^{p,q}$, $D = D(p,q;n)$, then*

$$H_n^{p,q}(\zeta, \eta) \equiv \sum_{j=1}^{D} f_j(\zeta) \overline{f_j(\eta)}$$

reproduces $\mathcal{H}^{p,q}$. That is, if $\phi \in \mathcal{H}^{p,q}, \zeta \in \Sigma$, then

$$\phi(\zeta) = \int_\Sigma H_n^{p,q}(\zeta, \eta) \phi(\eta) \, d\sigma(\eta).$$

(4) *The orthogonal projection $\pi_{p,q} : L^2(\Sigma) \to \mathcal{H}^{p,q}$ is given by*

$$\pi_{p,q}(f)(\zeta) = \int_\Sigma f(\eta) H_n^{p,q}(\zeta, \eta) \, d\sigma(\eta).$$

Proof. We leave the proofs of parts (1) and (2) as exercises.

To prove (3), notice that if $\phi \in \mathcal{H}^{p,q}$, then we may write $\phi = \sum_{j=1}^{D} a_j f_j$. Then

$$\int_{\Sigma} H_n^{p,q}(\zeta,\eta)\phi(\eta)\,d\sigma(\eta) = \int_{\Sigma}\left(\sum_{j=1}^{D} f_j(\zeta)\overline{f_j(\eta)}\right)\left(\sum_{j=1}^{D} a_k f_k(\eta)\right)d\sigma(\eta)$$

$$= \sum_{j,k=1}^{D} a_k f_j(\zeta)\int_{\Sigma}\overline{f_j(\eta)}f_k(\eta)\,d\sigma(\eta)$$

$$= \sum_{j=1}^{D} a_j f_j(\zeta)$$

$$= \phi(\zeta).$$

For (4), select $g \in L^2(\Sigma)$. Then

$$\pi_{p,q}g(\zeta) = \int_{\Sigma} g(\eta)\sum_{j=1}^{D}\overline{f_j(\eta)}f_j(\zeta)\,d\sigma(\eta)$$

$$= \sum_{j=1}^{D}\left(\int_{\Sigma} g(\eta)\overline{f_j(\eta)}\,d\sigma(\eta)\right)f_j(\zeta)$$

so that $\pi_{p,q}$ maps L^2 into $\mathcal{H}^{p,q}$. By (3) it follow that $\pi_{p,q}\circ\pi_{p,q} = \pi_{p,q}$. Finally, $\pi_{p,q}$ is plainly self-adjoint. Thus $\pi_{p,q}$ is the orthogonal projection onto $\mathcal{H}^{p,q}$. $\quad\square$

In order to present the solution of the Dirichlet problem for the Laplace–Beltrami operator $\triangle_B$, we need to define another special function. This one is defined by means of an ordinary differential equation.

Definition 4.3.3. Let $a,b \in \mathbb{R}$ and $c > 0$. The linear differential equation

$$x(1-x)y'' + [c-(a+b+1)x]y' - aby = 0 \qquad (4.3.3.1)$$

is called the *hypergeometric equation*.

If we divide the hypergeometric equation through by the leading factor $x(1-x)$, we see that this is an ordinary differential equation with a regular singularity at 0. It follows (see [COL]) that (4.3.3.1) has a solution of the form

$$x^\lambda \sum_{j=0}^{\infty} a_j x^j, \qquad (4.3.3.2)$$

where $a_0 \neq 0$ and the series converges for $|x| < 1$. Let us now sketch what transpires when the expression (4.3.3.2) is substituted into the differential equation (4.3.3.1).

We find that

$$\sum_{j=0}^{\infty} a_j x^{\lambda+j-1}(\lambda+j)(\lambda+j-1+c) - \sum_{j=0}^{\infty} a_j x^{\lambda+j}(\lambda+j+a)(\lambda+j+b) = 0,$$

which gives the following system of equations for determining the exponent λ and the coefficients a_j :

$$a_0\lambda(\lambda - 1 + c) = 0$$

$$a_j(\lambda + j)(\lambda + j - 1 + c) - a_{j-1}(\lambda + j - 1 + a)(\lambda + j - 1 + b) = 0 \quad , j \geq 1.$$

The first of these equations yields that either $\lambda = 0$ or $\lambda = 1 - c$.

First consider the case $\lambda = 0$. We find that

$$a_j = \frac{(j - 1 + a)(j - 1 + b)}{j(j - 1 + c)} a_{j-1} \qquad , j = 1, 2, \ldots.$$

Setting $a_0 = 1$ we obtain

$$a_j = \frac{a(a + 1) \cdots (a + j - 1)b(b + 1) \cdots (b + j - 1)}{j!c(c + 1) \cdots (c + j - 1)}$$

$$= \frac{\Gamma(a + j)\Gamma(b + j)\Gamma(c)}{j!\Gamma(a)\Gamma(b)\Gamma(c + j)},$$

where Γ is the classical gamma function (see [CCP]). Thus for $\lambda = 0$, a particular solution to (4.3.3.1) is

$$y(x) = F(a, b, c; x) \equiv \sum_{j=0}^{\infty} \frac{\Gamma(a + j)\Gamma(b + j)\Gamma(c)}{\Gamma(a)\Gamma(b)\Gamma(c + j)} \cdot \frac{x^j}{j!}.$$

Now consider the case $\lambda = 1 - c$. Arguing in the same manner, if $c \neq 2, 3, 4, \ldots$, we find (setting $a_0 = 1$ again) that a particular solution of the differential equation is given by

$$y(x) = F(1 - c + a, 1 - c + b, 2 - c; x)$$

$$\equiv x^{1-c} \sum_{j=0}^{\infty} \frac{\Gamma(1 - c + a + j)\Gamma(1 - c + b + j)\Gamma(2 - c)}{\Gamma(1 - c + a)\Gamma(1 - c + b)\Gamma(2 - c + j)} \cdot \frac{x^j}{j!}.$$

Remark 4.3.4. We leave as an exercise the following statement: By checking the asymptotic behavior of $F(1 - c + a, 1 - c + b, 2 - c; x)$ at the origin, one may see that this function is linearly independent from that found when $\lambda = 0$. The functions F are known as the *hypergeometric functions*. See [ERD] for more on these matters.

In the case that $c = 2, 3, 4, \ldots$, then a modification of the above calculations (again see [COL, p. 165]) gives rise to a solution with a logarithmic singularity at 0. Define

$$S_n^{p,q}(r) = r^{p+q} \frac{F(p, q, p + q + n; r^2)}{F(p, q, p + q + n; 1)}.$$

We want to show that $S_n^{p,q}$ is C^∞ on the interval $(-1, 1)$ and continuous on $[-1, 1]$. We will make use of the following classical summation tests for series. For more on these tests, see [STR].

Lemma 4.3.5 (Dini–Kummer). *For $j = 1, 2, \ldots$ let $a_j, b_j > 0$ and put*

$$D_j = b_j - b_{j+1} \frac{a_{j+1}}{a_j}.$$

If $\liminf_{j \to \infty} D_j > 0$, then the series $\sum_j a_j$ converges.

Remark 4.3.6. Notice that, if $b_j = 1$ for all j, then this test reduces to the ratio test. □

Proof. By hypothesis, we may find a $\beta > 0$ and an integer $j_0 > 0$ such that, if $j \geq j_0$, then $D_j > \beta$. Thus

$$\beta < b_j - b_{j+1} \frac{a_{j+1}}{a_j}$$

so that

$$0 < a_j < \frac{a_j b_j - b_{j+1} a_{j+1}}{\beta} \tag{4.3.5.1}$$

for $j \geq j_0$.
  Now

$$\sum_{j=j_0}^{\infty} \frac{1}{\beta}(a_j b_j - a_{j+1} b_{j+1}) = \lim_{J \to \infty} \frac{1}{\beta} \sum_{j=j_0}^{J} (a_j b_j - a_{j+1} b_{j+1}).$$

By our hypothesis, $a_j b_j > a_{j+1} b_{j+1} > 0$ for all $j \geq j_0$. Therefore we may set $\gamma = \lim_{j \to \infty} a_j b_j$. The number γ is finite and nonnegative. Using (4.3.5.1) we have

$$\sum_{j=j_0}^{\infty} a_j < \frac{1}{\beta} \sum_{j=j_0}^{\infty} (a_j b_j - a_{j+1} b_{j+1})$$

$$= \frac{1}{\beta}(a_{j_0} b_{j_0} - \gamma)$$

$$< \infty. \qquad \qquad \square$$

Corollary 4.3.7 (Raabe). *If $a_j > 0$ for $j = 1, 2, \ldots$, we set $Q_j = j(1 - a_{j+1}/a_j)$. If it holds that*

$$\liminf Q_j > 1, \tag{4.3.7.1}$$

then $\sum_j a_j$ converges.

Proof. Let $b_1 = 1$ and $b_j = j - 1$ for $j \geq 2$. Then

$$Q_j - 1 = j\left(1 - \frac{a_{j+1}}{a_j}\right) - 1$$

$$= (j - 1) - j\frac{a_{j+1}}{a_j}$$

$$= D_j,$$

where we are using the notation of the lemma. Then $\liminf_{j \to \infty} Q_j > 1$ if and only if $\liminf_{j \to \infty} D_j > 0$.   □

Proposition 4.3.8. *Take*

$$F(a, b, c; x) = \sum_{j=0}^{\infty} \frac{\Gamma(a + j)\Gamma(b + j)\Gamma(c)}{\Gamma(a)\Gamma(b)\Gamma(c + j)} \cdot \frac{x^j}{j!}$$

as usual. If $|x| = 1$ and $c > a + b$, then the series converges absolutely.

Proof. We want to apply Raabe's test. Thus we need to calculate the terms Q_j. Denote the absolute value of the jth summand by α_j. Then, since $|x| = 1$, we have

$$\frac{\alpha_{j+1}}{\alpha_j} = \frac{(a + j)(b + j)}{(j + 1)(c + j)}.$$

Set $c = a + b + \delta$, where this equality defines $\delta > 0$. Then

$$\frac{\alpha_{j+1}}{\alpha_j} = \frac{ab + aj + bj + j^2}{(j + 1)(a + b + \delta + j)}$$

$$= 1 - \frac{\delta j + a + b + \delta + j - ab}{(j + 1)(a + b + \delta + j)}$$

$$= 1 - \frac{(1 + \delta)j}{(j + 1)(a + b + \delta + j)} + O(1/j^2).$$

As a result,

$$Q_j \equiv j\left(1 - \frac{\alpha_{j+1}}{\alpha_j}\right)$$

$$= j\left(\frac{(1 + \delta)j}{(j + \delta)(a + b + \delta + j)} + O(1/j^2)\right)$$

and $\liminf_{j \to \infty} Q_j = 1 + \delta > 1$. Thus Raabe's test implies our result.   □

It follows from the proposition that $S_n^{p,q}$ is continuous on $[-1, 1]$ and C^∞ on $(-1, 1)$. We need to know when the function is in fact C^∞ up to the endpoints. If either $p = 0$ or $q = 0$, then the order zero term of the hypergeometric equation drops out. One may solve this hypergeometric equation for solutions of the form

$$\sum_{j=0}^{\infty} a_j (x - 1)^{j+\lambda} \tag{4.3.9}$$

and find that in fact the solutions are *real analytic* near 1; in particular they are smooth. On the other hand, if both p and q are not zero, then the hypergeometric equation never has real analytic solutions near 1 as we may learn by substituting (4.3.9) into the differential equation. In fact the solutions are never C^n, where n is the dimension of the complex space that we are studying.

Remark 4.3.10. Gauss found that

$$\lim_{x \to 1^-} F(a, b, c; x) = \frac{\Gamma(c)\Gamma(c - a - b)}{\Gamma(c - a)\Gamma(c - b)}.$$

Also one may substitute the function

$$y = \int_{[0,1]} (u - x)^{\xi-1} \phi(u) \, du,$$

where ξ is a constant to be selected, into the hypergeometric equation. Some calculations, together with standard uniqueness theorems for ordinary differential equations, lead to the formula

$$F(a, b, c; x) = \frac{\Gamma(c)}{\Gamma(b)\Gamma(a)} \int_0^1 t^{b-1} (1 - t)^{a-b-1} (1 - xt)^{-a} \, dt$$

for $0 < x < 1$. It is easy to see from this formula that F cannot be analytically continued past 1.   □

As a consequence of our last proposition,

$$S_n^{p,q}(r) = r^{p+q} \frac{F(p, q, p + q + n; r^2)}{F(p, q, p + q + n; 1)}$$

is well defined and C^∞ when $0 \leq r < 1$.

Theorem 4.3.11. *Let $f \in \mathcal{H}^{p,q}$. Then the solution of the Dirichlet problem*

$$\begin{cases} \triangle_B u = 0 \text{ on } B \\ u = f \quad \text{on } \partial B \equiv \Sigma \end{cases}$$

is given by

$$u(r\zeta) = f(\zeta)S_n^{p,q}(r)$$

for $\zeta \in \Sigma$ and $0 \le r \le 1$.

Proof. To simplify the calculations, we shall prove the theorem only in dimension $n = 2$.

Let $F_0(z) = z_1^p \bar{z}_2^q$ and $f_0 = F_0|_{\partial B}$. Then the ordinary Laplacian

$$\Delta \equiv 4\left(\frac{\partial^2}{\partial z_1 \partial \bar{z}_1} + \frac{\partial^2}{\partial z_2 \partial \bar{z}_2}\right)$$

annihilates F_0. Recall that $\mathcal{H}^{p,q}$ is irreducible for $U(2)$. This means that $\{f \circ \sigma\}_{\sigma \in U(2)}$ spans all of $\mathcal{H}^{p,q}$ (for if it did not, it would generate a nontrivial invariant subspace, and these do not exist by definition of irreducibility). Furthermore Δ_B commutes with $U(2)$, so if we prove the assertion for f_0, F_0, then the full result follows.

For $z \in B$ we set $r = |z|$. Then $r^2 = z_1\bar{z}_1 + z_2\bar{z}_2$. We seek a solution of our Dirichlet problem of the form

$$u(z) = g(r^2)z_1^p\bar{z}_2^q.$$

Recall that

$$\Delta_B = \frac{4}{n+1}(1 - |z|^2)\sum_{i,j=1}^{n}(\delta_{ij} - z_i\bar{z}_j)\frac{\partial^2}{\partial z_i \partial \bar{z}_j}.$$

We calculate $\Delta_B u$.

Now

$$\frac{\partial}{\partial \bar{z}_j}u = z_j g'(r^2)[z_1^p\bar{z}_2^q] + g(r^2)z_1^p\left(q\bar{z}_2^{q-1}\right)\delta_{2j}$$

and

$$\frac{\partial^2}{\partial z_i \bar{z}_j}u = g''(r^2)\bar{z}_i z_j\left[z_1^p\bar{z}_2^q\right] + \delta_{ij}g'(r^2)\left[z_1^p\bar{z}_2^q\right]$$

$$+ z_j g'(r^2)\left[pz_1^{p-1}\bar{z}_2^q\delta_{i1}\right]$$

$$+ \bar{z}_i g'(r^2)\left[z_1^p q\bar{z}_2^{q-1}\delta_{2j}\right]$$

$$+ g(r^2)\left[(pz_1^{p-1}\delta_{1i})(q\bar{z}_2^{q-1}\delta_{2j})\right].$$

Therefore

$$\sum_{i=1}^{2} \frac{\partial^2 u}{\partial z_i \, \partial \bar{z}_i} = \left\{ \sum_{i=1}^{2} [g''(r^2)|z_i|^2 + g'(r^2)]z_1^p \bar{z}_2^q \right\} + g'(r^2) p z_1^p \bar{z}_2^q + g'(r^2) q z_1^p \bar{z}_2^q$$

$$= z_1^p \bar{z}_2^q [g''(r^2)r^2 + (2 + p + q)g'(r^2)].$$

By a similar calculation we find that

$$\sum_{i,j=1}^{2} z_i \bar{z}_j \frac{\partial^2 u}{\partial z_i \bar{z}_j} = z_1^p \bar{z}_2^q [r^4 g''(r^2) + (p + q + 1)r^2 g'(r^2) + pq \, g(r^2)].$$

Substituting these two calculations into the equation $\triangle_B u = 0$ (and remembering that $n = 2$), we find that

$$0 = \triangle_B u = \frac{4}{2+1}(1 - r^2)z_1^p \bar{z}_2^q [g''(r^2)r^2 + (2 + p + q)g'(r^2)]$$

$$- \frac{4}{2+1}(1 - r^2)z_1^p \bar{z}_2^q [g''(r^2)r^4 + (p + q + 1)r^2 g'(r^2) + pq \, g(r^2)]$$

$$= \frac{4}{2+1}(1 - r^2)z_1^p \bar{z}_2^q \left\{ r^2(1 - r^2)g''(r^2) \right.$$

$$\left. + [(p + q + 2) - (p + q + 1)r^2]g'(r^2) - pq \, g(r^2) \right\}.$$

Therefore if a solution of our Dirichlet problem of the form of $u(z) = g(r^2)z_1^p \bar{z}_2^q$ exists, then g must satisfy the following ordinary differential equation:

$$r^2(1 - r^2)g''(r^2) + [(p + q + 2) - (p + q + 1)r^2]g'(r^2) - pq \, g(r^2) = 0.$$

We may bring the essential nature of this equation to the surface with the changes of variables $t = r^2, a = p, b = q$, and $c = p + q + 2$. Then the equation becomes

$$t(1 - t)g'' + [c - (a + b + 1)t]g' - ab \, g = 0.$$

This, of course, is a hypergeometric equation. Since u is the solution of an elliptic problem, it must be C^∞ on the interior. Thus g must be C^∞ on $[0, 1)$. Given the solutions that we have found of the hypergeometric equation, we conclude that

$$g(t) = F(p, q, p + q + n; t).$$

Consequently

$$u(z) = \frac{F(p,q,p+q+n;r^2)}{F(p,q,p+q+n;1)} z_1^p \bar{z}_2^q$$

$$= S_n^{p,q}(r) r^{p+q} f(\zeta). \qquad \square$$

Theorem 4.3.12. *Let* $0 \le r < 1$ *and* $\eta, \zeta \in \partial B$. *Then the Poisson–Szegő kernel for the ball* $B \subseteq \mathbb{C}^n$ *is given by the formula*

$$\mathcal{P}(r\eta, \zeta) = \sum_{p,q=0}^{\infty} S_n^{p,q}(r) H_n^{p,q}(\eta, \zeta).$$

Proof. Recall that, if $g \in C(\partial B)$, then

$$G(z) = \begin{cases} \int_{\partial B} \mathcal{P}(z,\zeta) g(\zeta) \, d\sigma(\zeta) & \text{on} \quad B \\ g(z) & \text{on} \ \partial B \end{cases}$$

solves the Dirichlet problem for $\triangle_B$ with data g. Recall also that $H_n^{p,q}(\eta, \zeta)$ is the zonal harmonic for $\mathcal{H}^{p,q}$.

Let us first prove that the series in the statement of the theorem converges. An argument similar to the one we gave for real spherical harmonics shows that

$$|H_n^{p,q}(\eta, \zeta)| \le C \cdot D(p,q;n).$$

Here $D(p,q;n)$ is the dimension of $\mathcal{H}^{p,q}$. Clearly

$$D(p,q;n) \le \dim \mathcal{H}_{2n}^{p+q} = \binom{2n + (p+q) - 1}{p+q} - \binom{2n + (p+q) - 3}{p+q-2}$$

$$\le C \cdot (p+q+1)^{2n}.$$

Recall that

$$S_n^{p,q}(r) = r^{p+q} \frac{F(p,q,p+q+n;r^2)}{F(p,q,p+q+n;1)}$$

and observe that $F(p,q,p+q+n;r^2)$ is an increasing function of r. Thus

$$S_n^{p,q}(r) \le r^{p+q} \cdot 1.$$

Putting together all of our estimates, we find that

$$S_n^{p,q}(r) \cdot H_n^{p,q}(\eta, \zeta) \le C \cdot r^{p+q} (p+q+1)^{2n}.$$

Summing on p and q for $0 \le r < 1$, we see that our series converges absolutely.

It remains to show that the sum of the series is actually the Poisson–Szegő kernel. What we will in fact show is that, for $\eta \in \partial B$ and $0 < r < 1$, we have

$$\int_{\partial B} \mathcal{P}(r\eta, \zeta) f(\zeta) \, d\sigma(\zeta) = \int_{\partial B} \sum_{p,q} S_n^{p,q}(r) H_n^{p,q}(\eta, \zeta) f(\zeta) \, d\sigma(\zeta)$$

for every $f \in C(\partial B)$. But we already know that this identity holds for $f \in \mathcal{H}^{p,q}$. Finite linear combinations of $\cup_{p,q} \mathcal{H}^{p,q}$ are dense in $C(\partial B)$. Hence the result follows. $\qquad\qquad\square$

Now we return to the question that has motivated all of our work: Namely, we want to understand the lack of boundary regularity for the Dirichlet problem for the Laplace–Beltrami operator on the ball. As a preliminary, we must introduce a new piece of terminology.

Definition 4.3.13. Let $U \subseteq \mathbb{C}^n$ be an open set and suppose that f is a continuous function defined on U. We say that f is *pluriharmonic* on U if, for every $a \in U$ and every $b \in \mathbb{C}^n$, it holds that the function

$$\zeta \mapsto f(a + \zeta b)$$

is harmonic on the open set (in $\mathbb{C}$) of those ζ such that $a + \zeta b \in U$.

A function is pluriharmonic if and only if it is harmonic in the classical sense on every complex line $\zeta \mapsto a + \zeta b$. Pluriharmonic functions arise naturally because they are (locally) the real parts of holomorphic functions of several complex variables (see [KRA1, Chap. 2] for a detailed treatment of these matters).

Remark that a C^2 function v is pluriharmonic if and only if we have $(\partial^2 / \partial z_j \, \partial \bar{z}_k) v \equiv 0$ for all j, k. In the notation of differential forms, this condition is conveniently written as $\partial \bar{\partial} v \equiv 0$.

Now we have

Theorem 4.3.14. Let $f \in C^\infty(\partial B)$. Consider the Dirichlet problem

$$\begin{cases} \triangle_B u = 0 \text{ on } B \\ u\big|_{\partial B} = f \text{ on } \partial B. \end{cases}$$

Suppose that the solution u of this problem (given in Theorem 4.3.11) lies in $C^\infty(\overline{B})$. Then u must be of the form

$$\sum_\alpha c_\alpha z^\alpha + \sum_\beta d_\beta \bar{z}^\beta.$$

That is, u must be pluriharmonic. The converse statement holds as well: If f is the boundary function of a pluriharmonic function u that is continuous on $\overline{B}$ and if f is C^∞ on the boundary, then $U \in C^\infty(\overline{B})$.

Proof. Now let $v \in C(\overline{B})$ and suppose that v is pluriharmonic on B. Let $v|_{\partial B} \equiv f$. Then the solution to the Dirichlet problem for $\triangle_B$ with data f is in fact the function v (exercise). But then v is also the ordinary Poisson integral of f. Thus if $f \in C^\infty(\partial B)$, then $v \in C^\infty(\overline{B})$. This proves the converse (the least interesting) direction of the theorem.

For the forward direction, let $f \in C^\infty(\partial B)$ and suppose that the solution u of the Dirichlet problem for $\triangle_B$ with data f is C^∞ on $\overline{B}$. We write

$$f = \sum_{p,q} Y_{p,q},$$

where each $Y_{p,q} \in \mathcal{H}^{p,q}$. We proved above that

$$P(r\eta, \zeta) = \sum_{p,q} S_n^{p,q}(r) H_n^{p,q}(\eta, \zeta)$$

and also that the solution to the Dirichlet problem for $\triangle_B$ is given by

$$
\begin{aligned}
u(r\eta) \quad &= \quad \int P(r\eta, \zeta) f(\zeta) \, d\sigma(\zeta) \\
&= \quad \sum_{p',q'} \sum_{p,q} \int S_n^{p',q'}(r) H_n^{p',q'}(\eta, \zeta) Y_{p,q}(\zeta) \, d\sigma(\zeta) \\
&\overset{\text{(orthogonality)}}{=} \sum_{p,q} S_n^{p,q}(r) \int H_n^{p,q}(\eta, \zeta) Y_{p,q}(\zeta) \, d\sigma(\zeta) \\
&= \quad \sum_{p,q} S_n^{p,q}(r) Y_{p,q}(\eta).
\end{aligned}
$$

Therefore if $\mathcal{P}f = u$ is smooth on $\overline{B}$, then we may define for each p, q the function

$$\mathcal{Q}_{p,q}(r) = \int_{\partial B} (\mathcal{P}f)(r\zeta) \overline{Y_{p,q}(\zeta)} \, d\sigma(\zeta)$$

$$= S_n^{p,q}(r) \|Y_{p,q}\|^2.$$

Thus if $\mathcal{P}f$ is C^∞ up to the boundary, then, by differentiation under the integral sign, $\mathcal{Q}(r)$ is C^∞ up to $r = 1$. But recall that

$$S_n^{p,q}(r) = r^{p+q} \frac{F(p, q, p+q+n; r^2)}{F(p, q, p+q+n; 1)}$$

is smooth at $r = 1$ if and only if either $p = 0$ or $q = 0$. So the only nonvanishing terms in the expansion of f are elements of $\mathcal{H}^{p,0}$ or $\mathcal{H}^{0,q}$. That is what we wanted to prove. $\square$

We leave it to the reader to prove the refined statement that if a solution u to the Dirichlet problem is C^n up to the closure, then u must be pluriharmonic.

The analysis of the Poisson–Szegő kernel using bigraded spherical harmonics is due to Folland [FOL]. We thank Folland for useful conversations and correspondence regarding this material.

An analysis of boundary regularity for the Dirichlet problem of the Laplace–Beltrami operator on strictly pseudoconvex domains began in [GRL]. Interestingly, these authors uncovered a difference between the cases of dimension 2 and dimensions 3 and higher.

4.4 An Application to the Bergman Projection

In recent years the Bergman projection $P : L^2(\Omega) \to A^2(\Omega)$ has been an object of intense study. The reason for this interest is primarily that Bell and Ligocka [BEL1, BEL2] have demonstrated that the boundary behavior of biholomorphic mappings of domains may be studied by means of the regularity theory of this projection mapping. Of central importance in these considerations is the following (see also our discussion in Sect. 2.1).

Definition 4.4.1 (Condition R). Let $\Omega \subseteq \mathbb{C}^n$ be a smoothly bounded domain. We say that Ω satisfies *Condition R* if P maps $C^\infty(\overline{\Omega})$ to $C^\infty(\overline{\Omega})$.

A representative theorem in the subject is the following.

Theorem 4.4.2 (Bell). *Let Ω_1, Ω_2 be smooth, pseudoconvex domains in $\mathbb{C}^n$. Let $\Phi : \Omega_1 \to \Omega_2$ be a biholomorphic mapping. If at least one of the two domains satisfies Condition R, then Φ extends to a C^∞ diffeomorphism of $\overline{\Omega}_1$ to $\overline{\Omega}_2$.*

There are roughly two known methods to establish Condition R for a domain. One is to use symmetries, as in [BAR1, BEB]. The more powerful method is to exploit the $\overline{\partial}$- Neumann problem. That is the technique that we treat here. Let us begin with some general discussion.

Let $\Omega \subset\subset \mathbb{C}^n$ be a fixed domain on which the equation $\overline{\partial} u = \alpha$ is always solvable when α is a $\overline{\partial}$- closed $(0, 1)$ form (i.e., a domain of holomorphy—in other words, a pseudoconvex domain). Let $P : L^2(\Omega) \to A^2(\Omega)$ be the Bergman projection. If u is any solution to $\overline{\partial} u = \alpha$, then $w = w_\alpha = u - Pu$ is the unique solution that is orthogonal to holomorphic functions. Thus w is well defined, independent of the choice of u. Define the mapping

$$T : \alpha \mapsto w_\alpha.$$

Then, for $f \in L^2(\Omega)$ it holds that

$$Pf = f - T(\overline{\partial} f). \tag{4.4.3}$$

To see this, first notice that $\overline{\partial}[f - T(\overline{\partial}f)] = \overline{\partial}f - \overline{\partial}f = 0$, where all derivatives are interpreted in the weak sense. Thus $f - T(\overline{\partial}f)$ is holomorphic. Also $f - [f - T(\overline{\partial}f)]$ is orthogonal to holomorphic functions by design. This establishes the identity (4.4.3). But we have a more useful way of expressing T : namely, $T = \overline{\partial}^* N$. Thus we have derived the following important result:

$$P = I - \overline{\partial}^* N \overline{\partial}. \tag{4.4.4}$$

This formula is usually attributed to J. J. Kohn.

Now suppose that our domain is strictly pseudoconvex. Then we know that N maps W^s to H^{s+1} for every s (see [FOK, KRA4]). Recall that $\overline{\partial}$ and $\overline{\partial}^*$ are first-order differential operators. Then a trivial calculation with (4.4.4) shows that

$$P : W^s \to H^{s-1}$$

for every s. By the Sobolev embedding theorem, a strictly pseudoconvex domain therefore satisfies Condition R. Thus thanks to the program of Bell and Ligocka (see [BEL1, KRA1]), we know that biholomorphic mappings of strictly pseudoconvex domains extend to be diffeomorphisms of their closures.

It is often convenient, and certainly aesthetically more pleasing, to be able to prove that $P : W^s \to W^s$. This is known to be true on strictly pseudoconvex domains. We now describe the proof, due to J. J. Kohn [KOH1], of this assertion.

Theorem 4.4.5. *Let Ω be a smoothly bounded, strictly pseudoconvex domain in $\mathbb{C}^n$. Then, for each $s \in \mathbb{R}$, there is a constant $C = C(s)$ such that*

$$\|Pf\|_{W^s} \le C \cdot \|f\|_{W^s}. \tag{4.4.5.1}$$

Remark 4.4.6. In fact the specific property of a strictly pseudoconvex domain that will be used is the following: For every $\epsilon > 0$ there is a $C(\epsilon) > 0$ so that the inequality

$$\|\phi\|^2 \le \epsilon Q(\phi, \phi) + C(\epsilon)\|\phi\|_{-1}^2 \tag{4.4.5.2}$$

for all $\phi \in \mathcal{D} \equiv \bigwedge^{0,1} \cap \operatorname{dom} \overline{\partial} \cap \operatorname{dom} \overline{\partial}^*$. We leave it as an exercise for the reader to check that property (4.4.5.2) is equivalent to the norm Q being compact in the following sense: If $\{\phi_j\}$ is bounded in the Q norm, then it has a convergent subsequence in the L^2 norm.

The theorem that we are about to prove is in fact true on any smoothly bounded domain with the property (4.4.5.2). Property (4.4.5.2) is known to hold for a large class of domains, including domains of finite type (see [CAT1, CAT2, DAN1, DAN2, DAN3]) and, in particular, domains with real analytic boundary [DIF2].

$\square$

Proof of the Theorem: We have already observed that the Bergman projection of a strictly pseudoconvex domain maps functions in $C^\infty(\overline{\Omega})$ to functions in $C^\infty(\overline{\Omega})$. Thus it suffices to prove our estimate (4.4.5.1) for $f \in C^\infty(\overline{\Omega})$.

Let r be a smooth defining function for Ω. Let $\zeta \in \partial\Omega$ and let $U \subseteq \mathbb{C}^n$ be a neighborhood of ζ. We may select a smooth function w on U such that $\omega^n \equiv w \cdot \partial r$ satisfies $|\omega^n| \equiv 1$ on U. We select $\omega^1, \ldots, \omega^{n-1}$ on U such that $\omega^1, \ldots, \omega^n$ forms an orthonormal basis of the $(1, 0)$ forms on U. Thus any $\phi \in \mathcal{D}^{0,1}$ can be expressed, on $\overline{\Omega} \cap U$, as a linear combination

$$\phi = \sum_j \phi_j \overline{\omega}^j.$$

Of course $\phi \in \mathcal{D}^{0,1}$ if and only if $\phi_n = 0$ on $\partial\Omega$.

Let Λ_t^s be the tangential Bessel potential of order s (see [FOK]). If η is any real-valued cutoff function supported in U, then, whenever $\phi \in \mathcal{D}^{0,1}$, we have $\eta\Lambda_t^s(\eta\phi) \in \mathcal{D}^{0,1}$ as well. The identity $Q(N\alpha, \psi) = \langle \alpha, \psi \rangle$, with $\alpha = \overline{\partial}f$ and $\psi = \eta^3\Lambda^{2s}\eta N\overline{\partial}f$, yields that

$$Q(N\overline{\partial}f, \eta^3\Lambda^{2s}\eta N\overline{\partial}f) = \langle \overline{\partial}f, \eta^3\Lambda^{2s}\eta N\overline{\partial}f \rangle. \qquad (4.4.5.3)$$

Now we apply the compactness inequality (4.4.5.2) with $\phi = \eta\Lambda_t^s(\eta N\overline{\partial}f)$ to obtain

$$\|\eta\Lambda_t^s(\eta N\overline{\partial}f)\|^2 \leq \epsilon Q(\eta\Lambda_t^s(\eta N\overline{\partial}f), \eta\Lambda_t^s(\eta N\overline{\partial}f)) + C(\epsilon)\|\eta\Lambda_t^s(\eta N\overline{\partial}f)\|_{-1}^2$$

$$\leq \epsilon Q(N\overline{\partial}f, \eta^3\Lambda_t^{2s}\eta N\overline{\partial}f) + \epsilon C\|N\overline{\partial}f\|_s^2 + C'(\epsilon)\|N\overline{\partial}f\|_{s-1}^2.$$

Of course in the last estimate, we have done two things: First, we have moved η and Λ_t^s across the inner product Q at the expense of creating certain acceptable error terms (which are controlled by the term $\epsilon C\|N\overline{\partial}f\|_s^2$). Second, we have used the fact that $\|\Lambda_t^s g\|_0^2 \leq \|g\|_s^2$ by definition. Now, using (4.4.5.3), we see that the last line is majorized by

$$\epsilon\langle f, \overline{\partial}^* \eta^3\Lambda_t^{2s}\eta N\overline{\partial}f \rangle + \epsilon C\|N\overline{\partial}f\|_s^2 + C'(\epsilon)\|N\overline{\partial}f\|_{s-1}^2. \qquad (4.4.5.4)$$

Now we may cover $\overline{\Omega}$ with boundary neighborhoods U as above plus an interior patch on which our problem is strongly elliptic. We obtain an estimate like (4.4.5.4) on each of these patches. We may sum the estimates, using (as we did in the solution of the $\overline{\partial}$- Neumann problem) the fact that $\partial\Omega$ is non-characteristic for Q, to obtain

$$\|N\overline{\partial}f\|_s^2 \leq \epsilon C\|\overline{\partial}^* N\overline{\partial}f\|_s^2 + C'(\epsilon)(\|f\|_s^2 + \|N\overline{\partial}f\|_{s-1}^2).$$

Applying this inequality, with s replaced by $s - 1$, to the last term on the right of (4.4.5.4), and then repeating, we may finally derive that

$$\|N\overline{\partial}f\|_s^2 \leq \epsilon C\|\overline{\partial}^* N\overline{\partial}f\|_s^2 + C'(\epsilon)(\|f\|_s^2 + \|N\overline{\partial}f\|_0^2). \qquad (4.4.5.5)$$

We know that $\overline{\partial}\overline{\partial}^* N\overline{\partial} f = \overline{\partial} f$. As a result,

$$\|\eta\Lambda_t^s\eta\overline{\partial}^* N\overline{\partial} f\|^2 = \langle N\overline{\partial} f, \eta^3\Lambda_t^{2s}\eta\overline{\partial}\overline{\partial}^* N\overline{\partial} f\rangle + \mathcal{O}(\|N\overline{\partial} f\|_s\|\eta\Lambda_t^s\eta\overline{\partial}^* N\overline{\partial} f\|)$$

$$= \langle N\overline{\partial} f, \eta^3\Lambda_t^{2s}\eta\overline{\partial} f\rangle + \mathcal{O}(\|N\overline{\partial} f\|_s\|\eta\Lambda_t^s\eta\overline{\partial}^* N\overline{\partial} f\|)$$

$$= \mathcal{O}\left(\left(\|N\overline{\partial} f\|_s + \|f\|_s\right)\|\eta\Lambda_t^s\eta\overline{\partial}^* N\overline{\partial} f\|\right).$$

Summing as before, we obtain the estimate

$$\|\overline{\partial}^* N\overline{\partial} f\|_s \leq C(\|N\overline{\partial} f\|_s + \|f\|_s).$$

Putting (4.4.5.5) into this last estimate gives

$$\|\overline{\partial}^* N\overline{\partial} f\|_s \leq \epsilon C\|\overline{\partial}^* N\overline{\partial} f\|_s^2 + C'(\epsilon)(\|f\|_s^2 + \|N\overline{\partial} f\|_0^2).$$

If we choose $\epsilon > 0$ small enough, then we may absorb the first term on the right into the left-hand side and obtain

$$\|\overline{\partial}^* N\overline{\partial} f\|_s \leq C \cdot \left(\|f\|_s + \|N\overline{\partial} f\|_0\right). \tag{4.4.5.6}$$

But the operator $\overline{\partial}^*$ is closed since the adjoint of a densely defined operator is always closed. It follows from the open mapping principle that

$$\|N\overline{\partial} f\| \leq C\|\overline{\partial}^* N\overline{\partial} f\|.$$

On the other hand, $\overline{\partial}^* N\overline{\partial}$ is projection onto the orthogonal complement of $A^2(\Omega)$. Thus it is bounded in L^2, and we see that

$$\|N\overline{\partial} f\| \leq C\|f\|_0.$$

Putting this information into (4.4.5.6) gives

$$\|\overline{\partial}^* N\overline{\partial} f\|_s \leq C\|f\|_s.$$

If we recall that $P = I - \overline{\partial}^* N\overline{\partial}$, then we may finally conclude that

$$\|Pf\|_s \leq C\|f\|_s.$$

That concludes the proof. $\qquad\qquad\square$

Exercises

1. Is there a maximum principle for functions that are annihilated by the invariant Laplacian on the ball?
2. The "invariant Laplacian" on the ball is in fact the Laplace–Beltrami operator for the Bergman metric. Use the Fefferman's ideas to produce an approximate formula for the Laplace–Beltrami operator for the Bergman metric on a strictly pseudoconvex domain.
3. Prove that, if a function is annihilated by the invariant Laplacian on the ball, then it must be real analytic.
4. Prove that the converse to Exercise 3 is false.
5. Prove that not every continuous function on the boundary of the ball is the boundary trace of a pluriharmonic function on B. [Note that this is in contrast to the situation for harmonic functions on the disc in the complex plane.]
6. Refer to Exercise 5. Bedford and Federbush [BEDF] have found a fourth-order partial differential equation that the boundary trace of a pluriharmonic function must satisfy. Look up that paper to learn what the differential equation is. Verify that the function $u(z) = 2\operatorname{Re} z_1 + z_2^2 + \bar{z}_2^2$ is indeed annihilated by this differential equation.
7. What is the bigraded spherical harmonic expansion of the function $u(z_1, z_2) = |z_1|^2 + |z_2|^2$?
8. A basic result of classical Fourier analysis is that if the Fourier series of a function sums to the identically zero function, then the function must be the zero function. Is there an analogous result for bigraded spherical harmonic expansions? How would you prove it?
9. A basic result of classical Fourier analysis is that the Fourier coefficients of an L^1 function must tend to zero (this is the Riemann–Lebesgue lemma). Is there an analogous result for bigraded spherical harmonic expansions? How would you prove it?
10. A basic result of classical Fourier analysis is that the Fourier coefficients of an L^2 function in fact form an L^2 sequence. Is there an analogous result for bigraded spherical harmonic expansions? How would you prove it?

Chapter 5
Further Geometric Explorations

5.1 Introductory Remarks

A *domain* Ω in $\mathbb{C}^n$ is a connected, open set. An *automorphism* of Ω is a biholomorphic self-map. The collection of automorphisms forms a group under the binary operation of composition of mappings. The standard topology on this group is uniform convergence on compact sets, or the compact-open topology. We denote the automorphism group by $\mathrm{Aut}(\Omega)$. When Ω is a bounded domain, the group $\mathrm{Aut}(\Omega)$ is a real (never a complex) Lie group.

Although domains with *transitive automorphism group* are of some interest, they are relatively rare (see [HEL, Sect. III.3]). A geometrically more natural condition to consider, and one that gives rise to a more robust and broader class of domains, is that of having *non-compact automorphism group*. Clearly a domain has non-compact automorphism group if there are automorphisms $\{\varphi_j\}$ which have no subsequence that converges to an automorphism. The following proposition of Henri Cartan is of particular utility in the study of these domains.

Proposition 5.1.1. *Let $\Omega \subseteq \mathbb{C}^n$ be a bounded domain. Then Ω has non-compact automorphism group if and only if there are a point $X \in \Omega$, a point $P \in \partial\Omega$, and automorphisms φ_j of Ω such that $\varphi_j(X) \to P$ as $j \to \infty$.*

In fact it is useful to put this result into a more general context:

Theorem 5.1.2. *Let Ω be a bounded domain in $\mathbb{C}^n$. Let $\{\varphi_j\} \subseteq \mathrm{Aut}(\Omega)$ be a sequence of automorphisms of Ω. Suppose that φ_j converges, uniformly on compact subsets of Ω, to a holomorphic mapping $\varphi : \Omega \to \mathbb{C}^n$. Then the following three properties are equivalent:*

(i) $\varphi \in \mathrm{Aut}(\Omega)$.
(ii) $\varphi(\Omega) \not\subseteq \partial\Omega$.
(iii) There exists a point $P \in \Omega$ such that the Jacobian matrix $(\partial\varphi_i/\partial z_j(P))$, for $\varphi = (\varphi_1, \varphi_2, \ldots, \varphi_n)$, has a nonzero determinant.

S.G. Krantz, *Geometric Analysis of the Bergman Kernel and Metric*,
Graduate Texts in Mathematics 268, DOI 10.1007/978-1-4614-7924-6_5,
© Springer Science+Business Media New York 2013

We leave it to the reader to determine why the proposition follows from the theorem. Now we shall concentrate on proving the theorem.

We begin with three preliminary lemmas.

Lemma 5.1.3. *Let $\Omega \subset \mathbb{C}^n$ and $f : \Omega \to \mathbb{C}^n$ be a holomorphic mapping. Suppose that $\det(\mathrm{d} f)_P \neq 0$ for some $P \in \Omega$. Then there exist neighborhoods U of P and V of $f(P)$ such that $f(U) \subset V$ and $f|U$ are analytic isomorphisms onto V.*

Of course this lemma is nothing other than a holomorphic version of the inverse function theorem in several complex variables. We have proved it in Theorem 1.1.12.

Lemma 5.1.4. *Let $\{g_j\}$ be a sequence of continuous open mappings of $\Omega \subset \mathbb{C}^n$ into $\mathbb{C}^n$. Suppose that the g_j converge, uniformly on compact sets, to a limit mapping $g : \Omega \to \mathbb{C}^n$.*

Further suppose that, for some point $P \in \Omega$, P is an isolated point of $g^{-1}(g(P))$. Then, for any neighborhood U of P, there is an index j_0 such that $g(P) \in g_j(U)$ for $j \geq j_0$.

Interestingly, this lemma is topological rather than analytic.

Proof of Lemma 5.1.4: Seeking a contradiction, we suppose the assertion to be false. Passing to a subsequence, we may suppose that U is such that

$$\overline{U} \text{ is compact, } g(P) \notin g_j(\overline{U}), \text{ for } j = 1, 2, \ldots, \text{ and } \overline{U} \cap g^{-1}(g(P)) = \{P\}.$$

Then $g(\partial U)$ is compact and $g(P) \notin g(\partial U)$. Therefore there is a neighborhood V of $g(\partial U)$ and a polydisc Q about $g(P)$ so that

$$Q \cap V = \emptyset. \tag{5.1.4.1}$$

Now, if j_0 is large enough, then $g_j(\partial U) \subset V$ for $j \geq j_0$. Since g_j is an open mapping, we see that $\partial g_j(U) \subset g_j(\partial U)$. Thus $g_j(U)$ is a relatively compact-open set in $\mathbb{C}^n$ with $\partial g_j(U) \subset V$.

We claim that $(\partial g_j(U)) \cap Q \neq \emptyset$ if j is large enough. [This would contradict (5.1.4.1) and end the proof.] Since $g(P) \in Q$ and $g_j(P) \to g(P)$, we see that if j is large then $g_j(P) \in Q$. But $g(P) \notin g_j(U)$ by hypothesis. If $\{\partial g_j(U)\} \cap Q = \emptyset$, then we would have

$$Q = \{g_j(U) \cap Q\} \cup \{(\mathbb{C}^n \setminus \overline{g_j(U)}) \cap Q\},$$

and each of the open sets in this last display is nonempty. [Note that $g_j(P)$ belongs to the first open set and $g(P)$ belongs to the second.] This contradicts the fact that Q is connected. Hence, $\{\partial g_j(U)\} \cap Q \neq \emptyset$, and the lemma is proved. □

Now our last lemma is this:

Lemma 5.1.5 (Hurwitz's theorem). *Let Ω be an open, connected set in $\mathbb{C}^n$ and let $\{f_j\}$ be a sequence of holomorphic functions on Ω. We assume that the f_j converge* *uniformly on compact sets to a holomorphic function f. Then, if $f_j(z) \neq 0$ for all indices j and all $z \in \Omega$, and if f is nonconstant, then $f(z) \neq 0$ for all $z \in \Omega$.*

Proof. Seeking a contradiction, we suppose that $f(P) = 0$ for some $P \in \Omega$. Let Q be a small open polydisc about P. Then $f \not\equiv 0$ on Q (since, if it were identically 0, then f would be identically 0 on Ω since Ω is connected). Let $R \in \Omega$ be a point where f does not vanish. Set

$$\mathcal{D} = \{\lambda \in \mathbb{C} : P + \lambda(R - P) \in Q\}.$$

Then $\mathcal{D}$ is a convex and hence a connected open set in $\mathbb{C}$. Set

$$\varphi_j(\lambda) = f_j(P + \lambda(R - P)),$$
$$\varphi(\lambda) = f(P + \lambda(R - P)).$$

Then $\varphi(0) = 0$, $\varphi(1) = f(R) \neq 0$. Hence, φ is nonconstant on $\mathcal{D}$. Thus for large values of the index j, φ_j is also nonconstant, hence an open mapping of $\mathcal{D}$ into $\mathbb{C}$. Thus by the preceding lemma, $f_j(\Omega) \supset \varphi_j(\mathcal{D}) \supset \{0\}$ if j is large. That is a contradiction. $\qquad\square$

Proof of Theorem 5.1.2

(i) $\Rightarrow$ (ii) Obvious.

(i) $\Rightarrow$ (iii) If $\varphi \in \text{Aut}(\Omega)$ and $P \in \Omega$ and if we set $\psi = \varphi^{-1} \in \text{Aut}(\Omega)$, then we have

$$\psi \circ \varphi = \text{identity}.$$

Thus $(d\psi)_{\varphi(P)} \circ (d\varphi)_P = \text{identity}$, so that $(d\varphi)_P$ is invertible.

(iii) $\Rightarrow$ (ii) If $(d\varphi)_P$ has nonzero determinant, then by our first lemma, $\varphi(\Omega)$ contains a nonempty neighborhood of $\varphi(P)$, hence $\varphi(\Omega) \not\subset \partial\Omega$.

(ii) $\Rightarrow$ (iii) Obviously $\varphi(\Omega) \subset \overline{\Omega}$. Thus if (ii) holds, then $\varphi(\Omega) \cap \Omega \neq \emptyset$. Let $P \in \Omega$ be a point such that $\varphi(P) = R \in \Omega$. Let $\psi_j = \varphi_j^{-1}$. Choose a subsequence j_k such that $\{\psi_{j_k}\}$ converges uniformly on compact subsets of Ω to a mapping $\psi : \Omega \to \mathbb{C}^n$. Now we have

$$\psi(R) = \lim_{k \to \infty} \varphi_{j_k}^{-1}(\varphi(P)).$$

Furthermore, if k is large enough, then $\varphi_{j_k}(P)$ is close to $\varphi(P)$. Hence, $\varphi_{j_k}(P)$ lies in a compact subset of Ω. Since $\varphi_{j_k}^{-1}$ converges uniformly on compact subsets of Ω, we may conclude that

$$\psi(R) = \lim_{k \to \infty} \varphi_{j_k}^{-1}(\varphi_{j_k}(P)) = \lim_{k \to \infty} P = P.$$

As a result, $\psi(R) = P \in \Omega$. Let V be a small neighborhood of R. Then $\psi(V)$ lies in a compact subset of Ω; hence, there is a K compact in Ω so that $\psi_{j_k}(V) \subset K$ for k large. Then, for $z \in V$, we have (since $g_{j_k}(V) \subset K$ and $\varphi_{j_k} \to \varphi$ uniformly on K) that

$$
\begin{aligned}
\varphi(\psi(z)) &= \lim_{k\to\infty} \varphi(\psi_{j_k}(z)) \\
&= \lim_{k\to\infty} \varphi_{j_k}(\psi_{j_k}(z)) \\
&= z.
\end{aligned}
$$

As a result, $(d\varphi)_{\psi(z)} \circ (d\psi)_z = $ identity for $z \in V$. In particular, $\det((d\varphi)_w) \neq 0$ for $w \in \psi(V)$, which proves (iii).

(iii) $\Rightarrow$ (i) Note that the function $g_j(z) = \det(d\varphi_j)_z$ is holomorphic on Ω and converges to $g(z) = \det(d\varphi)_z$ uniformly on compact subsets of Ω. If (iii) holds, then $g(z) \neq 0$. Also $g_j(z) \neq 0$ for all j and all z since $\varphi_j \in \text{Aut}(\Omega)$ [refer to the proof that (i) $\Rightarrow$ (iii)]. If $g(z)$ is constant, then $g(z)$ is obviously never 0, and if $g(z)$ is nonconstant, then it is again never 0 by the third lemma above. In either case, $g(z) \neq 0$ for all $z \in \Omega$. By the first lemma, $\varphi : \Omega \to \mathbb{C}^n$ is an open mapping, and any $z \in \Omega$ is isolated in $\varphi^{-1}(\varphi(z))$. It follows from the second lemma that $\varphi(\Omega) \subset \cup \varphi_j(\Omega) = \Omega$.

Let $\{j_k\}$ be a subsequence of $\{j\}$ so that ψ_{j_k} converges uniformly on compact subsets of Ω. Then, for $z \in \Omega$, $\{\varphi_{j_k}(z)\}$ converges to $\varphi(z) \in \Omega$, hence lies in a compact subset of Ω. Therefore

$$
\psi(\varphi(z)) = \lim_{k\to\infty} \psi_{j_k}(\varphi_{j_k}(z)) = z
$$

for all $z \in \Omega$. In particular, $\det(d\psi)_w \neq 0$ for $w \in \varphi(\Omega)$. As a result, repeating the preceding argument, we conclude that $\psi(\Omega) \subset \Omega$. Hence, $\psi_{j_k}(z)$ lies in a compact subset of Ω for any $z \in \Omega$. We conclude that

$$
\varphi(\psi(z)) = \lim_{k\to\infty} \varphi_{j_k}(\psi_{j_k}(z)) = z.
$$

Therefore $\varphi \circ \psi = $ identity, $\psi \circ \varphi = $ identity, and we find that $\varphi \in \text{Aut}(\Omega)$. $\square$

We say that a domain $\Omega \subseteq \mathbb{C}^n$ has C^k boundary, $k \geq 1$ an integer, if it is possible to write

$$
\Omega = \{z \in \mathbb{C}^n : \rho(z) < 0\}
$$

for a function ρ that is C^k and which satisfies $\nabla \rho \neq 0$ on $\partial \Omega$. This definition is equivalent to a number of other natural definitions of C^k boundary for a domain (see the Appendices in [KRA1]). Below we shall define a topology on the collection of domains with C^k boundary.

Domains with compact automorphism group exhibit certain rigidities which are of interest for our studies. We begin this section by showing that, for certain smoothly bounded domains with compact automorphism group, the convergence of automorphisms will take place in a much stronger topology than the standard one specified in the first paragraph. This fact has intrinsic interest but is also of considerable use for further studies in complex function theory. It is even new in the context of one complex variable.

As an application of the ideas in the last paragraph, we offer a new result about the semicontinuity of the automorphism group under perturbation of the underlying domain. This generalizes results of [GRK1]. We also offer a direct generalization of the result of [GRK1, Theorem 0.1] to finite-type domains. Some of the proof techniques presented here are new.

5.2 Semicontinuity of Automorphism Groups

Symmetry is easily destroyed but not so easily created. To turn a symmetric object into an asymmetric one requires only an arbitrarily small effort, while to turn an asymmetric object into a symmetric one requires a distinct push.

These intuitions that symmetry is unstable but an increase in symmetry requires a substantial change hold with precision in a variety of circumstances. The goal of this section is a result of this type for the automorphism groups of C^∞ strictly pseudoconvex domains. This result will depend for its proof on a theorem in the context of the compact Riemannian manifolds that is similar in spirit. [EBI1, EBI2]:

Theorem 5.2.1 (Ebin). *If (M, g_0) is a C^∞ compact Riemannian manifold, then there is a neighborhood $\mathcal{G}$ of g_0 in the C^∞ topology on the C^∞ Riemannian metrics such that: If $g \in \mathcal{G}$ then there is a diffeomorphism $F : M \to M$ (C^∞ close to the identity) such that the set*

$$\{F \circ \alpha \circ F^{-1} : \ \alpha : M \to M \text{ is an isometry for } g\}$$

is a subset of and hence a subgroup of

$$\{\beta : \ \beta : M \to M \text{ is an isometry for } g_0\} \ .$$

In particular, the group of isometries of M relative to g is isomorphic to a subgroup of the group of isometries of g_0.

Ebin's original proof of the theorem just stated involved infinite-dimensional manifolds and the construction of "slices" in the Lie group sense for the action of the diffeomorphism group on the manifold M. However, the result can in fact be established by finite-dimensional methods and ordinary Lie group theory. See [KIMYW] for the details.

The possibility of averaging over compact groups gives a useful corollary about group actions as such. For the statement of the corollary, we say that a sequence of C^∞ group actions $G_j \times M \to M$ sub-converges in the C^∞ topology to an action $G_0 \times M \to M$ if every sequence α_j of G_j-action elements has a subsequence α_{j_k} which converges in the C^∞ topology to a G_0-action element.

Corollary 5.2.2. *If $G_j \times M \to M$ is a sequence of actions on a compact manifold M by compact Lie groups G_j and if the G_j-actions sub-converge in the C^∞ topology to a compact Lie group action $G_0 \times M \to M$, then for all j sufficiently large, there is a diffeomorphism $F_j : M \to M$ such that the conjugation by F_j of the G_j-action is a subgroup of the G_0-action. Moreover, the F_j may be chosen to converge to the identity map of M in the C^∞ topology.*

This corollary follows from the proof of Ebin's theorem (Theorem 5.2.1) by averaging a fixed Riemannian metric over the group actions to produce G_j-invariant metrics g_j converging in the C^∞ topology to a G_0-invariant metric g_0.

Generically, that is for a dense open set of metrics, the isometry group is in fact the identity alone (see [EBI1, EBI2]). Our interest here, however, is in the metrics which have a nontrivial isometry group.

The main goal of this section is to prove the statement analogous to Ebin's theorem (Theorem 5.2.1) for C^∞, strictly pseudoconvex domains:

Theorem 5.2.3 ([GRK2]). *If Ω_0 is a bounded, C^∞, strictly pseudoconvex domain in $\mathbb{C}^n$ that is not biholomorphic to the ball, then there is a neighborhood $\mathcal{U}$ of Ω_0 in the C^∞ topology (on bounded domains with the C^∞ boundary) such that if $\Omega \in \mathcal{U}$, then there is a real diffeomorphism $F : \Omega \to \Omega_0$ such that F is C^∞ close to the identity and*

$$\{F \circ \alpha \circ F^{-1} : \alpha \in Aut(\Omega)\} \subset Aut(\Omega_0).$$

In particular, $Aut(\Omega)$ is isomorphic to a subgroup of $Aut(\Omega_0)$.

The essential idea of the proof of this theorem is to note, from the Lu Qi-Keng theorem (Theorem 5.8.1), that the Bergman metric of Ω_0 does not have constant holomorphic sectional curvature, while at the same time the holomorphic sectional curvature is asymptotically constant at the boundary. So far, this is just a recapitulation of the curvature proof of Bun Wong's theorem (Sect. 3.4). Noting further that these curvature estimates are stable under C^∞ perturbations of $\partial\Omega_0$, one expects to find that the smooth extension to the closure $cl(\Omega_0)$ of $Aut(\Omega_0)$, guaranteed by Fefferman's result on smoothness to the boundary, [FEF1, Part I] will also be stable under perturbation of $\partial\Omega_0$ in the following sense: If Ω is C^∞ close to Ω_0, then $Aut(\Omega)$ on $cl(\Omega)$ is C^∞ close to $Aut(\Omega_0)$ on $cl(\Omega_0)$ in the sense that each element of $Aut(\Omega)$ belongs to some pre-chosen C^∞ neighborhood of $Aut(\Omega)$ on $cl(\Omega_0)$. Of course $cl(\Omega_0)$ is a compact manifold with boundary so that Ebin's theorem (Theorem 5.2.1) as just stated and proved (for manifolds without boundary) does not apply as such. But, by passing to the "metric double" and introducing suitable automorphism-invariant metrics, we can apply Ebin's theorem on manifolds without boundary. We now turn to a more detailed version of the outline just given.

The detailed proof will be based on two propositions:

Proposition 5.2.4. *If Ω_0 is a C^∞ strictly pseudoconvex domain and if Ω_0 is not biholomorphic to the unit ball, then there is a point p in Ω_0, a compact set $K_0 \subset \Omega_0$, and a C^∞ neighborhood V of Ω_0 in the C^∞ topology on domains such that, if $\Omega \in V$, then $\Omega \supset K_0 \cup \{p\}$ and the $Aut(\Omega)$-orbit of p lies in K_0.*

Proposition 5.2.5. *If Ω_0 is a C^∞ strictly pseudoconvex domain not biholomorphic to the unit ball then, for each $\ell = 1, 2, \ldots$, there is a C^∞ neighborhood V of Ω_0 and a positive constant C_ℓ such that, for each $\Omega \in V$ and each $f \in Aut(\Omega)$, the Euclidean derivatives of order $\leq \ell$ of f at points $p \in \Omega$ have absolute value $\leq C_\ell$.*

For brevity, we shall summarize this last statement by saying that

The derivatives of order $\leq \ell$ of elements in $Aut(\Omega)$ are stably uniformly bounded (where "stably" refers to variation of Ω near Ω_0 and "uniformly" refers to variation over the points of the domain Ω).

This proposition, which is in effect a stable version of smoothness-to-the-boundary theorem by Fefferman, will be established later.

Armed with these propositions, we can now establish the following lemma of normal families type.

Lemma 5.2.6. *If Ω_j, $j = 1, 2, \ldots$, converge in the C^∞ topology to Ω_0 (with Ω_0 being C^∞, strictly pseudoconvex, and not biholomorphic to the ball), and if $g_j \in Aut(\Omega_j)$, then there are subsequences $\Omega_{j_k}, g_{j_k}, k = 1, 2, \ldots$, such that g_{j_k} converges in the C^∞ topology to an element $g_0 \in Aut(\Omega_0)$.*

Hereinafter, we write $G_j = Aut(\Omega_j)$ and $G_0 = Aut(\Omega_0)$. The lemma then says in effect that, for j large, the action of each element of G_j is close to the action of an element of G_0.

Proof of the Lemma: Fix a point p and a compact set K_0 as in Proposition 5.2.4. Then, for j large, $g_j(p) \in K_0 \subset \Omega_j$. By normal families, there is a subsequence g_{j_k} which converges uniformly on each compact subset of Ω_0. And the limit of this subsequence is an element g_0 of G_0 (this follows from a straightforward modification of Theorem 4, page 78 of [NAR]). A standard result of Cartan then implies the C^∞ convergence of $\{g_{j_k}\}$ on $cl(\Omega_{j_k})$ [respectively to g_0 on $cl(\Omega_0)$].

To check this last assertion in detail, it suffices to show that $\{g_{j_k}\}$ on $cl(\Omega_{j_k})$ is a Cauchy sequence in the $C^{\ell+1}$ norm for each fixed $\ell = 1, 2, \ldots$. For this, suppose that $\epsilon > 0$ is given. Choose a compact set $K \subset \Omega_0$ such that, for all Ω which are C^∞ close enough to Ω_0 and $x \in \partial\Omega$, there is a polygonal arc in Ω, of length not exceeding $\epsilon/[3C_{\ell+1}]$, from some point $s \in K$ to the point x. [Here $C_{\ell+1}$ is the constant from Proposition 5.2.5.] The possibility of choosing K in this fashion is elementary: Simply let the set K be the $\epsilon/[4C_\ell]$ normal "push-in" of Ω_0.

Now choose k_0 so large that (from Cauchy estimates), $g_{j_{K_{\Omega_1}}} - g_0$ and $g_{j_{K_{\Omega_2}}} - g_0$ have C^ℓ norm on K bounded above by $\epsilon/3$ if $K_{\Omega_1}, K_{\Omega_2} \geq k_0$. For such $K_{\Omega_1}, K_{\Omega_2}$, the C^ℓ norm of the difference $g_{jK_{\Omega_1}} - g_{jK_{\Omega_2}}$ is $\leq \epsilon$ on $\mathrm{cl}(\Omega_{K_{\Omega_1}})$, $\mathrm{cl}(\Omega_{K_{\Omega_2}})$ provided that K_{Ω_1}, K_{Ω_2} are also required to be so large that $\Omega_{K_{\Omega_1}}$, $\Omega_{K_{\Omega_2}}$ are sufficiently C^∞ close to Ω_0 and hence to each other.   □

Lemma 5.2.7. *There is a neighborhood $\mathcal{V}$ of Ω_0 in the C^∞ topology on domains and a family g_Ω, $\Omega \in \mathcal{V}$, with g_Ω a C^∞ Riemannian metric on $\mathrm{cl}(\Omega)$ such that (1) if $\mathrm{Aut}(\Omega)$ acts isometrically on g_Ω and (2) if $\{\Omega_j\}$ is a sequence in $\mathcal{V}$ converging C^∞ to Ω_0, then $\{g_{\Omega_j}\}$ converges C^∞ to g_{Ω_0}.*

Proof. Set g_{Ω_0} equal to the average with respect to $\mathrm{Aut}(\Omega_0)$ of the Euclidean metric on $\mathrm{cl}(\Omega_0)$. For each $\Omega \neq \Omega_0$, choose diffeomorphisms $F_\Omega : \mathrm{cl}(\Omega) \to \mathrm{cl}(\Omega_0)$ such that F_Ω converges as Ω tends to Ω_0 in the C^∞ topology. Set g_Ω equal to the average over the compact (for $\mathcal{V}$ small enough) group $\mathrm{Aut}(\Omega)$ of the pullback metric $F_\Omega^* g_{\Omega_0}$. By arguments in [GRK3], each element of $\mathrm{Aut}(\Omega)$ acts nearly isometrically on $F_\Omega^* g_{\Omega_0}$, in the C^∞ sense of "nearly," on $\mathrm{cl}(\Omega)$. This is because g_{Ω_0} is $\mathrm{Aut}(\Omega_0)$-invariant and each element of $\mathrm{Aut}(\Omega)$ is C^∞ close to an element of $\mathrm{Aut}(\Omega_0)$. The convergence conclusion of the lemma follows.   □

Lemma 5.2.8. *The metrics g_Ω in Lemma 5.2.7 can be chosen to be product metrics near the boundary.*

Here "the product metric" near the boundary of Ω means precisely that, for each boundary point x of $\mathrm{cl}(\Omega)$, there is a real local coordinate system $(x_1, x_2, \ldots, x_{2n})$ in a neighborhood of x with:

- The boundary $\mathrm{cl}(\Omega) \setminus \Omega$ equaling $\{(x_1, x_2, \ldots, x_{2n-1}, 0)\}$
- The points of Ω in the neighborhood of x satisfying $x_{2n} < 0$ (and vice versa)
- The metric in the given neighborhood having at $(x_1, x_2, \ldots, x_{2n})$ the form

$$dx_{2n}^2 + \big(\text{a positive definite quadratic form}$$

$$\text{in } dx_1, dx_2, \ldots, dx_{2n-1} \text{ with coefficients}$$

$$\text{depending only on } (x_1, x_2, \ldots, x_{2n-1})\big)$$

Proof of Lemma 5.2.8: An $\mathrm{Aut}(\Omega)$ product metric of this sort at and near the boundary is easily obtained using the map

$$\partial\Omega \times [0, \delta) \to \Omega$$

defined by

$$(b, t) \mapsto \exp_p(tN),$$

where N is the inward-pointing normal at b relative to the previous g_Ω metric and $\exp_p$ is the g_Ω-exponential map. Choose δ so small that the map is a diffeomorphism and define the metric by declaring this diffeomorphism to be isometric for [(the metric on $\partial\Omega$) + dt^2]. This construction is Aut(Ω)-invariant. Using an Aut(Ω)-invariant partition of unity to make a transition to the previous g_Ω will provide all properties: The partition of unity function is taken to depend only on the t variable.   □

The proof of Theorem 5.2.3 can now be completed as follows: With the metrics g_Ω chosen as in Lemma 5.2.8, in particular as product metrics near the boundary, we form the compact Riemannian manifolds $(\hat{\Omega}, \hat{g}_\Omega)$ by taking $\hat{\Omega}$ to be the manifold "double" of Ω and $\hat{g}_\Omega$ to be the natural metric on $\hat{\Omega}$, equal to g_Ω on each copy of Ω and fitting together to form a C^∞ metric across the (one copy of) $\partial\Omega$ on account of the product metric. Let G_Ω be the group generated by Aut(Ω) and the interchange operation I_Ω that interchanges the two copies of Ω that are "glued" to form $\hat{\Omega}$. We now apply Ebin's theorem (Theorem 5.2.1) to deduce that the isometry group of $\hat{\Omega}$ is diffeomorphism-conjugate (via a diffeomorphism close to the identity) to a subgroup H_Ω of the isometry group of $\hat{\Omega}_0$. Now, by our previous analysis via normal families, H_Ω lies in a small neighborhood of G_{Ω_0} in the isometry group of $\hat{\Omega}_0$. This isometry group is a compact Lie group and G_{Ω_0} is a compact, hence closed, subgroup and H_Ω is also compact and therefore closed. Standard Lie group theory yields that H_Ω is conjugate to a subgroup of $G_{\hat{\Omega}_0}$ by way of an isometry of $\hat{\Omega}_0$ close to the identity. Thus the diffeomorphism conjugation together with this second conjugation gives a close-to-the-identity diffeomorphism $F : \hat{\Omega} \to \hat{\Omega}_0$ conjugating $G_{\hat{\Omega}}$ to $G_{\hat{\Omega}_0}$.

Now $G_{\hat{\Omega}_0}$ contains I_{Ω_0}. Also the only possible fixed points of an element of $G_{\hat{\Omega}}$ that is not preserving each copy of Ω are lying in $\partial\Omega$. It follows that F in fact maps $\partial\Omega$ diffeomorphically to $\partial\Omega$, and thus F, being close to the identity, maps Ω to Ω_0. As a result,

$$F\big|_{\mathrm{cl}(\Omega)} : \mathrm{cl}(\Omega) \to \mathrm{cl}(\Omega_0)$$

is the conjugating diffeomorphism called for in the theorem.   □

The reader with a mind towards maximum generality will have noticed that complex analysis really played no role in the latter part of this proof. In particular, the proof technique gives rise to the following results:

Theorem 5.2.9 (Ebin's Theorem for Manifolds with Boundary). *If (M, g_0) is a compact, the C^∞ Riemannian manifold with boundary, then there is a neighborhood $\mathcal{U}$ of g_0 in the C^∞ topology on the Riemannian metrics such that, for each $g \in \mathcal{U}$, there is a diffeomorphism $F : M \to M$ (which can be chosen to be C^∞ close to the identity) such that, for each g-isometry $f : M \to M$, the mapping $F^{-1} \circ f \circ F$ is a g_0 isometry.*

Theorem 5.2.10. *If G_0 is a compact subgroup of the diffeomorphism group of a compact manifold (possibly with boundary), then there is a neighborhood $\mathcal{V}$ of G_0 in the C^∞ topology on the diffeomorphism group such that every compact subgroup*

G of the diffeomorphism group, with $G \subset \mathcal{V}$, is conjugate to a subgroup of G_0 via a diffeomorphism (which may be taken C^∞ close to the identity).

The proofs of these results are obtained by extracting suitable portions of the proof of Theorem 5.2.2.

5.3 Convergence of Holomorphic Mappings

Throughout this section, and in subsequent parts of the chapter, we shall use the concept of finite type as developed by Kohn–Catlin–D'Angelo. See [KRA1, Sect. 11.5] for an explication of these ideas. For completeness we supply the relevant definitions here.

Next we shall develop in full generality both the geometric and the analytic notions of "type" for domains in complex dimension 2. In this low-dimensional context, the whole idea of type is rather clean and simple (misleadingly so). In retrospect we shall see that the reason for this is that the varieties of maximal dimension that can be tangent to the boundary (i.e., one-dimensional complex analytic varieties) have no interesting subvarieties (the subvarieties are all zero dimensional). Put another way, any irreducible one-dimensional complex analytic variety V has a holomorphic parametrization $\phi : D \to V$. Nothing of the kind is true for higher-dimensional varieties. 8

5.3.1 Finite Type in Dimension Two

We begin with the formal definitions of geometric type and of analytic type for a point in the boundary of a smoothly bounded domain $\Omega \subseteq \mathbb{C}^2$. The main result of this subsection will be that the two notions are equivalent. We will then describe, but not prove, some sharp regularity results for the $\bar{\partial}$ problem on a finite-type domain. Good references for this material are [KOH2, BLG, KRA1].

Remark 5.3.3: The idea of commutator is an essential concept from symplectic geometry. You will see that it fits the context here very nicely.

Definition 5.3.3. A *first-order commutator* of vector fields is an expression of the form

$$[L, M] \equiv LM - ML.$$

Here the right-hand side is understood according to its action on C^∞ functions:

$$[L, M](\phi) \equiv (LM - ML)(\phi) \equiv L(M(\phi)) - M(L(\phi)).$$

Inductively, an mth order commutator is the commutator of an $(m-1)$st order commutator and a vector field N. The commutator of two vector fields is again a vector field.

Definition 5.3.4. A *holomorphic vector field* is any linear combination of the expressions

$$\frac{\partial}{\partial z_1} \quad , \quad \frac{\partial}{\partial z_2}$$

with coefficients in the ring of C^∞ functions.

A *conjugate holomorphic vector field* is any linear combination of the expressions

$$\frac{\partial}{\partial \bar{z}_1} \quad , \quad \frac{\partial}{\partial \bar{z}_2}$$

with coefficients in the ring of C^∞ functions.

Definition 5.3.5. Let M be a vector field defined on the boundary of $\Omega = \{z \in \mathbb{C}^2 : \rho(z) < 0\}$. We say that M is *tangential* if $M\rho = 0$ at each point of $\partial\Omega$.

Now we define a gradation of vector fields which will be the basis for our definition of analytic type. Throughout this section $\Omega = \{z \in \mathbb{C}^2 : \rho(z) < 0\}$ and ρ is C^∞. If $P \in \partial\Omega$ then we may make a change of coordinates so that $\partial\rho/\partial z_2(P) \neq 0$. Define the holomorphic vector field

$$L = \frac{\partial\rho}{\partial z_1}\frac{\partial}{\partial z_2} - \frac{\partial\rho}{\partial z_2}\frac{\partial}{\partial z_1}$$

and the conjugate holomorphic vector field

$$\bar{L} = \frac{\partial\rho}{\partial \bar{z}_1}\frac{\partial}{\partial \bar{z}_2} - \frac{\partial\rho}{\partial \bar{z}_2}\frac{\partial}{\partial \bar{z}_1}.$$

Both L and $\bar{L}$ are tangent to the boundary because $L\rho = 0$ and $\bar{L}\rho = 0$. They are both nonvanishing near P by our normalization of coordinates.

The real and imaginary parts of L (equivalently of $\bar{L}$) generate (over the ground field $\mathbb{R}$) the complex tangent space to $\partial\Omega$ at all points near P. The vector field L alone generates the space of all holomorphic tangent vector fields and $\bar{L}$ alone generates the space of all conjugate holomorphic vector fields. Again we stress that we are working in complex dimension two.

Definition 5.3.6. Let $\mathcal{L}_1$ denote the module, over the ring of C^∞ functions, generated by L and $\bar{L}$. Inductively, $\mathcal{L}_\mu$ denotes the module generated by $\mathcal{L}_{\mu-1}$ and all commutators of the form $[F, G]$ where $F \in \mathcal{L}_1$ and $G \in \mathcal{L}_{\mu-1}$.

Clearly $\mathcal{L}_1 \subseteq \mathcal{L}_2 \subseteq \cdots$. Each $\mathcal{L}_\mu$ is closed under conjugation. *It is not generally the case that $\cup_\mu \mathcal{L}_\mu$ is the entire three-dimensional tangent space at each point of the boundary.* A counterexample is provided by

$$\Omega = \{z \in \mathbb{C}^2 : |z_1|^2 + 2e^{-1/|z_2|^2} < 1\}$$

and the point $P = (1, 0)$. We invite the reader to supply details of this assertion.

Definition 5.3.7. Let $\Omega = \{\rho < 0\}$ be a smoothly bounded domain in $\mathbb{C}^2$, and let $P \in \partial\Omega$. Let $m \geq 2$ be an integer. We say that $\partial\Omega$ is of *finite analytic type* m at P if $\langle \partial\rho(P), F(P) \rangle = 0$ for all $F \in \mathcal{L}_{m-1}$ while $\langle \partial\rho(P), G(P) \rangle \neq 0$ for some $G \in \mathcal{L}_m$. In this circumstance we call P a point of type m.

Remark 5.3.8. A point is of finite analytic type m if it requires the commutation of m vector fields to obtain a component in the complex normal direction. Such a commutator lies in $\mathcal{L}_m$.

If P is a point of type m, then the module $\mathcal{L}_m$ is precisely equal to the full, 3-dimensional real tangent space at P. This is because there is an element G with a nontrivial complex normal component.

There is an important epistemological observation that needs to be made at this time. Complex tangential vector fields do not, after being commuted with each other finitely many times, suddenly "pop out" into the complex normal direction. What is really being discussed in this definition is an order of vanishing of coefficients.

For instance, suppose that, at the point P, the complex normal direction is the z_2 direction. A vector field

$$F(z) = a(z)\frac{\partial}{\partial z_1} + b(z)\frac{\partial}{\partial z_2},$$

such that b vanishes to some finite positive order at P and $a(P) \neq 0$ will be tangential at P. But when we commute vector fields, *we differentiate their coefficients*. Thus if F is commuted with the appropriate vector fields finitely many times, then b will be differentiated (lowering the order of vanishing by one each time) until the coefficient of $\partial/\partial z_2$ vanishes to order 0. This means that, after finitely many commutations, the coefficient of $\partial/\partial z_2$ does not vanish at P. In other words, after finitely many commutations, the resulting vector field has a component in the normal direction at P.   □

Notice that the condition $\langle \partial\rho(P), G(P) \rangle \neq 0$ is just an elegant way of saying that the vector $G(P)$ has nonzero component in the complex normal direction. Any point of the boundary of the unit ball is of finite analytic type 2. Any point of the form $(e^{i\theta}, 0)$ in the boundary of $\{(z_1, z_2) : |z_1|^2 + |z_2|^{2m} < 1\}$ is of finite analytic type $2m$. Any point of the form $(e^{i\theta}, 0)$ in the boundary of $\Omega = \{z \in \mathbb{C}^2 : |z_1|^2 + 2e^{-1/|z_2|^2} < 1\}$ is *not* of finite analytic type. We say that such a point is of *infinite analytic type*.

Now we shift gears and turn to a precise definition of finite geometric type. If P is a point in the boundary of a smoothly bounded domain, then we say that an analytic disc $\phi : D \to \mathbb{C}^2$ is a *nonsingular disc tangent to* $\partial\Omega$ at P if $\phi(0) = P$, $\phi'(0) \neq 0$, and $(\rho \circ \phi)'(0) = 0$:

Definition 5.3.9. Let $\Omega = \{\rho < 0\}$ be a smoothly bounded domain and $P \in \partial\Omega$. Let m be a nonnegative integer. We say that $\partial\Omega$ is of *finite geometric type m at P* if the following condition holds: there is a nonsingular disc ϕ tangent to $\partial\Omega$ at P such that for small ζ,

$$|\rho \circ \phi(\zeta)| \leq C|\zeta|^m$$

BUT there is no nonsingular disc ψ tangent to $\partial\Omega$ at P such that, for small values of ζ,

$$|\rho \circ \phi(\zeta)| \leq C|\zeta|^{(m+1)}.$$

In this circumstance we call P a point of finite geometric type m.

We invite the reader to reformulate the definition of geometric finite type in terms of the order of vanishing of ρ restricted to the image of ϕ. The principal result of this section is the following theorem:

Theorem 5.3.10. *Let* $\Omega = \{\rho < 0\} \subseteq \mathbb{C}^2$ *be smoothly bounded and* $P \in \partial\Omega$. *The point P is of finite geometric type* $m \geq 2$ *if and only if it is of finite analytic type m.*

Proof. We may assume that $P = 0$. Write ρ in the form

$$\rho(z) = 2\operatorname{Re} z_2 + f(z_1) + \mathcal{O}(|z_1 z_2| + |z_2|^2).$$

We do this of course by examining the Taylor expansion of ρ and using the theorem of E. Borel (see [HIR]) to manufacture f from the terms that depend on z_1 only. Notice that

$$L = \frac{\partial f}{\partial z_1}\frac{\partial}{\partial z_2} - \frac{\partial}{\partial z_1} + (\text{error terms}).$$

Here the error terms arise from differentiating $\mathcal{O}(|z_1 z_2| + |z_2|^2)$. Now it is a simple matter to notice that the best order of contact of a one-dimensional nonsingular complex variety with $\partial\Omega$ at 0 equals the order of contact of the variety $\zeta \mapsto (\zeta, 0)$ with $\partial\Omega$ at 0 which is just the order of vanishing of f at 0.

On the other hand,

$$[L, \bar{L}] = \left[-\frac{\partial^2 \bar{f}}{\partial z_1 \partial \bar{z}_1} \frac{\partial}{\partial \bar{z}_2} \right] - \left[-\frac{\partial^2 f}{\partial \bar{z}_1 \partial z_1} \frac{\partial}{\partial z_2} \right]$$

$$+ (\text{error terms})$$

$$= 2i \operatorname{Im} \left[\frac{\partial^2 f}{\partial \bar{z}_1 \partial z_1} \frac{\partial}{\partial z_2} \right]$$

$$+ (\text{error terms}).$$

Inductively, one sees that a commutator of m vector fields chosen from $L, \bar{L}$ will consist of (real or imaginary parts of) mth order of derivatives of f times $\partial/\partial z_2$ plus the usual error terms. And the pairing of such a commutator with $\partial \rho$ at 0 is just the pairing of that commutator with dz_2; in other words it is just the coefficient of $\partial/\partial z_2$. We see that this number is nonvanishing as soon as the corresponding derivative of f is nonvanishing. Thus the analytic type of 0 is just the order of vanishing of f at 0.

Since both notions of type correspond to the order of vanishing of f, we are done. □

Remark 5.3.11. It is worth noting that a strictly pseudoconvex boundary point (such as in the boundary of the unit ball B) is of finite type 2. This is easily calculated with *either* definition of finite type, and we encourage the reader to do so.

Likewise a boundary point $P = (1, 0)$ of the domain $\Omega = \{(z_1, z_2) : |z_1|^2 + |z_2|^{2m} < 1\}$ is of finite type $2m$. This is easily calculated with *either* definition of finite type, and we encourage the reader to do so. □

Remark 5.3.12. The value of this last theorem may be appreciated in the following context. It is useful to know that if $P \in \partial \Omega$ is a point of finite type, then nearby boundary points are also of finite type. Further, one would like to be able to say something about the type of those nearby points in terms of the type of P. These goals are quite difficult to achieve in the language of geometric finite type. But, in the argot of analytic finite type, they are quite easy. For, if G is an element of $\mathcal{L}_m$ so that $\langle \partial \rho(P), G(P) \rangle \neq 0$, then also $\langle \partial \rho(P), G(\zeta) \rangle \neq 0$ for ζ near P. Thus such ζ will be of finite type *at most m*.

The example

$$\Omega = \{(z_1, z_2) \in \mathbb{C}^2 : |z_1|^2 + |z_2|^{2m} < 1\} , \quad 0 < m \in \mathbb{Z},$$

illustrates the idea. Consider the boundary point $P = (1, 0)$. It is of course of finite type $2m$. The nearby boundary point $(e^{i\epsilon}, 0)$ for ϵ small is also of finite type $2m$. But the nearby boundary point $(1 - \epsilon, \sqrt[2m]{2\epsilon - \epsilon^2})$, $\epsilon > 0$ small, is strictly pseudoconvex, hence of finite type 2. □

From now on, when we say "finite type" (in dimension two), we can mean either the geometric or the analytic definition.

We say that a domain $\Omega \subseteq \mathbb{C}^2$ is of *finite type* if there is a number M such that every boundary point is of finite type not exceeding M.

The definition of finite type in higher dimensions (due to J. P. D'Angelo) is more complex. We give it in three steps.

Definition 5.3.13. Let f be a scalar-valued holomorphic function of a complex variable and P a point of its domain. The *multiplicity* of f at P is defined to be the least positive integer k such that the kth derivative of f does not vanish at P. If m is that multiplicity then we write $v_P(f) = v(f) = m$.

If ϕ is instead a vector-valued holomorphic function of a complex variable, then its multiplicity at P is defined to be the minimum of the multiplicities of its entries. If that minimum is m, then we write $v_P(\phi) = v(\phi) = m$:

Definition 5.3.14. Let $\phi : D \to \mathbb{C}^n$ be a holomorphic curve and ρ the defining function for a smoothly bounded domain Ω. Then the pullback of ρ under ϕ is the function $\phi^* \rho(\zeta) = \rho \circ \phi(\zeta)$.

Definition 5.3.15. Let Ω be a smoothly bounded domain in $\mathbb{C}^n$ and $\partial \Omega$ its boundary. Let $P \in \partial \Omega$. Let ρ be a defining function for Ω in a neighborhood of P. We say that P is a point of finite type (or finite 1-type) if there is a constant $C > 0$ such that

$$\frac{v(\phi^* \rho)}{v(\phi)} \le C$$

whenever ϕ is a nonconstant (possibly singular) one-dimensional holomorphic curve through P such that $\phi(0) = P$.

The infimum of all such constants C is called the type (or 1-type) of P. It is denoted by $\Delta(M, P) = \Delta_1(M, P)$.

Again, the reference [KRA1, Sect. 11.5] provides a thorough treatment, with examples, of the concept of point of finite type.

It is a basic fact—see, for instance, [BEL1, Main Theorem, p. 103] and the discussion in [KRA1, Sect. 11.5]—that any automorphism of a smoothly bounded, finite-type domain Ω extends to be a C^∞ diffeomorphism of the closure of the domain Ω to itself.[1] Thus it is natural in the present context to equip the automorphism group with a different topology which we shall call the C^k topology. Fix k a positive integer. Let $\epsilon > 0$. If $\varphi_0 \in \mathrm{Aut}(\Omega)$, then a subbasic neighborhood of φ_0 is one of the form

---

[1]In fact the standard condition to guarantee such an extension to a diffeomorphism of the closures is Bell's Condition R—see [KRA1, Sect. 11.5] and also our Sect. 4.4. Condition R is guaranteed by a subelliptic estimate for the $\bar{\partial}$-Neumann problem, and that condition is known to hold on domains of finite type.

$$\mathcal{U}_{k,\epsilon}(\varphi_0) \equiv \left\{ \varphi \in \mathrm{Aut}(\Omega) : \left| \frac{\partial^\alpha}{\partial z^\alpha} (\varphi - \varphi_0)(z) \right| < \epsilon \text{ for all } z \in \Omega \right.$$

$$\left. \text{and all multi-indices } \alpha \text{ with } |\alpha| \le k \right\}.$$

It is easy to see that, with this topology, $\mathrm{Aut}(\Omega)$ is still a real Lie group (see [KOB2, Sect. V.2]) when Ω is a bounded domain.

Our first result of this section is as follows:

Proposition 5.3.16. *Let $\Omega \subseteq \mathbb{C}^n$ be a bounded domain with compact automorphism group in the C^k topology, $k > 0$ an integer. Let α be a multi-index such that $|\alpha| \le k$. Then there is a positive, finite constant K_α such that*

$$\sup_{z \in \Omega} \left| \frac{\partial^\alpha}{\partial z^\alpha} \varphi(z) \right| \le K_\alpha \tag{5.3.16.1}$$

for all $\varphi \in \mathrm{Aut}(\Omega)$.

The point here is that we have a uniform bound on the αth derivative of *all* automorphisms of Ω, that bound being valid *up to the boundary*. A result of this kind was proved in [GRK1, Proposition 5.1] for the automorphism group of a *strictly pseudoconvex* domain considered in the compact-open topology. That proof was rather complicated, using the Fefferman's asymptotic expansion for the Bergman kernel of a strictly pseudoconvex domain [FEF1, Theorem 2] as well as the concept of the Bergman representative coordinates [GKK, Sect. 4.2]. The proof presented here—for the C^k topology—is much simpler and works in considerably greater generality:

Proof of Proposition 5.3.16: Suppose to the contrary that, for some fixed multi-index α, there is no bound K_α. Then there are a sequence φ_j of automorphisms of Ω and points $P_j \in \Omega$ such that

$$\left| \frac{\partial^\alpha}{\partial z^\alpha} \varphi_j(P_j) \right| \to +\infty.$$

But $\mathrm{Aut}(\Omega)$ is compact, so there is a subsequence φ_{j_k} that converges in the C^k topology to a limit automorphism φ_0. Let

$$L_0 \equiv \sup_{z \in \Omega} \left| \frac{\partial^\alpha}{\partial z^\alpha} \varphi_0(z) \right|.$$

Let $\epsilon > 0$. Choose K so large that

$$\left| \frac{\partial^\alpha}{\partial z^\alpha} \varphi_{j_k}(P_{j_k}) \right| > L_0 + 2\epsilon$$

for $k > K$. Choose M so large that

$$\left| \frac{\partial^\alpha}{\partial z^\alpha} \varphi_{j_m}(z) - \frac{\partial^\alpha}{\partial z^\alpha} \varphi_0(z) \right| < \epsilon$$

for all $m > M, z \in \Omega$. It then follows that, for $\ell > \max(K, M)$,

$$\left| \frac{\partial^\alpha}{\partial z^\alpha} \varphi_0(P_{j_\ell}) \right| > L_0 + \epsilon \,.$$

This is impossible. $\qquad\qquad\qquad\qquad\qquad\qquad\qquad\qquad\qquad\qquad \square$

The next result relates our different topologies on the automorphism group in an important new way.

Proposition 5.3.17. *Let k be a positive integer. Let Ω be a smoothly bounded domain on which*

$$\left| \frac{\partial^\alpha}{\partial z^\alpha} \varphi(z) \right| \le K_\alpha \qquad\qquad (5.3.17.1)$$

for all $\varphi \in \mathrm{Aut}(\Omega)$, all $z \in \Omega$, and all multi-indices α such that $|\alpha| \le k$. Then any sequence φ_j of automorphisms that converges uniformly on compact sets to a limit automorphism φ_0 in fact converges in the C^{k-1} topology to φ_0.

Remark 5.3.18. As the previous result shows, the converse of this proposition is true as well. $\qquad\qquad\qquad\qquad\qquad\qquad\qquad\qquad\qquad\qquad\qquad\qquad\qquad \square$

Proof of the Proposition: From (5.3.17.1), there is a constant K_{Ω_1} so that

$$\left| \nabla \varphi_j(z) \right| \le K_{\Omega_1}$$

for all $\varphi \in \mathrm{Aut}(\Omega)$, all j, and all $z \in \Omega$. Let $\epsilon > 0$. Choose a compact set $K \subseteq \Omega$ so large that if $w \in \Omega \setminus K$, then there is a line segment ℓ_w connecting w to an element $k_w \in K$ (and parametrized by $\gamma_w(t) = (1-t)w + tk_w$) which has a length less than ϵ / K_{Ω_1}.

Now choose j so large that

$$\left| \varphi_j(z) - \varphi_0(z) \right| < \epsilon \qquad\qquad (5.3.17.2)$$

for all $z \in K$. Choose a point $w \in \Omega \setminus K$. Then

$$\left| \varphi_j(w) - \varphi_0(w) \right| \le \left| \varphi_j(w) - \varphi_j(k_w) \right| + \left| \varphi_j(k_w) - \varphi_0(k_w) \right| + \left| \varphi_0(k_w) - \varphi_0(w) \right|$$
$$\equiv I + II + III \,.$$

Now we know that $II < \epsilon$ by (5.3.17.2). For I, notice that

$$\left| \varphi_j(w) - \varphi_j(k_w) \right| = \left| \int_0^1 \frac{\mathrm{d}}{\mathrm{d}t} \left[\varphi_j \circ \ell_w(t) \right] \, \mathrm{d}t \right|$$

$$\leq K_{\Omega_1} \cdot \frac{\epsilon}{K_{\Omega_1}}$$

$$= \epsilon .$$

A similar estimate obtains for III.

In summary,

$$\left| \varphi_j(w) - \varphi_0(w) \right| < 3\epsilon .$$

This gives the uniform convergence estimate that we want for all points of Ω. That proves the result for $k = 1$.

Of course similar estimates may be applied to $|(\partial^\alpha / \partial z^\alpha)\varphi_j(w) - (\partial^\alpha / \partial z^\alpha)\varphi_0(w)|$ for any $|\alpha| < k$. Thus we get convergence in the C^{k-1} topology. $\qquad\square$

Corollary 5.3.19. *Let $\Omega \subseteq \mathbb{C}^n$ be a smoothly bounded domain on which automorphisms satisfy uniform bounds on derivatives as in (5.3.17.1). Let $\varphi_j \in \mathrm{Aut}(\Omega)$ be a sequence of automorphisms that converges uniformly on compact sets to a limit automorphism φ_0. Then in fact $\varphi_j \to \varphi_0$ uniformly on $\overline{\Omega}$.*

Proof. This is a special case of the preceding result. $\qquad\square$

Remark 5.3.20. Let Ω be a strictly pseudoconvex domain with real analytic boundary which is not biholomorphic to the ball. Then the results on uniform bounds of derivatives of automorphisms are particularly easy to prove. For $\mathrm{Aut}(\Omega)$ must be compact (see [WON, Main Theorem, p. 253]). It is further known—see [GRK3]— that there is an open neighborhood U of $\overline{\Omega}$ such that every automorphism (and its inverse, of course) analytically continues to U. It then follows directly from Cauchy estimates that, if α is a multi-index, then

$$\left| \frac{\partial^\alpha}{\partial z^\alpha} \varphi(z) \right| \leq K_\alpha$$

for all $\varphi \in \mathrm{Aut}(\Omega)$ and all $z \in \Omega$. $\qquad\square$

It is possible to use the Bergman representative coordinates (see our Sect. 3.1) in a new fashion to obtain the uniform-bounds-on-derivatives result for finite type domains in $\mathbb{C}^2$ *in the compact-open topology.* More precisely,

Theorem 5.3.21. *Let $\Omega \subseteq \mathbb{C}^2$ be a smoothly bounded, finite-type domain in $\mathbb{C}^2$ with compact automorphism group in the compact-open topology. Let α be a multi-index. Then there is a constant $K_\alpha > 0$ so that*

$$\left| \frac{\partial^\alpha}{\partial z^\alpha} \varphi(z) \right| \leq K_\alpha$$

for all $\varphi \in \mathrm{Aut}(\Omega)$ and all $z \in \Omega$.

Proof. For a fixed $w \in \Omega$, let δ_w denote the Dirac delta mass at w. Then of course

$$K(z, w) = P(\delta_w)(z) \tag{5.3.21.1}$$

for all $z \in \Omega$, where K is the Bergman kernel for Ω and P the Bergman projection. Now, by a well-known formula of Kohn (see Sect. 6.6),

$$P = I - \bar{\partial}^* N \bar{\partial} .$$

Here N is the $\bar{\partial}$-Neumann operator. It follows that P is pseudolocal up to the boundary (again see [KRA2, Sects. 7.8 and 7.9]).

Let U be a tubular neighborhood of $\partial\Omega$. Let $L \subset\subset \Omega$ be a compact set so that $\partial L \subseteq U$. Now pick $w \in \partial L$. So there will be an $r > 0$, with r greater than the radius of U, so that $K(\cdot, w)$ is smooth on $\overline{\Omega} \cap B(w, r)$.

Now assume that $w \in U \cap \Omega$. Let $\tilde{w}$ be the point of $\partial\Omega$ that is nearest to w. Then, because we are in complex dimension 2 (see [BEF1, Theorem 3.1]), there is a holomorphic peak function[2] $f_{\tilde{w}}$ for $\tilde{w}$. We may replace $f_{\tilde{w}}(z)$ with $[9 + f_{\tilde{w}}(z)]/10$ so that our peak function does not vanish on $\overline{\Omega}$. Continue to denote the peak function by $f_{\tilde{w}}$. Then we may write

$$K(z, w) = P(\delta_w)(z)$$

$$= \int_\Omega K(z, \zeta) \delta_w(\zeta)$$

$$= \int_\Omega K(z, \zeta) \sum_j (\alpha_j) \cdot \left(\frac{1}{\eta_j^4 \cdot \Omega_4} \right) \cdot \chi_{B(w, \eta_j)}(\zeta) \, dV(\zeta) .$$

Here χ_S denotes the characteristic function of the set S. In the right-hand part of this last sequence of equalities, the α_j are positive numbers that sum to 1, and Ω_4 is the volume of the unit ball in $\mathbb{R}^4 \approx \mathbb{C}^2$ (see [KRA1, Sect. 1.4]). [We are simply invoking here the mean-value property of a holomorphic function on balls.] Also the η_j is an increasing sequence of finitely many positive radii with the largest of them equaling the distance τ of w to $\partial\Omega$.

Now this last equals

$$\int_\Omega K(z, \zeta) c \cdot f_{\tilde{w}}^j(\zeta) \, dV(\zeta) + \mathcal{E}(z, w) = f_{\tilde{w}}^j(z) + \mathcal{E}(z, w) ,$$

---

[2]The construction of peaking functions in [BEF1, Theorem 3.1] is quite difficult and technical. It amounts to a delicate scaling procedure. An alternative approach to the matter, using entire functions that grow at a certain rate at infinity, appears in [FOM]. The paper [BEF1] proves the peak point result for domains with real analytic boundary. The paper [FOM] proves the result for finite-type domains.

where $c > 0$ is a constant, j (interpreted as a *power*) is a suitably chosen positive integer, and $\mathcal{E}(z, w)$ is an error term. Now we know that the first term in this last displayed expression *does not vanish* on Ω intersect a ball about w that has radius larger than τ and the error term is negligible in this regard—because the Bergman projection of $\sum_j \alpha_j \left(\frac{1}{\eta_j^4 \cdot \Omega_4} \right) \chi_{B(w, \eta_j)}(\zeta)$ is, by inspection, approximated closely in the uniform topology by the dilated peaking function.

Thus the Bergman representative coordinates (see [GKK, Sect. 4.2] for this concept), which are given by

$$b_{j,w}(z) = \frac{\partial}{\partial \bar{\zeta}_j} \log \frac{K(z, \zeta)}{K(\zeta, \zeta)} \bigg|_{\zeta = w},$$

are well defined on $\beta_w \cap \Omega$ with $\beta_w = B(w, \tau')$ for some $\tau' > \tau$. And the size of $b_{j,w}$ may be taken to be uniformly bounded, independent of w, just by the noted regularity properties of the Bergman kernel. Of course $L \cap \beta_w \neq \emptyset$.

Now fix a multi-index α. Then certainly $|(\partial^\alpha / \partial z^\alpha)\varphi(z)|$ is bounded by some M_α for all $\varphi \in \mathrm{Aut}(\Omega)$ and all $z \in L$. But then the Bergman representative coordinates enable us to realize each automorphism as a linear map (namely, the Jacobian—again see [GKK, Sect. 4.2]) on $\beta_w \cap \Omega$. And the size of the coefficients of these linear maps depends only on the Jacobian of the automorphism at the center of the ball. Of course the center of the ball lies in a compact subset of Ω, so these Jacobians have uniformly bounded coefficients. The conclusion then is that $|(\partial^\alpha / \partial z^\alpha)\varphi(z)|$ is uniformly bounded on $L \cup \beta_w$. And the bound is independent of w. Remembering that w is an arbitrary element of ∂L, we see that $|(\partial^\alpha / \partial z^\alpha)\varphi(z)|$ is uniformly bounded on all of Ω, uniformly for all $\varphi \in \mathrm{Aut}(\Omega)$. $\square$

5.4   The Semicontinuity Theorem

Now one of the main results of this section is the following:

Theorem 5.4.1. *Let Ω be a smoothly bounded, finite-type domain in $\mathbb{C}^2$ which has compact automorphism group in the compact-open topology. Let k an integer be sufficiently large. Then there is an $\epsilon > 0$ so that if Ω' is a smoothly bounded, finite-type domain with C^k distance less than ϵ from Ω, then $\mathrm{Aut}(\Omega')$ can be realized as a subgroup of $\mathrm{Aut}(\Omega)$. By this we mean that there is a smooth diffeomorphism $\Phi : \Omega' \to \Omega$ so that*

$$\varphi \longmapsto \Phi \circ \varphi \circ \Phi^{-1}$$

is a univalent homomorphism of $\mathrm{Aut}(\Omega')$ into $\mathrm{Aut}(\Omega)$.

Proof. The proof of this result has been indicated earlier in this chapter (see also [GRK3, Theorem 0.1]), so we only sketch the steps.

Step 1: There is a Riemannian metric, smooth on $\overline{\Omega}$, which is invariant under any automorphism of Ω. We construct this metric simply by averaging the Euclidean metric with respect to Haar measure on the automorphism group of Ω. In order for the resulting metric to be smooth to the boundary, we must invoke the uniform bounds on automorphism derivatives that we proved in Sect. 5.1.

Step 2: The metric in Step 1 can be modified so that it is a product metric near the boundary, and still invariant. This is a standard construction from the Riemannian geometry, and we omit the details.

Step 3: We may form the metric double $\hat{\Omega}$ of $\overline{\Omega}$, and the resulting metric is smooth on $\hat{\Omega}$.

Step 4: Any automorphism of Ω can now be realized as an isometry of $\hat{\Omega}$.

Step 5: By a classical result of David Ebin [EBI1, Sect. 1], there is a semicontinuity result for isometries of compact Riemannian manifolds. We may apply this result to the isometry group of $\hat{\Omega}$. In particular, any smooth deformation Ω' of Ω gives rise to a smooth deformation $\widehat{\Omega'}$ of $\hat{\Omega}$ and hence to a deformation of the invariant metric on $\hat{\Omega}$. Thus, we may compare the isometry group of the perturbed metric to the isometry group of the original metric.

Step 6: We may unravel the construction to see that Step 5 may be interpreted to say that the automorphism group of Ω' is a subgroup of the automorphism group of Ω and we may extract the conjugation map Φ from the conjugation map provided by Ebin's theorem.

That completes the argument. $\qquad\square$

Since we introduced the C^k metric for the space of automorphisms, it is worthwhile to formulate a result for that topology. We have:

Theorem 5.4.2. *Let Ω be a smoothly bounded, finite-type domain in $\mathbb{C}^2$. Equip $Aut(\Omega)$ with the C^k topology, some integer $k \geq 0$. Assume that Ω has compact automorphism group in the C^k topology. Then there is an $\epsilon > 0$ so that if Ω' is a smoothly bounded, finite-type domain with C^m distance less than ϵ from Ω (with $m \leq k$), then $Aut(\Omega')$ can be realized as a subgroup of $Aut(\Omega)$. By this we mean that there is a smooth mapping $\Phi : \Omega' \to \Omega$ so that*

$$\varphi \longmapsto \Phi \circ \varphi \circ \Phi^{-1}$$

is a univalent homomorphism of $Aut(\Omega')$ into $Aut(\Omega)$.

Proof. The proof is just the same as that for the last theorem. The main point is to have a uniform bound for derivatives of automorphisms (Proposition 5.2.5), so that the smooth-to-the-boundary invariant metric can be constructed. $\qquad\square$

5.5 Some Examples

In this section we provide some examples which bear on the context of Theorems 5.4.1 and 5.4.2.

Example 5.5.1. Let

$$\Omega = B(0, 2) \setminus \overline{B}(0, 1) .$$

Then Ω is a bounded domain, but it is not pseudoconvex.

Of course any automorphism of Ω continues analytically to $B(0, 2)$. But it also must preserve $S_1 \equiv \{z : |z| = 1\}$ and $S_2 \equiv \{z : |z| = 2\}$. It follows that $\mathrm{Aut}(\Omega) = U(n)$. Now an obvious Lie subgroup of $U(n)$ is $SU(n)$. But $SU(n)$ has precisely the same orbits as $U(n)$—in fact the orbit of any point in S_2 is S_2 itself and the orbit of any point in S_1 is S_1 itself. It follows that there is no domain that is "near" to Ω in any C^k topology and with automorphism group that is precisely $SU(n)$. Therefore an obvious sort of converse to Theorem 5.2.3 fails in this case. That is to say, not every closed subgroup of the automorphism group of Ω arises as the automorphism group of a nearby domain.

We note, however, that with suitable hypotheses (including strong pseudoconvexity), there is a sort of converse to Theorem 5.2.3—see [MIN, Sect. 1].

Example 5.5.2. If we do not mandate that the domain Ω have smooth boundary, then Theorem 5.2.3 need not be true. As a simple example, consider

$$\Omega = \{z \in \mathbb{C}^n : 0 < |z| < 1\} .$$

Of course this Ω is not pseudoconvex and does not have a smooth defining function (so does not have smooth boundary by our reckoning). The automorphism group of Ω is $U(n)$. A "small" perturbation of Ω is $\Omega' = B = \{z \in \mathbb{C}^n : |z| < 1\}$. But the automorphism group of Ω' is much larger than $U(n)$ (it includes $U(n)$, but it also includes the Möbius transformations). So semicontinuity of automorphism groups fails.

5.6 Further Remarks

The idea of semicontinuity for automorphism groups is an important paradigm that has far-reaching applicability in geometry. In any situation where symmetries are considered, one may formulate the idea of semicontinuity. The basic idea is that symmetry is hard to create but easy to destroy: small perturbations can and will reduce symmetry, but it takes a large perturbation to create symmetry.

In the present discussion we have taken a fundamental theorem of [GRK1, Theorem 0.1] in the strictly pseudoconvex setting and extended it in various ways to the finite-type setting. It would be interesting to know whether the result is true in complete generality. Even more interesting would be an example—say in the infinite-type context—in which semicontinuity fails.

5.7 The Lu Qi-Keng Conjecture

In the paper [BOA1, BOA2], Harold Boas gave an example of a strictly pseudoconvex, topologically trivial domain on which the Bergman kernel has zeroes (off the diagonal, of course). This is a counterexample to an old question of Lu Qi-Keng. We provide the details here.

It should first be noted that, in one complex dimension, any topologically trivial domain (except for the entire complex plane itself) is conformally equivalent to the unit disc. Since the kernel for the disc certainly never vanishes, we may conclude that the kernel for the topologically trivial domain also does not vanish. Such an analysis is not valid in higher dimensions, for in that context there is no Riemann mapping theorem.

We note that in [GRK1] it is proved that, in the class of smoothly bounded strictly pseudoconvex domains, the collection of domains for which the Bergman kernel function is bounded away from zero is open. Also the collection of domains for which the Bergman kernel has no zeros is closed. Boas's example shows that the second collection of domains is strictly larger than the first.

It is clear that the Bergman kernel for the ball $B \subseteq \mathbb{C}^n$,

$$K_B(z, \zeta) = c_n \cdot \frac{1}{(1 - z \cdot \bar{\zeta})^{n+1}},$$

has no zeros. This is a useful feature, as many geometric constructions (see [GOL]) entail division by the Bergman kernel. For instance, construction of Bergman representative coordinates entails division by the Bergman kernel. Thus in 1966, Lu Qi-Keng conjectured that the Bergman kernel of any bounded domain does not vanish.

It was shown, by direct calculation, by Skwarczyński [SKW] and Rosenthal [ROSE] that the Lu Qi-Keng conjecture fails on the annulus in the plane. By contrast, it is clear from conformal mapping, and the usual transformation formula for the Bergman kernel, that the Bergman kernel of any *simply connected* domain in the plane will be nonvanishing.

The operative form of the Lu Qi-King conjecture then became: The Bergman kernel of any topologically trivial domain in $\mathbb{C}^n$ has no zeros. Greene and Krantz [GRK1, GRK2] proved that the Bergman kernel of a strictly pseudoconvex domain deforms stably under perturbation of the domain. Hence, if $\Omega_0 \subseteq \mathbb{C}^n$ is a domain on which the Lu Qi-Keng conjecture is true, then the conjecture will remain true on "nearby" domains, where the sense of nearby must be interpreted in a suitable topology on domains. Hurwitz's theorem also shows that the limit of a sequence of domains (again, in a suitable topology) on which the Lu Qi-Keng conjecture holds will also have the Lu Qi-Keng property.

Thus it was a bit of a surprise when Boas [BOA1] proved that the Lu Qi-Keng conjecture is false. We provide the details here. Indeed, consider the unbounded, logarithmically convex, complete Reinhardt domain

$$\Omega = \{(z, w) \in \mathbb{C}^2 : |w| < (1 + |z|)^{-1}\}.$$

Of course Ω is a complete circular domain, so any holomorphic function on Ω will have a globally convergent power series. Which holomorphic monomials are square integrable on Ω? Consider the calculation

$$\iint_{\Omega} |z^j w^k|^2 \, dV(z, w) = \int_z \int_0^{2\pi} \int_0^{(1+|z|)^{-1}} |z|^{2j} r^{2k} \cdot r \, dr d\theta dA(z)$$

$$= 2\pi \int_z \frac{r^{2k}}{2k+2} \bigg|_0^{(1+|z|)^{-1}} \cdot |z|^{2j} \, dA(z)$$

$$= 2\pi \cdot \int_z \frac{(1+|z|)^{-2k-2}}{2k+2} \cdot |z|^{2j} \, dA(z).$$

Now it is plain that the integrand is integrable near the origin. As $|z| \to +\infty$, the size of the integrand is $|z|^{2j-2k-2}$. In order for this to be integrable at infinity, we must have $k > j$ (just use polar coordinates). But then we see that any holomorphic, square-integrable function on Ω will have power series expansion, in terms of monomials $z^j w^k$, with k at least 1. Hence, the function will vanish at the origin.

The domain that we have just constructed fails the Lu Qi-Keng property, but it is unbounded. To find a bounded example, let $\Omega_R = \Omega \cap B(0, R)$. Ramadanov's theorem (see also [KRA8]) now tells us that the Bergman kernel K_{Ω_R} for Ω_R converges uniformly on compact sets to the Bergman kernel K_Ω for Ω. By Hurwitz's theorem, we may conclude that when R is large enough, the kernels K_{Ω_R} will vanish.

The domain Ω_R is a bounded, topologically trivial domain on which the Bergman kernel has zeros. To obtain a *smoothly bounded* example, we simply exhaust Ω_R by smooth, strictly logarithmically convex complete Reinhardt domains. See [KRA1, Chap. 1] for the details of this process. Then Hurwitz's theorem and Ramadanov's theorem give the result.

In the paper [BOA2], Boas shows that the Lu Qi-Keng conjecture fails "generically" in the following precise sense. Let $\mathcal{H}$ denote the Hausdorff distance on domains. To describe this idea, first note that if $S \subseteq \mathbb{R}^N$ and $x \in \mathbb{R}^N$, then

$$\text{dist}(x, S) = \inf_{s \in S} |s - x|.$$

Now, if Ω_1, Ω_2 are domains in $\mathbb{C}^n$, then set

$$\mathcal{H}(\Omega_1, \Omega_2) = \max \left\{ \sup_{z^1 \in \Omega_1} \operatorname{dist}(z^1, \Omega_2), \ \sup_{z^2 \in \Omega_2} \operatorname{dist}(z^2, \Omega_1) \right\}.$$

Following the notation and language of Boas, we now set

$$\rho_1(\Omega_1, \Omega_2) = \mathcal{H}(\overline{\Omega}_1, \overline{\Omega}_2) + \mathcal{H}(\partial \Omega_1, \partial \Omega).$$

The main result in [BOA2] is as follows:

Theorem 5.7.1. *The domains in $\mathbb{C}^n$ having the Lu Qi-Keng property are nowhere dense in the collection of all domains with any of the following secondary properties:*

(a) bounded, pseudoconvex
(b) bounded, strictly pseudoconvex

The proof of this result would take us far afield, and we omit it. Details may be found in [BOA2].

5.8 The Lu Qi-Keng Theorem

Next we would like to present an application of Bergman representative coordinates to the proof of a remarkable theorem of Lu Qi-Keng on domains with a Bergman metric of constant holomorphic sectional curvature. Our presentation here owes a debt to [GKK].

Theorem 5.8.1 (Lu Qi-Keng). *Let Ω be a bounded domain in $\mathbb{C}^n$. Assume that the Bergman metric is complete and has constant holomorphic sectional curvature. Then Ω is biholomorphic to the unit ball.*

Notice that this result is certainly specific to the Bergman metric. For example, the annulus $\{\zeta \in \mathbb{C} : 1 < |\zeta| < R\}$, $R > 1$ admits a complete metric of constant (holomorphic sectional) curvature. But it is not even homeomorphic to the unit disc, much less biholomorphic to it.

Proof of Theorem 5.8.1:

Background Information

If the curvature c were positive, then Ω would be a complete Riemannian manifold with all sectional curvatures greater than or equal to $c/4 > 0$. [This is because of the formula for Riemannian sectional curvature in case the holomorphic sectional curvature is constant.] Hence, Ω would be compact by standard Riemannian geometry. [Myers's theorem—A complete Riemannian manifold with sectional curvature everywhere $\geq \epsilon > 0$ has diameter $\leq \pi/\sqrt{\epsilon}$ and is hence compact [PET].]

If instead the curvature c were zero, then the universal cover of Ω would be a complete, simply connected Kähler manifold of sectional curvature 0 and hence would be biholomorphically isometric to $\mathbb{C}^n$. But then, since Ω is bounded, the covering map into Ω would be constant by Liouville's theorem. This would contradict surjectivity of the covering map.

The only remaining case is that the curvature c is negative. If g_Ω is the Bergman metric of Ω (with constant negative holomorphic sectional curvature c), then the metric

$$g \equiv -\frac{c(n+1)}{4} g_\Omega$$

has constant (negative) holomorphic sectional curvature $-4/(n+1)$. [Here g_Ω is, as usual, the Bergman metric.] Thus the simply connected covering space $\hat{\Omega}$ of Ω with the pullback $\hat{g}$ of the metric g is a complete, simply connected Kähler manifold with constant holomorphic sectional curvature $-4/(n+1)$. By standard Kähler geometry (cf., [KON]), $(\hat{\Omega}, \hat{g})$ is biholomorphically isometric to B^n with its Bergman metric. Thus we obtain a holomorphic covering map $F : B^n \to \Omega$ which is locally isometric for the Bergman metric on B^n and g on Ω, respectively.

To prove the theorem, we need only show that F is in fact injective.

The Heart of the Proof

For this let $q = F(0)$. Since F is a covering map, it is locally invertible. Namely, there exists an open neighborhood U of q and a neighborhood V of 0 such that $F|_V : V \to U$ is a biholomorphism. Denote by H_0 the inverse of $F|_V$.

With $z, w \in U$, let

$$K_0(z, w) \equiv K_{B^n}(H_0(z), H_0(w)).$$

Then

$$\frac{\partial^2}{\partial z_j \, \partial \bar{z}_k} \log K_0(z, z) = g_{j\bar{k}} = \lambda \, g_{\Omega \, j\bar{k}}$$

by the condition on F above, where $\lambda = -\frac{c(n+1)}{4}$. This implies that

$$\frac{\partial^2}{\partial z_j \, \partial \bar{z}_k} \log K_0(z, z) - \frac{\partial^2}{\partial z_j \, \partial \bar{z}_k} \lambda \log K_\Omega(z, z) = 0$$

for every $z \in U$, and furthermore that

$$\log K_0(z, w) - \lambda \log K_\Omega(z, w) = \varphi(z) + \overline{\varphi(w)}$$

for every $z, w \in U$, for some holomorphic function $\varphi : U \to \mathbb{C}$. For this one may need to replace U by a smaller, simply connected neighborhood. Consequently one obtains

$$\frac{\partial}{\partial \overline{w}_j} \log \frac{K_0(z, w)}{K_0(w, w)} - \frac{\partial}{\partial \overline{w}_j} \lambda \log \frac{K_\Omega(z, w)}{K_\Omega(w, w)} = 0$$

for every $z, w \in U$.

This gives rise to a direct computation with the Bergman representative coordinate systems $b^1 : V \to \mathbb{C}^n$ and $b^2 : U \to \mathbb{C}^n$. One obtains that

$$H_0(\zeta) = (F|_V)^{-1} = (b^1)^{-1} \circ A \circ b^2(\zeta) \qquad (5.8.1.1)$$

for every $\zeta \in U$. Here A is the linear map represented by the matrix with the (j, k)-th entry

$$\lambda \frac{\partial F_k}{\partial z_j}\bigg|_0.$$

Now look at the expressions in (5.8.1.1). The map b^1 is in fact a constant multiple of the Euclidean coordinate system. Therefore it extends to all of $\mathbb{C}^n$ holomorphically. So does the linear map A. The map $\zeta \to b^2(\zeta)$ extends to a holomorphic mapping of $\Omega \setminus Z_q$, where

$$Z_q = \{\zeta \in \Omega \mid K_\Omega(\zeta, q) = 0\}.$$

Since $K_\Omega(\cdot, q)$ is a holomorphic function on Ω with $K_\Omega(q, q) \neq 0$, the set Z_q is an analytic variety whose complex codimension in Ω is 1. Hence, $\Omega \setminus Z_q$ is a connected, dense, and open subset of Ω. Therefore using the expression of H_0 in (5.8.1.1), the map H_0 extends to a holomorphic mapping of $\Omega \setminus Z_q$ into $\mathbb{C}^n$. Let H denote this extension.

Now, let $X \equiv F^{-1}(Z_q)$. Then one immediately sees that

$$X = \{z \in B^n \mid K_\Omega(F(z), q) = 0\}.$$

Since $K_\Omega(F(0), q) = K_\Omega(q, q) \neq 0$, we see that X is again a complex analytic subvariety of B^n with complex codimension 1. Thus $B^n \setminus X$ is a connected, dense, and open subset of B^n. Furthermore, $H \circ F : B^n \setminus X \to \mathbb{C}^n$ is holomorphic with $H \circ F(z) = z$ for every $z \in V$, as $H = H_0$ on V. This means that $H \circ F(z) = z$ for every $z \in B^n \setminus X$. Now, for every $\zeta \in \Omega \setminus Z_q$, choose $x \in B^n$ such that $F(x) = \zeta$. Then

$$H(\zeta) = H(F(x)) = x.$$

This implies that $H(\Omega \setminus Z_q) \subset B^n$.

We see that H is holomorphic on $\Omega \setminus Z_q$. The removable singularity theorem for bounded holomorphic maps (the Riemann removable singularities theorem) yields that H extends to a holomorphic mapping of Ω into $\mathbb{C}^n$. Since H continues to play the role of left inverse of F, it follows easily that F must be injective. That completes the proof.   □

5.9   The Dimension of the Bergman Space

The material in this section has intrinsic interest, but it is also offered as an introduction to the next section.

It is a fact (which we shall prove below) that a domain in $\mathbb{C}$ will have either an ∞-dimensional Bergman space or a 0-dimensional Bergman space. As an instance, the disc or the annulus have infinite-dimensional Bergman space; the entire complex plane has zero-dimensional Bergman space.

Matters are different in several complex variables. Jan Wiegerinck [WIE] showed that there is a domain in $\mathbb{C}^2$ with positive, finite-dimensional Bergman space. We provide the details here.

We may note that the Bergman space for a *bounded* domain is immediately infinite dimensional, for all the polynomials will be in the Bergman space. The same assertion is true for any domain that is biholomorphically equivalent to a bounded domain. So Wiegerinck's example will perforce be unbounded. We have:

Theorem 5.9.1. *For every integer $k > 0$ there is a Reinhardt domain in $\mathbb{C}^2$ with k-dimensional Bergman space.*

Proof. We work in $\mathbb{C}^2$ (although there are analogues of these results in any dimension greater than one). Consider the domains

$$X_1 = \{(z, w) \in \mathbb{C}^2 : |w| < 1/(|z| \log |z|) \,, \, |z| > e\}$$

and

$$X_2 = \{(z, w) \in \mathbb{C}^2 : |z| < 1/(|w| \log |w|) \,, \, |w| > e\} \,.$$

Set

$$\Omega = X_1 \cup X_2 \cup \{(z, w) \in \mathbb{C}^2 : |z| < 2e, |w| < 2e\} \,.$$

Lemma 5.9.2. *The monomials in $A^2(\Omega)$ are precisely*

$$c z^k w^k, \qquad k = 0, 1, 2, \ldots .$$

Here c is some complex constant.

Proof: We calculate

$$\int_{X_1} |z|^{2p} |w|^{2p} \, dV = (2\pi)^2 \int_{r_1=e}^{\infty} \int_{r_2=0}^{1/r_1 \log r_1} r_1^{2p+1} r_2^{2q+1} \, dr_1 dr_2$$

$$= (2\pi)^2 \int_{e}^{\infty} \frac{r_1^{2p-2q-1}}{(\log r_1)^{2q+2}} \, dr_1 .$$

This last integral is finite if and only if $q \geq p$. Thus

$$c z^p w^q \in A^2(X_1) \iff q \geq p . \tag{5.9.2.1}$$

A similar calculation shows that

$$c z^p w^q \in A^2(X_2) \iff q \leq p . \tag{5.9.2.2}$$

The result follows from (5.9.2.1) and (5.9.2.2). □

Now we shall enlarge Ω in the direction of $|z| = |w|$ in order to get rid of the monomials with high exponents. Set

$$B_m = \left\{ (z, w) \in \mathbb{C}^2 : |z| > 1, |w| > 1, ||z| - |w|| < \frac{1}{(|z| + |w|)^m} \right\} .$$

Also define

$$\Omega_k = \Omega \cup B_{4k} , \qquad k = 1, 2, \ldots .$$

See Fig. 5.1.

Our goal is to show that Ω_k has k-dimensional Bergman space.

Lemma 5.9.3. *For $k = 0, 1, 2, \ldots$ we have*

$$z^p w^q \in A^2(\Omega_k) \iff p = q < k .$$

Fig. 5.1 The domain Ω_k

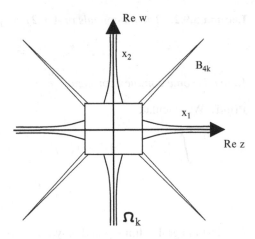

Proof. We calculate, using the change of variables $r_1 + r_2 = t, r_1 - r_2 = s$, that

$$\|z^p w^q\|^2_{A^2(B_m)} = \int_{B_m} |z^{2p} w^{2p}| \, dV$$

$$= (2\pi)^2 \int_{\substack{|r_1 - r_2| < 1/(r_1 + r_2)^m \\ r_1 > 1, r_2 > 1}} (r_1 r_2)^{2p+1} \, dr_1 dr_2$$

$$\approx \int_2^\infty (2\pi)^2 \int_{-(1/t)^m}^{(1/t)^m} \frac{1}{2} \left(\frac{t^2 - s^2}{4} \right)^{2p+1} dsdt .$$

This last integral converges if and only if $m = 4k > 4p + 3$. Then Lemma 5.9.2 and the last calculation give Lemma 5.9.3.   □

Proof of the Theorem: The logarithmically convex hull of the Reinhardt domain Ω_k is $\mathbb{C}^2$. Therefore every $f \in A^2(\Omega_k)$ has a holomorphic continuation to $\mathbb{C}^2$. As a result, f has a power series expansion

$$\sum_{m,n} a_{m,n} z^m w^n$$

that converges to f uniformly on bounded subsets of $\mathbb{C}^2$. As usual $B(0, R)$ denotes the ball in $\mathbb{C}^2$ with center 0 and radius R. Then

$$\|f\|_{A^2(\Omega_k)} \geq \int_{\Omega_k \cap B(0,R)} |f|^2 \, dV = \sum |a_{mn}|^2 \int_{\Omega_k \cap B(0,R)} |z^{2m} w^{2n}| \, dV . \quad (5.9.1.1)$$

This last equality follows from the fact that the monomials $z^p w^q$ form an orthogonal set on every bounded Reinhardt domain, as is easily seen using polar coordinates.

With Lemma 5.9.2, and letting $R \to \infty$ in (5.9.1.1), we conclude that $a_{mn} = 0$ unless $n = m < k$. Thus

$$A^2(\Omega_k) = \text{span}\{1, zw, z^2w^2, \ldots, z^{k-1}w^{k-1}\}. \qquad \square$$

Now, to complete the discussion, we prove the following result:

Theorem 5.9.4. *Let Ω be a domain in $\mathbb{C}$. Then the dimension of $A^2(\Omega)$ is either 0 or ∞.*

Proof: After a Möbius transformation, we may assume that Ω contains the point ∞. Let $f \in A^2(\Omega)$ and suppose that f is not identically zero. There are now two cases:

Case 1: *The function f is rational.* In this case we note that

$$\int_{\mathbb{C}} |f|^2 \, dxdy = \infty, \qquad \int_{\Omega} |f|^2 \, dxdy < \infty.$$

As a result, the complement $^c\Omega$ of Ω in $\mathbb{C}$ has positive 2-dimensional Lebesgue measure. It is known (see [GAR]) that, in this situation, the Cauchy transform

$$g(z) = \int_{^c\Omega} \frac{1}{\zeta - z} \, d\xi d\eta, \qquad \zeta = \xi + i\eta,$$

is a nonconstant, bounded, holomorphic function on Ω satisfying $g(\infty) = 0$. Clearly $g^2, g^3, \ldots$ will be in $L^2(\Omega)$. Hence, $\dim(A^2(\Omega)) = \infty$.

Case 2: *The function f is nonrational.* We expand f in a Laurent series around the point ∞:

$$f(z) = \sum_{p}^{\infty} c_k z^{-k}, \qquad p \geq 2, \quad c_p \neq 0.$$

We shall construct a function $g \in A^2(\Omega)$ such that $g \neq 0$ and the Laurent series of g has no terms with $z^{-1}, z^{-2}, \ldots, z^{-p}$. Let $z^1, z^2, \ldots, z^{p+1}$ be distinct points in Ω. Take g to be of the form

$$g(z) = \sum_{j=1}^{p+1} \frac{b_j(f(z) - f(z_j))}{z - z_j}, \qquad b_j \in \mathbb{C}.$$

Expanding around ∞ we see that

$$g(z) = \sum_{k=1}^{\infty} a_k z^{-k} \,,$$

with coefficients a_k depending on b_j, z_j, and f. Indeed, for $k = 1, 2, \ldots, p$, we have

$$a_k = \sum_{j=1}^{p+1} -b_j f(z_j) z_j^{k-1} \,.$$

Now select the constants b_j to be a nontrivial solution of the homogeneous linear equations

$$a_k = 0 \,, \quad k = 1, 2, \ldots, p \,.$$

Such solutions always exist.

Note that $a_1 = 0$ and $f \in A^2(\Omega)$ imply that $g \in A^2(\Omega)$, and we also see that g is nonrational (otherwise f would be rational). In particular, $g \not\equiv 0$. Thus for some $q > p$, $a_q \neq 0$. We can continue this reasoning with g instead of f and see that $\dim(A^2(\Omega))$ is infinite. $\square$

It is known (see [CAR, GAR]) that $A^2(\Omega)$ for $\Omega \subseteq \mathbb{C}$ has nontrivial functions if and only if ${}^c\Omega$ has positive logarithmic capacity.

5.10   The Bergman Theory on a Manifold

The paper [KOB1] gives a nice presentation of the theory of the Bergman space, the Bergman kernel, and the Bergman metric on a complex manifold. We present the key ideas here.

5.10.1   *Kernel Forms*

Let M be an n-dimensional complex manifold. Let $A^2(M)$ be the set of holomorphic n-forms on M such that

$$\left| \int_M f \wedge \overline{f} \right| < \infty \,.$$

The vector space $A^2(M)$ is a separable, complex Hilbert space with inner product given by

$$\langle f_1, f_2 \rangle \equiv (-1)^{n^2/2} \int_M f_1 \wedge \overline{f}_2 .$$

We remark that the space A^2 could be finite dimensional. This happens, for instance, when M is compact (see the elementary Hodge theory in [KON]). Also refer to our Sect. 5.9 in which we give the example of Wiegerinck of a Reinhardt domain in $\mathbb{C}^2$ on which the classical Bergman space is finite dimensional.

With this inner product, let $h_0, h_1, h_2, \ldots$ be a complete orthonormal basis for $A^2(M)$. Then

$$K(z, \overline{w}) \equiv \sum_{j=0}^{\infty} h_j(z) \wedge \overline{h}_j(\overline{w})$$

is a holomorphic $2n$-form on $M \times \overline{M}$, where $\overline{M}$ is the complex manifold conjugate to M. It is a fact that the definition of K is independent of the choice of orthonormal basis. We call K the (Bergman) kernel form for M.

If $\overline{z}$ is the point of $\overline{M}$ corresponding to a point $z \in M$, then the set of points $(z, \overline{z}) \in M \times \overline{M}$ is identified, in a natural fashion, with M. Thus $K(z, \overline{z})$ can be thought of as a $2n$-form on M.

Theorem 5.10.1. *The form $K(z, \overline{z})$ is invariant under the group of holomorphic transformations of M.*

Remark 5.10.2. One nice feature of the differential form formalism is that the Jacobian of the mapping is built in. Whereas our Theorem 1.1.14 had to specify the Jacobian (twice!) explicitly, we now need not worry about it. $\quad\square$

Proof of the Theorem: Let Φ be any one-to-one, onto, holomorphic transformation of M to itself. If $h_0, h_1, h_2, \ldots$ is a complete orthonormal basis for $A^2(M)$, then so is $\Phi^* h_0, \Phi^* h_1, \Phi^* h_2, \ldots$ a complete orthonormal basis (calculation for the reader). Since the kernel form is independent of the choice of basis, we see immediately that

$$K(z, \overline{z}) = \sum \Phi^* h_j(z) \wedge \Phi^* \overline{h}_j(\overline{z}) = \Phi^* K(z, \overline{z}) .$$

That completes the proof. $\quad\square$

If $f_1, f_2 \in A^2(M)$ then, for any point $z \in M$, there are real numbers $(c_1, c_2) \neq (0, 0)$ such that

$$c_1 f_1 \wedge \overline{f}_1(z) = c_2 f_2 \wedge \overline{f}_2(z) .$$

If $c_1 = 0$ or $c_2/c_1 \geq 1$, then we say that $f_1 \wedge \overline{f}_1 \geq f_2 \wedge \overline{f}_2$ at z.

Theorem 5.10.3. *We have that*

$$K(z, \bar{z}) = \max_{\langle f, f \rangle = 1} f(z) \wedge \overline{f}(z).$$

If $K(z, \bar{z}) \neq 0$, then an n-form $f \in A^2(M)$ satisfying the above identity is unique up to a constant factor c with $|c| = 1$ and it is characterized by the following two properties:

(a) $\langle f, f \rangle = 1$.
(b) $\langle f, f' \rangle = 0$ for all $f' \in A^2(M)$ which vanish at z.

Proof. Fix a point $z \in M$ and let $A^2(M)'$ be the set of n-forms $f' \in A^2(M)$ which vanish at z. If $A^2(M)' = A^2(M)$, then our result is trivial.

Suppose instead that $A^2(M)' \neq A^2(M)$. Let h_0 be an element of $A^2(M)$ which is orthogonal to $A^2(M)'$ and has unit length. Let g be any element of $A^2(M)$ such that $g(z) \neq 0$. Let c be a complex number such that $g(z) = ch_0(z)$. Then $g - ch_0$ is in $A^2(M)'$. Therefore $A^2(M)$ is spanned by $A^2(M)'$ and h_0. Given $z \in M$, we can choose therefore a complete orthonormal basis $h_0, h_1, h_2, \ldots$ for $A^2(M)$ so that $h_0(z) \neq 0$ and $h_1(z) = h_2(z) = \cdots = 0$. This immediately implies the theorem. $\square$

Now Theorem 5.10.3 implies the following:

Theorem 5.10.4. *Let M' be a domain (i.e., a connected open subset) in M, and let K_M and $K_{M'}$ be the kernel forms for M and M', respectively. Then*

$$K_M \leq K_{M'}$$

in the sense that there exists a function c on M' such that $K_M = cK_{M'}$ and $0 \leq c \leq 1$ on M'.

Remark 5.10.5. If $M \setminus M'$ contains a nonempty, open set in M, then either $K_{M'}(z, \bar{z}) > K_M(z, \bar{z})$ or else $K_{M'}(z, \bar{z}) = K_M(z, \bar{z}) = 0$ for every $z \in M'$. $\square$

We have in addition the following result:

Theorem 5.10.6. *If M' is a domain in M and if $M \setminus M'$ is an analytic subvariety of M with complex dimension $\leq n - 1$, then*

$$K_{M'}(z, \bar{z}) = K_M(z, \bar{z}) \qquad \text{for all } z \text{ in } M'.$$

Proof. Let f be a square-integrable holomorphic n-form on M'. We shall show that f can be continued analytically to M. Let z^0 be any nonsingular point of the variety $M \setminus M'$. Let $z_1, z_2, \ldots, z_n$ be local coordinates on a neighborhood U of z^0 in M so that $(M \setminus M') \cap U$ is given by $z_1 = z_2 = \cdots = z_k = 0$ and z^0 is the origin of the coordinate system.

Let f be written in $M' \cap U$ as

$$f = f^* dz_1 \wedge \cdots \wedge dz_n,$$

where f^* is a function holomorphic in $M' \cap U$. Let V be the plane defined by

$$z_2 = z_3 = \cdots = z_n = 0.$$

Then f^* is a function holomorphic on $V \setminus \{z^0\}$ and can be expanded in a Laurent series in z_1 about the origin. Since f is square-integrable, the quantity

$$\int_{V \setminus \{z^0\}} f^* dz_1 \wedge d\bar{z}_1$$

must be finite. [It is of course essential here that f^* be holomorphic.] From this fact we may conclude that f^* is a power series in z_1 about the origin. [See the reasoning on p. 363 of [BRE].] Thus f can be analytically continued to M'', where M'' is the union of M' and the nonsingular points of $M \setminus M'$. Since $M \setminus M'$ is a subvariety of M and $\dim(M \setminus M'') < \dim(M \setminus M')$, we conclude by induction that our theorem is true. $\qquad\square$

Theorem 5.10.7. *Let M and M' be complex manifolds of complex dimensions n and n', respectively. Then*

$$K_{M \times M'} = (-1)^{nn'} K_M \wedge K_{M'}.$$

Proof. In the displayed formula in the theorem, the projections from $M \times M'$ onto M and onto M' are omitted. We can think of K_M and $K_{M'}$ as forms on $M \times M'$ in a natural manner.

Let $z^0 \in M$ and $z^{0'} \in M'$. From Theorem 5.10.3, we see that there exist forms h, h', and h'' on M, M', and $M \times M'$, respectively, so that

$$\langle h, h \rangle = \langle h', h' \rangle = \langle h'', h'' \rangle = 1,$$
$$K_M = h \wedge \bar{h} \qquad \text{at } z^0,$$
$$K_{M'} = h' \wedge h' \qquad \text{at } z^{0'},$$
$$K_{M \times M'} = h'' \wedge h'' \qquad \text{at } (z^0, z^{0'}).$$

If $K_{M \times M'}$ vanishes at $(z^0, z^{0'})$, then our result is trivial. If it does not vanish at $(z^0, z^{0'})$, then h'' is characterized by these properties:

$$\langle h'', h'' \rangle = 1 \text{ and } \langle h'', f'' \rangle = 0 \text{ for every } f'' \text{ vanishing at } (z^0, z^{0'}).$$

More precisely, h'' defined by these last two properties is unique up to a constant factor c with $|c| = 1$. However, $h'' \wedge h''$ is unique. It is easy to see that the form

$(-1)^{nn'} h \wedge h'$ on $M \times M'$ possesses these two properties. The proof is therefore complete.   □

5.10.2   The Invariant Metric

Let M be a complex manifold of complex dimension n. Suppose that

A1 *Given any point z of M, there is a square-integrable holomorphic n-form f such that $f(z) \neq 0$. In other words, the kernel form $K(z, \bar{z})$ of M is different from 0 at every point of M.*

Let $z_1, \ldots, z_n$ be a local coordinate system on M. Let

$$K(z, \bar{z}) = K^*(z, \bar{z}) dz_1, \wedge \cdots \wedge dz_n \wedge d\bar{z}_1 \wedge \cdots \wedge d\bar{z}_n ,$$

where $K^*(z, \bar{z})$ is a locally defined function. Define a quadratic Hermitian differential form ds^2 by

$$ds^2 = \sum_{\alpha, \beta} \frac{\partial^2 \log K^*}{\partial z_\alpha, \partial \bar{z}_\beta} dz_\alpha d\bar{z}_\beta .$$

It is easily shown that ds^2 is independent of the choice of coordinates.

Theorem 5.10.8. *The quadratic form ds^2 is positive semidefinite and invariant under the holomorphic automorphisms of M.*

Proof. Let z be any point of M and let $z_1, \ldots, z_n$ be local coordinates around z. Let

$$h_j = h_j^* dz_1 \wedge \cdots \wedge dz_n , \quad j = 0, 1, 2, \ldots,$$

be an orthonormal basis for $A^2(M)$ such that

$$h_0(z) \neq 0, \quad h_1(z) = h_2(z) = \cdots = 0 .$$

Then, from the identity $K^* = \sum H_j^* \bar{h}_j^*$, it follows that

$$ds^2 = \frac{\left(\sum_{k=1}^{\infty} dh_k^* \cdot d\bar{h}_k^* \right)}{K^*} \quad \text{at } z.$$

This shows that ds^2 is positive definite.

To prove that ds^2 is invariant under a holomorphic automorphism Φ of M, let $w_1, w_2, \ldots, w_n$ be a coordinate system around the point $\Phi(z)$. So

$$w_i = \Phi_i(z_1, \ldots, z_n) \quad , \quad j = 1, 2, \ldots, n.$$

Let

$$K = K^* dz_1 \wedge \cdots \wedge d\bar{z}_n = L^*(\Phi(z), \overline{\Phi}(\bar{z})) \cdot J\overline{J},$$

where J is the Jacobian $\partial\Phi_i/\partial z_k$. Now the invariance of ds^2 follows from the definition of ds^2 and the analyticity of J. $\qquad\square$

The last proof shows that ds^2 is positive definite if and only if the following hypothesis is satisfied:

A2 *For every holomorphic vector Z at z, there exists a square-integrable holomorphic n-form f such that $f(z) = 0$ and $Z(f^*) \neq 0$, where $f = f^* dz_1 \wedge \cdots \wedge dz_n$. By a holomorphic vector Z at z, we mean simply a complex tangent vector of the form*

$$Z = \sum \zeta^j (\partial/\partial z_j)_z,$$

where the $\zeta_j s$ are complex numbers.

The metric ds^2 obtained in this fashion is Kählerian and is in fact the *Bergman metric* of M.

As an immediate consequence of Theorem 5.10.8, we see the following:

Theorem 5.10.9. *Let M and M' be complex manifolds satisfying A1 and A2. Let ds^2 and ds'^2 be the Bergman metrics of M and M', respectively. Then the Bergman metric of $M \times M'$ is $ds^2 + ds'^2$.*

From Theorem 5.10.6 we obtain:

Theorem 5.10.10. *Let M' be a domain in M such that $M \setminus M'$ is an analytic subvariety of M with complex dimension $\leq n - 1$ (where $n = \dim M$). Then the Bergman metric of M' is the restriction to M' of the Bergman metric of M.*

Let $\text{Aut}(M)$ be the automorphism group of M and assume hypotheses A1 and A2. Then $\text{Aut}(M)$ is a closed subgroup of the group of isometries of M equipped with the Bergman metric. Since it is a well-known result of Myers and Steenrod (see [KON, vol. 1]) that the group of isometries of a Riemannian manifold is a Lie group with compact isotropy group at each point, we now have the following result:

Theorem 5.10.11. *Let M be a complex manifold satisfying A1 and A2. Then the automorphism group $\text{Aut}(M)$ is a (real) Lie group. Furthermore, the isotropy group of $\text{Aut}(M)$ at each point of M is compact.*

Now let us specialize down to the case that M is a bounded domain in $\mathbb{C}^n$. As we usually do, call it Ω. Fix a point $P \in \Omega$. Then the mapping

$$\mathrm{Aut}(\Omega) \longrightarrow \mathbb{C}^n \times \mathbb{C}^{n^2}$$

$$\varphi \longmapsto \left(\varphi(P), \nabla\varphi(P)\right)$$

is univalent (because an automorphism, being an isometry of the Bergman metric, is completely determined by its first-order data at a point—see [KON, vol. 1]). This shows that the automorphism group is locally Euclidean. By the solution of Hilbert's fifth problem, we may conclude from this reasoning that $\mathrm{Aut}(\Omega)$ is a Lie group. The compactness of an isotropy group follows from Montel's theorem and Cartan's classical result about limits of sequences of automorphisms. This gives an alternative means of thinking about Theorem 5.10.11.

5.11 Boundary Behavior of the Bergman Metric

The Bergman metric on the disc D is given by

$$g_{jk} = \frac{\partial}{\partial z}\frac{\partial}{\partial \bar{z}} \log K(z,z) = \frac{\partial}{\partial z}\frac{\partial}{\partial \bar{z}} \left[-\log \pi - 2\log(1-|z|^2)\right] = \frac{2}{(1-|z|^2)^2}.$$

Thus the length of a curve $\gamma : [0,1] \to D$ is given by

$$\ell_B(\gamma) = \int_0^1 \frac{2\|\gamma'(t)\|^2}{(1-|\gamma(t)|^2)^2}\, dt.$$

It is natural to wonder what one can say about the Bergman metric on a more general class of domains. Let $\Omega \subseteq \mathbb{C}$ be a bounded domain with C^2 boundary, and suppose for the moment that Ω is simply connected. Then the Riemann mapping theorem tells us that there is a conformal mapping $\Phi : \Omega \to D$. Of course then we know that

$$K_\Omega(z,z) = |\Phi'(z)|^2 K_D(\Phi(z), \Phi(z)) = \frac{|\Phi'(z)|^2}{\pi \cdot (1-|\Phi(z)|^2)^2}.$$

Then the Bergman metric on Ω is given by

$$g_{jk} = \frac{\partial}{\partial z}\frac{\partial}{\partial \bar{z}} \log \left[\frac{|\Phi'(z)|^2}{\pi \cdot (1-|\Phi(z)|^2)^2}\right]$$

$$= \frac{\partial}{\partial z}\frac{\partial}{\partial \bar{z}} \left[\log|\Phi'(z)|^2 - \log \pi - 2\log(1-|\Phi(z)|^2)\right].$$

Of course $|\Phi'(z)|$ is bounded and bounded from 0 (see [BEK]). So the second derivative of $\log |\Phi'(z)|^2$ is a bounded term. The second derivative of $\log \pi$ is of course 0.

The second derivative of the remaining (and most interesting) term may be calculated to be

$$\frac{\partial^2}{\partial z \partial \bar{z}} \log K_\Omega(z,z) = \frac{\partial^2}{\partial \bar{z} \partial z}\left[-2 \log(1 - |\Phi(z)|^2)\right]$$

$$= \frac{\partial}{\partial \bar{z}}\left(\frac{-2}{1 - |\Phi(z)|^2} \cdot \left[-\Phi'(z)\overline{\Phi}(z)\right]\right)$$

$$= \frac{2}{(1 - |\Phi(z)|^2)^2} \cdot \left[-\overline{\Phi}'(z)\Phi(z)\right] \cdot \left[-\Phi'(z)\overline{\Phi}(z)\right]$$

$$- \frac{2}{1 - |\Phi(z)|^2} \cdot \left[-\Phi'(z)\overline{\Phi}'(z)\right].$$

This in turn, after some simplification, equals

$$\frac{2|\Phi'(z)|^2}{(1 - |\Phi(z)|^2)^2}. \tag{5.11.1}$$

As previously noted, the numerator is bounded and bounded from 0. Hopf's lemma (see [KRA1, Chap. 11] tells us that $(1 - |\Phi(z)|^2) \approx \text{dist}_{\partial\Omega}(z)$. So that the Bergman metric on Ω blows up like the reciprocal of the square of the distance to the boundary—just as on the disc.

In the case that Ω has C^2 boundary and is finitely connected—not necessarily simply connected—then one may use the Ahlfors map (see [KRA5]) instead of the Riemann mapping and obtain a result similar to that in (5.11.1). We omit the details.

Exercises

1. Formulate a semicontinuity theorem for automorphism groups of domains in the complex plane $\mathbb{C}^1$. Could one prove such a result using normal families?
2. Use our results on the Bergman kernel of the annulus in the plane to calculate an asymptotic formula for the Bergman metric on the annulus.
3. Calculate the Bergman metric on the bidisc. What is the asymptotic behavior of this metric as z tends to a point $(e^{i\theta}, e^{i\theta})$ of the distinguished boundary? What about the asymptotic behavior as z tends to a boundary point $(e^{i\theta}, 0)$ that is *not* in the distinguished boundary?
4. The Bergman kernel for the annulus has zeros. Where are those zeros located?

5. Suppose that we say that a sequence of domains Ω_j approaches a domain Ω if

$$\|\chi_{\Omega_j} - \chi_\Omega\| \to 0$$

as $j \to \infty$. Here χ_E is the characteristic function of the set E. Is there a semicontinuity theorem for this topology on domains? Why or why not?

6. The Bergman space for the planar domain

$$\Omega = \mathbb{C} \setminus \{\zeta \in \mathbb{C} : |\zeta| \le 1\}$$

is infinite dimensional. Write down an infinite, linearly independent, set of elements of the Bergman space on this Ω.

7. What is the invariant Laplacian on the unit disc in the plane? How is it related to the usual Laplacian?

8. What can you say about the Lu Qi-Keng conjecture for the Bergman kernel on a Riemann surface?

9. Is there a semicontinuity theorem for automorphism groups of Riemann surfaces? How would you formulate it? How might one prove it?

10. Consider open regions in the Euclidean plane. The automorphism group of a domain is the collection of rigid motions. Can you formulate a semicontinuity theorem for this context? How would you prove it?

11. Calculate the Bergman metric for the interior of the unit square in the plane. What is the asymptotic boundary behavior as you approach a corner of the square? How does this differ from the asymptotic boundary behavior as you approach an interior point of an edge?

12. What can you say about the asymptotic boundary behavior of the Bergman metric as you approach the boundary point $(1, 0)$ of the domain

$$\Omega = \{(z_1, z_2) : |z_1|^2 + |z_2|^{2m} < 1\} , \quad 1 < m \in \mathbb{Z}?$$

How does this differ from the boundary behavior as you approach the boundary point $(0, 1)$. [**Hint:** Think about the fact that Ω holomorphically covers the ball.]

Chapter 6
Additional Analytic Topics

6.1 The Diederich–Fornæss Worm Domain

The concept of "domain of holomorphy" is central to the function theory of several complex variables. The celebrated solution of the Levi problem tells us that a connected open set (a *domain*) is a domain of holomorphy if and only if it is pseudoconvex. For us, in the present book, pseudoconvexity is *Levi* pseudoconvexity; this is defined in terms of the positive semi-definiteness of the Levi form. This notion requires the boundary of the domain to be at least C^2. When the boundary is not C^2 we can still define a notion of pseudoconvexity (called *Hartogs pseudoconvexity*) that coincides with the Levi pseudoconvexity in the C^2-case. See [KRA1, Chap. 3] for the details. When the Levi form is positive *definite* then we say that the domain is *strictly* pseudoconvex. The geometry of pseudoconvex domains has become an integral part of the study of several complex variables. (See [KRA1] for basic ideas about analysis in several complex variables.)

Consider a pseudoconvex domain $\Omega \subseteq \mathbb{C}^n$. Any such domain has an exhaustion $U_1 \subset\subset U_2 \subset\subset U_3 \subset\subset \cdots \Omega$ with $\cup_j U_j = \Omega$ by smoothly bounded, strictly pseudoconvex domains. This information was fundamental to the solution of the Levi problem (see [BERS] for this classical approach) and is an important part of the geometric foundations of the theory of pseudoconvex domains.

It is natural to ask whether there is a dual result for the exterior of Ω. Specifically, given a pseudoconvex domain Ω, are there smoothly bounded, pseudoconvex domains $W_1 \supset\supset W_2 \supset\supset W_3 \supset\supset \cdots \supset\supset \cdots \overline{\Omega}$ such that $\cap_j W_j = \overline{\Omega}$? A domain having this property is said to have a *Stein neighborhood basis*. A domain failing this property is said to have *nontrivial Nebenhülle*.

Early on, in 1906, F. Hartogs produced the following counterexample (which has come to be known as the *Hartogs triangle*): Let $\Omega = \{(z_1, z_2) \in \mathbb{C}^2 : 0 < |z_1| < |z_2| < 1\}$.

Theorem 6.1.1. *Any function holomorphic on a neighborhood of $\overline{\Omega}$ actually continues analytically to $D^2(0, 1) \equiv D \times D$. Thus $\overline{\Omega}$ cannot have a neighborhood basis of pseudoconvex domains. Instead, it has a nontrivial Nebenhülle.*

S.G. Krantz, *Geometric Analysis of the Bergman Kernel and Metric*,
Graduate Texts in Mathematics 268, DOI 10.1007/978-1-4614-7924-6_6,
© Springer Science+Business Media New York 2013

Fig. 6.1 The function η

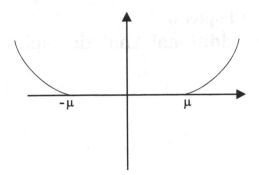

Proof: Let U be a neighborhood of $\overline{\Omega}$. For $|z_1| < 1$, the analytic discs

$$\zeta \mapsto (z_1, \zeta \cdot |z_1|)$$

have boundary lying in U. But, for $|z_1|$ sufficiently small, the entire disc lies in U. Thus a standard argument (as in the proof of the Hartogs extension phenomenon—see [KRA1, Chap. 3]), sliding the discs for increasing $|z_1|$, shows that a holomorphic function on U will analytically continue to $D(0, 1) \times D(0, 1)$. That proves the result.
$\square$

It was, however, believed for many years that the Hartogs example worked only because the boundary of Ω is not smooth—it is only Lipschitz away from the origin, and the origin is a non-manifold point at which the boundary is a finite union of Lipschitz surfaces. Thus for over 70 years, mathematicians sought a proof that a smoothly bounded pseudoconvex domain *will* have a Stein neighborhood basis. In 1977, it came as quite a surprise when Diederich and Fornæss [DIF1] produced a smoothly bounded domain—now known as the *worm*—which is pseudoconvex and which does *not* have a Stein neighborhood basis. In fact the Diederich–Fornæss example is the following:

Definition 6.1.2. Let $\mathcal{W}$ denote the domain

$$\mathcal{W} = \left\{ (z_1, z_2) \in \mathbb{C}^2 : \left| z_1 - e^{i \log |z_2|^2} \right|^2 < 1 - \eta(\log |z_2|^2) \right\},$$

where

  (i) $\eta \geq 0$, η is even, η is convex.
  (ii) $\eta^{-1}(0) = I_\mu \equiv [-\mu, \mu]$.
  (iii) There exists a number $a > 0$ such that $\eta(x) > 1$ if $|x| > a$.
  (iv) $\eta'(x) \neq 0$ if $\eta(x) = 1$.

We illustrate the function η in Fig. 6.1.

Notice that the slices of $\mathcal{W}$ for z_2 fixed are discs centered on the unit circle with centers that wind μ/π times about that circle as $|z_2|$ traverses the range of values for which $\eta(\log|z_2|^2) < 1$.

It is worth commenting here on the parameter μ in the definition of $\mathcal{W}$. The number μ in some contexts is selected to be greater than $\pi/2$. The number $\nu = \pi/2\mu$ is half the reciprocal of the number of times that the centers of the circles that make up the worm traverse their circular path.

Many authors use the original choice of parameter β, where $\mu = \beta - \pi/2$ (see, e.g., [BAR2, CHS, KRP1]). Here, we have preferred to use the notation μ, in accord with the sources [CHR1, CHR2].

Proposition 6.1.3. *The domain $\mathcal{W}$ is smoothly bounded and pseudoconvex. Moreover, its boundary is strictly pseudoconvex except at the boundary points $(0, z_2)$ for $|\log|z_2|^2| \leq \mu$. These points constitute an annulus in $\partial\mathcal{W}$.*

Proposition 6.1.4. *The smooth worm domain $\mathcal{W}$ has nontrivial Nebenhülle.*

The proofs of these propositions are deferred to Sect. 6.2.

As Diederich and Fornæss [DIF1] showed, the worm provides a counterexample to a number of interesting questions in the geometric function theory of several complex variables. As an instance, the worm gives an example of a smoothly bounded, pseudoconvex domain which lacks a global plurisubharmonic defining function. It also provides counterexamples in holomorphic approximation theory. Clearly, the worm showed considerable potential for a central role in the function theory of several complex variables. But in point of fact the subject of the worm lay dormant for nearly 15 years after the appearance of [DIF1]. It was the remarkable article of Kiselman [KIS] that reestablished the importance and centrality of the worm.

In order to put Kiselman's work into context, we must provide a digression on the subject of biholomorphic mappings of pseudoconvex domains. In the present discussion, all domains Ω are smoothly bounded. We are interested in one-to-one, onto, invertible mappings (i.e., biholomorphic mappings or biholomorphisms) of domains

$$\Phi : \Omega_1 \longrightarrow \Omega_2 \,.$$

Thanks to a classical theorem of Liouville (see [KRPA1]), there are no conformal mappings, other than trivial ones, in higher-dimensional complex Euclidean space. Thus biholomorphic mappings are studied instead. It is well known that the Riemann mapping theorem fails in several complex variables (see [KRA1, GRK2, GRK9, ISK]). It is thus a matter of considerable interest to find means to classify domains up to biholomorphic equivalence.

The Poincaré program for such a classification consisted of two steps: (1) to prove that a biholomorphic mapping of smoothly bounded pseudoconvex domains extends smoothly to a diffeomorphism of the closures of the domains and (2) to then calculate biholomorphic differential invariants on the boundary. His program was stymied for more than 60 years because the machinery did not exist to tackle step (1). The breakthrough came in 1974 with Fefferman's seminal article [FEF1]. In it he used remarkable techniques of differential geometry and partial differential equations to prove that a biholomorphic mapping of smoothly bounded, *strictly pseudoconvex* domains will extend to a diffeomorphism of the closures.

The Fefferman's proof was quite long and difficult and left open the question of (a) whether there was a more accessible and more natural approach to the question and (b) whether there were techniques that could be applied to a more general class of domains. Steven Bell [BEL1] as well as Bell and Ewa Ligocka [BELL] provided a compelling answer.

Let Ω be a fixed, bounded domain in $\mathbb{C}^n$. Let $A^2(\Omega)$ be the square-integrable holomorphic functions on Ω. Then $A^2(\Omega)$ is a closed subspace of $L^2(\Omega)$. The Hilbert space projection

$$P : L^2(\Omega) \longrightarrow A^2(\Omega)$$

can be represented by an integration formula

$$Pf(z) = \int_\Omega K(z, \zeta) f(\zeta) \, dV(\zeta).$$

The kernel $K(z, \zeta) = K_\Omega(z, \zeta)$ is called the *Bergman kernel*. It is an important biholomorphic invariant. See [BERS, CHS, KRA1, Chap. 1] for all the basic ideas concerning the Bergman kernel.

Clearly, the Bergman projection P is bounded on $L^2(\Omega)$. Notice that, if Ω is smoothly bounded, then $C^\infty(\overline{\Omega})$ is dense in $L^2(\Omega)$. In fact, more is true: If Ω is a Levi pseudoconvex and smoothly bounded, then $C^\infty(\overline{\Omega}) \cap \{\text{holomorphic functions}\}$ is dense in $A^2(\Omega)$ (see [CAT3]).

Bell [BEL1] has formulated the notion of Condition R for the domain Ω. We say that Ω satisfies *Condition R* if $P : C^\infty(\overline{\Omega}) \longrightarrow C^\infty(\overline{\Omega})$. It is known, thanks to the theory of the $\overline{\partial}$-Neumann problem (see Sect. 6.6), that strictly pseudoconvex domains satisfy Condition R. Deep work of Diederich–Fornæss [DIF3] and Catlin [CAT1, CAT2] shows that domains with real analytic boundary, and also finite type domains, satisfy Condition R. An important formula of Kohn, which we shall discuss in Sect. 6.6, relates the $\overline{\partial}$-Neumann operator to the Bergman projection in a useful way (see also [KRA4, Chap. 8]). The fundamental result of Bell and Bell–Ligocka is as follows.

Theorem 6.1.5. *Let $\Omega_j \subset \mathbb{C}^n$ be smoothly bounded, Levi pseudoconvex domains. Suppose that one of the two domains satisfies Condition R. If $\Phi : \Omega_1 \longrightarrow \Omega_2$ is a biholomorphic mapping, then Φ extends to be a C^∞ diffeomorphism of $\overline{\Omega}_1$ to $\overline{\Omega}_2$.*

This result established the centrality of Condition R. The techniques of proof are so natural and accessible that it seems that Condition R is certainly the "right" approach to questions of boundary regularity of biholomorphic mappings. Work of Boas–Straube in [BOS2] shows that Condition R is virtually equivalent to natural regularity conditions on the $\bar{\partial}$-Neumann operator.[1]

For later reference, and for its importance in its own right, we mention here that the above Theorem 6.1.5 can be "localized." To be precise, we say that a given smoothly bounded domain Ω satisfies the *Local Condition R* at a point $p_0 \in \partial\Omega$ if there exists a neighborhood U of p_0 such that $P : C^\infty(\overline{\Omega}) \longrightarrow L^2(\Omega) \cap C^\infty(\overline{\Omega} \cap U)$. Then Bell's local result is as follows; see [BEL2].

Theorem 6.1.6. *Let $\Omega_j \subset \subset \mathbb{C}^n$ be smoothly bounded, pseudoconvex domains, $j = 1, 2$. Suppose that Ω_1 satisfies Local Condition R at $p_0 \in \partial\Omega_1$. If $\Phi : \Omega_1 \longrightarrow \Omega_2$ is a biholomorphic mapping, then there exists a neighborhood U of p_0 such that Φ extends to be a C^∞ diffeomorphism of $\overline{\Omega}_1 \cap U$ onto its image.*

We might mention, as important background information, a result of David Barrett [BAR1] from 1984. This considerably predates the work on which we now concentrate. It does not concern the worm, but it does concern the regularity of the Bergman projection. We present the details in Sect. 6.13 below.

Theorem 6.1.7. *There exists a smoothly bounded, non-pseudoconvex domain $\Omega \subseteq \mathbb{C}^2$ on which Condition R fails.*

Although Barrett's result is *not* on a pseudoconvex domain, it provides some insight into the trouble that can be caused by rapidly varying normals to the boundary. See [BAR3] for some pioneering work on this idea.

As indicated above, it was Kiselman [KIS] who established an important connection between the worm domain and Condition R. He proved that, for a certain non-smooth version of the worm (see below), a form of Condition R fails.

For $s > 0$, let $W^s(\Omega)$ denote the usual Sobolev space on the domain Ω (see, for instance, [HOR2, KRA4]). Building on Kiselman's idea, Barrett [BAR2] used an exhaustion argument to show that the Bergman projection fails to preserve the Sobolev spaces of sufficiently high order on the *smooth* worm.

Theorem 6.1.8. *For $\mu > 0$, let W be the smooth worm, defined as in Definition 6.1.2, and let $v = \pi/2\mu$. Then the Bergman projection P on W does not map $W^s(W)$ to $W^s(W)$ when $s \geq v$.*

The capstone of results, up until 1996, concerning analysis on the worm domain is the seminal article of M. Christ. Christ finally showed that Condition R fails on the smooth worm. Precisely, his result is the following (see [CHR1]):

Theorem 6.1.9. *Let W be the smooth worm. Then there is a function $f \in C^\infty(\overline{W})$ such that its Bergman projection Pf is not in $C^\infty(\overline{W})$.*

---

[1]Here the $\bar{\partial}$-Neumann operator N is the natural right inverse to the $\bar{\partial}$-Laplacian $\Box = \bar{\partial}^*\bar{\partial} + \bar{\partial}\bar{\partial}^*$.

We note explicitly that the result of this theorem is closely tied to, indeed is virtually equivalent to, the assertion that the $\bar{\partial}$-Neumann problem is not hypoelliptic on the smooth worm, [BOS2].

In the work of Kiselman et al., the geometry of the boundary of the worm plays a fundamental role in the analysis. In particular, the fact that for large μ the normal does considerable winding is fundamental to all of the negative results. It is of interest to develop a deeper understanding of the geometric analysis of the worm domain, because it will clearly play a seminal role in future work in the analysis of several complex variables.

We conclude this discussion of biholomorphic mappings with a consideration of biholomorphic mappings of the worm. It is at this time unknown whether a biholomorphic mapping of the smooth worm $\mathcal{W}$ to another smoothly bounded, pseudoconvex domain will extend to a diffeomorphism of the closures. Of course the worm does *not* satisfy Condition R, so the obvious tools for addressing this question are not available.

Our discussion is organized as follows:

Section 6.2 gives particulars of the Diederich–Fornæss worm. Specifically, we prove that the worm is a Levi pseudoconvex, and we establish that there is no global plurisubharmonic defining function. We also examine the Diederich–Fornæss bounded plurisubharmonic exhaustion function on the smooth worm.

Section 6.3 considers non-smooth versions of the worm (these originated with Kiselman). We outline some of Kiselman's results.

Section 6.4 discusses the irregularity of the Bergman projection on the worm. In particular, we reproduce some of Kiselman's and Barrett's analysis.

Sections 6.4 and 6.5 discuss the failure of Condition R on the worm domains.

6.2 More on the Worm

We now present the details of the first basic properties of the Diederich–Fornæss worm domain $\mathcal{W}$. Recall that $\mathcal{W}$ is defined in Definition 6.1.2. Some material of this section can also be found in the excellent monograph [CHS].

We begin by proving Proposition 6.1.3.

Proof of Proposition 6.1.3. Property (iii) of the worm shows immediately that the worm domain is bounded. Let

$$\rho(z_1, z_2) = \left| z_1 + e^{i \log |z_2|^2} \right|^2 - 1 + \eta(\log |z_2|^2) \,.$$

Then ρ is (potentially) a defining function for $\mathcal{W}$. If we can show that $\nabla \rho \neq 0$ at each point of $\partial \mathcal{W}$, then the implicit function theorem guarantees that $\partial \mathcal{W}$ is smooth.

If it happens that $\partial \rho / \partial z_1(p) = 0$ at some boundary point $p = (p_1, p_2)$, then we find that

$$\frac{\partial \rho}{\partial z_1}(p) = \overline{p}_1 + e^{-i \log |p_2|^2} = 0$$

Now let us look at $\partial \rho / \partial z_2$ at the point p. The first term in ρ differentiates to 0, and we find that

$$\frac{\partial \rho}{\partial z_2}(p) = \eta'(\log |p_2|^2) \cdot \frac{\overline{p}_2}{|p_2|^2}.$$

Since $\rho(p) = 0$, we have that $\eta(\log |p_2|^2) = 1$. Hence, by property of η (iv), $\eta'(\log |p_2|^2) \neq 0$. It follows that $\partial \rho / \partial z_2(p) \neq 0$. We conclude that $\nabla \rho(z) \neq 0$ for every boundary point z.

For the pseudoconvexity, we write

$$\rho(z) = |z_1|^2 + 2\text{Re}\left(z_1 e^{-i \log |z_2|^2}\right) + \eta(\log |z_2|^2).$$

Multiplying through by $e^{\arg z_2^2}$, we have that locally $\mathcal{W}$ is given by

$$|z_1|^2 e^{\arg z_2^2} + 2\text{Re}\left(z_1 e^{-i \log z_2^2}\right) + \eta(\log |z_2|^2)e^{\arg z_2^2} < 0.$$

The function $e^{-i \log z_2^2}$ is locally well defined and holomorphic, and its modulus is $e^{\arg z_2^2}$. Thus the first two terms are plurisubharmonic. Therefore we must check that the last term is plurisubharmonic. Since it only depends on z_2, we merely have to calculate its Laplacian. We have, arguing as before, that

$$\Delta\left(\eta(\log |z_2|^2)e^{\arg z_2^2}\right) = \left(\Delta\eta(\log |z_2|^2)\right)e^{\arg z_2^2} + \eta(\log |z_2|^2)\Delta e^{\arg z_2^2} \geq 0.$$

Because η is convex and nonnegative [property (i)], the nonnegativity of this last expression follows. This shows that $\mathcal{W}$ is smoothly bounded and pseudoconvex.

It may be worth noting explicitly that we have proved that the (locally defined) defining function for the worm is plurisubharmonic. This does not contradict the fact (proved by Diederich and Fornæss) that the worm has no *globally defined* plurisubharmonic defining function.

In order to describe the locus of weakly pseudoconvex points, we consider again the local defining function

$$\rho(z_1, z_2) = |z_1|^2 e^{\arg z_2^2} + 2\text{Re}\left(z_1 e^{-i \log z_2^2}\right) + \eta(\log |z_2|^2)e^{\arg z_2^2}.$$

This function is strictly plurisubharmonic at all points (z_1, z_2) with $z_1 \neq 0$ because of the first two terms or where $\left| \log |z_2|^2 \right| > \mu$, because of the last term. Thus consider the annulus $\mathcal{A} \subset \partial \mathcal{W}$ given by

$$\mathcal{A} = \left\{(z_1, z_2) \in \partial \mathcal{W} : z_1 = 0 \text{ and } \left| \log |z_2|^2 \right| \leq \mu \right\}.$$

A direct calculation shows that the complex Hessian for ρ at a point $z \in \mathcal{A}$ acting on $v = (v_1, v_2) \in \mathbb{C}^2$ is given by

$$|v_1|^2 + 2\mathrm{Re}\left(v_1 \bar{v}_2 \frac{e^{i \log |z_2|^2}}{z_2}\right).$$

By pseudoconvexity, such an expression must be nonnegative for all complex tangential vectors v at z. But such vectors are of the form $v = (0, v_2)$, so that the Levi form $\mathcal{L}_\rho \equiv 0$ on $\mathcal{A}$. This proves the result.   □

It is appropriate now to give the proof of Diederich and Fornæss that the worm has nontrivial Nebenhülle. What is of interest here, and what distinguishes the worm from the older example of the Hartogs triangle, is that the worm is a bounded, pseudoconvex domain with *smooth boundary*.

We now show that $\overline{\mathcal{W}}$ does not have a Stein neighborhood basis.

Proof: What we actually show is that if U is *any* neighborhood of $\overline{\mathcal{W}}$, then U will contain

$$K = \{(0, z_2) : -\pi \leq \log |z_2|^2 \leq \pi\} \cup \{(z_1, z_2) : \log |z_2|^2 = \pi \text{ or } -\pi \text{ and } |z_1 - 1| < 1\}.$$

In fact, this assertion is immediate by inspection.

By the usual Hartogs extension phenomenon argument, it then follows immediately that, if U is pseudoconvex, then U must contain

$$\hat{K} = \{(0, z_2) : -\pi \leq \log |z_2|^2 \leq \pi \text{ and } |z_1 - 1| < 1\}.$$

Thus there can be no Stein neighborhood basis.   □

We turn next to a few properties of the smooth worm $\mathcal{W}$ connected with potential theory. The significance of the next result stems from the article [BOS1]. In that article, Boas and Straube established the following:

Theorem 6.2.1. *Let Ω be a smoothly bounded pseudoconvex domain that admits a defining function that is plurisubharmonic on the boundary. Then, for every $s > 0$,*

$$P : W^s(\Omega) \to W^s(\Omega)$$

is bounded. In particular, Ω satisfies Condition R.

For the sake of completeness, we mention here that if the Bergman projection P on a domain Ω is such that $P : C^\infty(\overline{\Omega}) \longrightarrow C^\infty(\overline{\Omega})$ is bounded (i.e., Ω satisfies Condition R), then P is said to be *regular*, while if $P : W^s(\Omega) \to W^s(\Omega)$ for every $s > 0$ (and hence, Ω satisfies Condition R a fortiori), P is said to be *exactly regular*.

Thanks to the result of Christ [CHR1], we now know that W does not satisfy Condition R; hence, a fortiori it cannot admit a defining function which is plurisubharmonic on the boundary. However, it is simpler to give a direct proof of this fact.

Proposition 6.2.2. *There exists no defining function $\tilde{\rho}$ for W that is plurisubharmonic on the entire boundary.*

Proof: Suppose that such a defining $\tilde{\rho}$ exist. Then, there exists a smooth positive function h such that $\tilde{\rho} = h\rho$. A direct calculation shows that the complex Hessian for $\tilde{\rho}$ at a point $z \in \mathcal{A}$ acting on $v = (v_1, v_2) \in \mathbb{C}^2$ is given by

$$\mathcal{L}_{\tilde{\rho}}(z; (v_1, v_2)) = 2\mathrm{Re}\left[\bar{v}_1 v_2 \left(\frac{ih}{z_2} + \partial_{z_2}h\right)e^{i\log|z_2|^2}\right] + \left[h + 2\mathrm{Re}\left(\partial_{z_1}h \cdot e^{i\log|z_2|^2}\right)\right]|v_1|^2 .$$

Since this expression is assumed to be always nonnegative, we must have

$$\left(\frac{ih}{z_2} + \partial_{z_2}h\right)e^{i\log|z_2|^2} = \overline{\partial_{\bar{z}_2}\left(he^{-i\log|z_2|^2}\right)} \equiv 0 ,$$

on $\mathcal{A}$. Therefore the function $g(z_2) = h(0, z_2)e^{-i\log|z_2|^2}$ is a holomorphic function on $\mathcal{A}$. Hence, $g(z_2)e^{i\log|z_2|^2} = h(0, z_2)e^{2\arg z_2}$ is locally a holomorphic function. Thus it must be locally a constant, hence a constant c on all of $\mathcal{A}$.

Therefore on $\mathcal{A}$,

$$h(0, z_2) = ce^{-2\arg z_2}$$

which is impossible. This proves the result. $\square$

We conclude this section with another important result about the Diederich–Fornæss worm domain W. In what follows, we say that λ is a *bounded plurisubharmonic exhaustion function* for a domain Ω if:

(a) λ is continuous on $\overline{\Omega}$.
(b) λ is strictly plurisubharmonic on Ω.
(c) $\lambda = 0$ on $\partial\Omega$.
(d) $\lambda < 0$ on Ω.
(e) For any $c < 0$, the set $\Omega_c = \{z \in \Omega : \lambda(z) < c\}$ is relatively compact in Ω.

A bounded plurisubharmonic exhaustion function carries important geometric information about the domain Ω.

Now Diederich–Fornæss have proved the following [DIF3] (see also [RAN] for a simpler proof when the boundary is smoother):

Theorem 6.2.3. *Let Ω be any smoothly bounded pseudoconvex domain, $\Omega = \{z \in \mathbb{C} : \varrho(z) < 0\}$. Then there exists $\delta, 0 < \delta \leq 1$, and a defining function $\tilde{\varrho}$ for Ω such that $-(-\tilde{\varrho})^\delta$ is a bounded strictly plurisubharmonic exhaustion function for Ω.*

The importance of this result in the setting of the regularity of the Bergman projection appears in the following related result, proved by Berndtsson–Charpentier [BEC] and [KOH1], respectively.

Theorem 6.2.4. *Let Ω be a smoothly bounded pseudoconvex domain, and let P denotes its Bergman projection. Let $\tilde{\rho}$ be a smooth defining function for Ω such that $-(-\tilde{\rho})^\delta$ is strictly plurisubharmonic. Then there exists $s_0 = s_0(\Omega, \delta)$ such that*

$$P : W^s(\Omega) \to W^s(\Omega)$$

is continuous for all $0 \le s < s_0$.

Remark 6.2.5. The sharp value of s_0 is not known, and most likely the exact determination of such a value might prove a very difficult task. The two sources [BEC] and [KOH1] present completely different approaches and descriptions of s_0, that is of the range $[0, s_0)$ for which P is bounded on W^s, with $s \in [0, s_0)$. In [BEC] it is proved that such a range is at least $[0, \delta/2)$, i.e., they show that $s_0 \ge \delta/2$, while in [KOH1] the parameter s_0 is not so explicit, but it tends to infinity as $\delta \to 1$. The value found in [BEC] has the advantage of providing an explicit lower bound for the regularity of the Bergman projection on a given domain, while the value given in [KOH1] is sharp in the sense given by Boas and Straube's Theorem 6.2.1. $\square$

The domain $\mathcal{W}$ serves as an example that the exponent δ may be arbitrarily small. To illustrate this point, the following result is essentially proved in [DIF1]. Here we add the precise estimate that such an exponent δ is less than the value ν.

Theorem 6.2.6. *Let $\delta_0 > 0$ be fixed. Then there exists $\mu_0 > 0$ such that, for all $\mu \ge \mu_0$, the following holds. If $\tilde{\varrho}$ is a defining function for $\mathcal{W} = \mathcal{W}_\mu$, with $\mu \ge \mu_0$ and $\delta > 0$ is such that $-(-\tilde{\varrho})^\delta$ is a bounded plurisubharmonic exhaustion function for $\mathcal{W}$, then $\delta < \delta_0$.*
More precisely, we show that, in the notation above, $\delta < \nu = \pi/2\mu$.

Proof: We may assume that $\tilde{\rho} = h\rho$, where $\rho = \rho_\mu$ is defined in Proposition 6.2.2 and h is a smooth positive function on $\overline{\mathcal{W}}$. Then, by hypothesis $-h^\delta(-\rho)^\delta$ is strictly plurisubharmonic on $\mathcal{W}$.

Let

$$\sigma(z_1, z_2) = -\frac{1}{2\pi} \int_0^{2\pi} h^\delta(z_1, e^{i\theta} z_2) \big(-\rho(z_1, e^{i\theta} z_2)\big)^\delta \, d\theta$$

$$= -\frac{1}{2\pi} \int_0^{2\pi} h^\delta(z_1, e^{i\theta} z_2) \, d\theta \big(-\rho(z_1, z_2)\big)^\delta$$

$$= -\tilde{h}(z_1, z_2) \big(-\rho(z_1, z_2)\big)^\delta.$$

Obviously, σ is also strictly plurisubharmonic on $\mathcal{W}$, and $\tilde{h}$ is strictly positive and smooth on $\mathcal{W}$. We can also write $\tilde{h}(z_1, z_2) = h^\#(z_1, |z_2|^2)$, where $h^\#$ is defined for

$(z_1, t) \in \mathbb{C} \times \mathbb{R}^+$ such that if $|z_2|^2 = t$ then $(z_1, z_2) \in \mathcal{W}$. For simplicity of notation, we rename such a function h again.

Thus we have that

$$\sigma(z_1, z_2) = -h(z_1, |z_2|^2)\big(-\rho(z_1, z_2)\big)^\delta$$

is strictly plurisubharmonic on $\mathcal{W}$.

Now consider the points in $\mathcal{W}$ of the form $p = (z_1, z_2) = (\varepsilon e^{i \log |z_2|^2}, z_2)$ with $e^{-\mu/2} \le |z_2| \le e^{\mu/2}$. For these points one has that

$$\partial\rho(p) = \big((1 - \varepsilon)e^{i \log |z_2|^2}, 0\big) .$$

A straightforward computation shows that, at such points $p \equiv (\varepsilon e^{i \log |z_2|^2}, z_2)$, the Levi form $\mathcal{L}_\sigma$ of σ calculated at vectors $v = (v_1, v_2) \in \mathbb{C}^2$ equals (all the functions are evaluated at the point p and we write ζ in place of $e^{i \log |z_2|^2}$)

$$
\begin{aligned}
\mathcal{L}_\sigma\big(p; (v_1, v_2)\big) = \varepsilon^{\delta-2}(2 - \varepsilon)^{\delta-2} \Big\{ & (2 - \varepsilon)\Big(-\varepsilon^2(2 - \varepsilon)\partial^2_{z_1 \bar{z}_1} h \\
& + 2\delta\varepsilon(1 - \varepsilon)\mathrm{Re}\left(\zeta\partial_{z_1} h\right) + \delta\varepsilon h \\
& + \delta(1 - \delta)\frac{(1 - \varepsilon)^2}{2 - \varepsilon}h\Big)|v_1|^2 \\
& + 2w\varepsilon(2 - \varepsilon)\mathrm{Re}\Big[\big(-\varepsilon(2 - \varepsilon)\partial^2_{z_1 \bar{z}_2} h \\
& + \delta(1 - \varepsilon)\partial_{z_2} h + \delta\frac{i\zeta}{z_2}h\big)v_1\bar{v}_2\Big] \\
& + \delta^2(2 - \delta)\Big(-(2 - \delta)\partial^2_{z_2 \bar{z}_2} h + \frac{2\delta}{|z_2|^2}h\Big)|v_2|^2 \Big\} .
\end{aligned}
$$

Next, we evaluate the above Levi form at vectors of the form $(v_1, v_2) = (u_1, \varepsilon u_2)$. Making the obvious simplification, we see that the necessary condition in order for σ to be strictly plurisubharmonic is that and $0 < \varepsilon < 1$ and, for all $(u_1, u_2) \in \mathbb{C}^2$,

$$
\begin{aligned}
\Big(-\varepsilon^2(2 - \varepsilon)\partial^2_{z_1 \bar{z}_1} h &+ 2\delta\varepsilon(1 - \varepsilon)\mathrm{Re}\left(\zeta\partial_{z_2} h\right) + \delta\varepsilon h \\
&+ \delta(1 - \delta)\frac{(1 - \varepsilon)^2}{2 - \varepsilon}h\Big)|u_1|^2 \\
&+ 2\mathrm{Re}\Big[\big(-\varepsilon(2 - \varepsilon)\partial^2_{z_1 \bar{z}_2} h + \delta(1 - \varepsilon)\partial_{z_2} h + \delta\frac{i\zeta}{z_2}h\big)u_1\bar{u}_2\Big] \\
&+ \Big(-(2 - \varepsilon)\partial^2_{z_2 \bar{z}_2} h + \frac{2\delta}{|z_2|^2}h\Big)|u_2|^2 \ge 0 .
\end{aligned}
$$

$$\tag{6.2.6.1}$$

Since $h \in C^\infty(\overline{\mathcal{W}})$ this inequality must hold also for $\varepsilon = 0$ and $(0, z_2) \in \mathcal{A}$. Then we have

$$\left(\frac{1}{2}\delta(1-\delta)h\right)|u_1|^2 + 2\operatorname{Re}\left[(\delta\zeta\partial_{z_2}h + \delta\frac{i\zeta}{z_2}h)u_1\bar{u}_2\right]$$

$$+ \left(-(2-\varepsilon)\partial^2_{z_2\bar{z}_2}h + \frac{2\delta}{|z_2|^2}\right)|u_2|^2 \geq 0. \qquad (6.2.6.2)$$

Next, we substitute for h the function $\tilde{h}$ defined on $\mathbb{C} \times \mathbb{R}^+$ such that $h(z_1, z_2) = \tilde{h}(z_1, |z_2|^2)$. Then

$$\partial_{z_2}h(0, z_2) = \bar{z}_2\partial_t\tilde{h}(0, |z_2|^2) \text{ and } \partial^2_{z_2\bar{z}_2}h(0, z_2) = |z_2|^2\partial^2_t\tilde{h}(0, |z_2|^2) + \partial_t\tilde{h}(0, |z_2|^2).$$

Plugging these into (6.2.6.1) we then obtain the differential inequality for the function $\tilde{h}$:

$$\frac{1}{2}\delta(1-\delta)\tilde{h}|u_1|^2 + 2\operatorname{Re}\left[\delta\zeta(\partial_t\tilde{h} + \frac{i}{|z_2|^2}\tilde{h})u_1\bar{u}_2\right]$$

$$+ \left(-2|z_2|^2\partial^2_t\tilde{h} - 2\partial_t\tilde{h} + \frac{2\delta}{|z_2|^2}\tilde{h}\right)|u_2|^2 \geq 0$$

for all $(u_1, u_2) \in \mathbb{C}^2$, $e^{-\mu/2} \leq |z_2| \leq e^{\mu/2}$ (and the function $\tilde{h}$ being evaluated at the point $(0, |z_2|^2)$). Now if we choose (u_1, u_2) of the form $(2e^{i\theta}/|z_2|, 1)$ in such a way that the second term in the above display becomes nonpositive, we obtain that the function σ is plurisubharmonic only if

$$\frac{\delta(1-\delta)}{|z_2|^2}\tilde{h} - 2\delta\left((\partial_t\tilde{h})^2 + \frac{\tilde{h}^2}{|z_2|^4}\right)^{1/2} - |z_2|^2\partial^2_t\tilde{h} - \partial_t\tilde{h} + \frac{\delta}{|z_2|^2}\tilde{h} \geq 0$$

which in turn gives

$$- \delta^2\tilde{h} - |z_2|^4\partial^2_t\tilde{h} - |z_2|^2\partial_t\tilde{h} \geq 0 \qquad (6.2.6.3)$$

for all points $(0, |z_2|^2)$ with $e^{-\mu/2} \leq |z_2| \leq e^{\mu/2}$.

We now set $g(s) = \tilde{h}(0, e^s)$ for $s \in [-\mu, \mu]$. Notice that $|z_2|^2 = e^s$ and that

$$g' = e^s\partial_t\tilde{h} \quad \text{and} \quad g'' = e^s\partial_t\tilde{h} + e^{2s}\partial^2_t\tilde{h}. \qquad (6.2.6.4)$$

From (6.2.6.2) we obtain the differential inequality

$$g'' + \delta^2 g \leq 0,$$

for $s \in [-\mu, \mu]$, where g is a smooth strictly positive function. From the strict positivity of g, it follows that, for all $0 < \delta' < \delta$, it must be that

$$g'' + \delta'^2 g < 0 ,$$

again for all $s \in [-\mu, \mu]$. Setting $\tilde{g}(s) = g(s/\delta')$ the differential inequality above can be rewritten as

$$\tilde{g}'' + \tilde{g} < 0$$

for all $s \in [-\mu\delta', \mu\delta']$. Finally, by translation (calling the new function g again), i.e., setting $g(s) = \tilde{g}(s + \mu\delta')$, we obtain that

$$g'' + g < 0 \tag{6.2.6.5}$$

for a smooth strictly positive function g, for all $s \in [0, 2\mu\delta']$.

We now claim that there exists a smooth strictly positive function φ such that

$$\varphi'' + \varphi < 0 \qquad \text{and} \qquad \varphi' < 0 \tag{6.2.6.6}$$

for $s \in [0, \mu\delta']$. For notice that if g as above is such that $g'(a) < 0$, then $g'(s) < 0$ for $s \in [a, 2\mu\delta']$, while if instead $g'(a) \geq 0$, then $g'(s) > 0$ for $s \in [0, a)$, since $g'' < 0$ on $[0, 2\mu\delta']$. In this latter case, making the substitution $s \mapsto 2\mu\delta' - s$ that preserves (6.2.6.6), we obtain a function with negative derivative on $[a, 2\mu\delta')$. By the arbitrariness of $\delta' < \delta$ we establish the claim.

Now, the argument at the end of the proof of Theorem 6 in [DIF1] shows that the differential inequalities (6.2.6.3) above are possible only if $\mu\delta' < \pi/2$, i.e.,

$$\delta' < \frac{\pi}{2\mu} = \nu .$$

This proves the result.   □

6.3   Non-Smooth Versions of the Worm Domain

In order to perform certain analyses on $\mathcal{W}$, some simplifications of the domain turn out to be particularly useful.

In the first instance, one can simplify the expression of the defining function ρ for $\mathcal{W}$ by taking η to be 1 minus the characteristic function of the interval $[-\mu, \mu]$. This has the effect of truncating the two caps and destroying in part the smoothness of the boundary. Precisely, we can define

$$\mathcal{W}' = \left\{ (z_1, z_2) \in \mathbb{C}^2 : \left| z_1 - e^{i \log |z_2|^2} \right|^2 < 1, \ \left| \log |z_2|^2 \right| < \mu \right\}.$$

We remark that $\mathcal{W}'$ is a bounded pseudoconvex domain with boundary that is C^∞ except at points that satisfy

1. $|z_2| = e^{\mu/2}$ and $|z_1 - e^{-i \log |z_2|^2}| = 1$.
2. $|z_2| = e^{-\mu/2}$ and $|z_1 - e^{-i \log |z_2|^2}| = 1$.

Of interest are also two non-smooth, *unbounded* worms. Here, in order to be consistent with the results obtained in [KRP1], we change the notation a bit. (In practice, we set $\mu = \beta - \pi/2$.)

For $\beta > \pi/2$ we define

$$D_\beta = \left\{ \zeta \in \mathbb{C}^2 : \text{Re}\left(\zeta_1 e^{-i \log |\zeta_2|^2}\right) > 0, \; \left| \log |\zeta_2|^2 \right| < \beta - \frac{\pi}{2} \right\}$$

and

$$D'_\beta = \left\{ z \in \mathbb{C}^2 : \left| \text{Im}\, z_1 - \log |z_2|^2 \right| < \frac{\pi}{2}, \; \left| \log |z_2|^2 \right| < \beta - \frac{\pi}{2} \right\}.$$

It should be noted that these latter two domains are biholomorphically equivalent via the mapping

$$(z_1, z_2) \ni D'_\beta \mapsto (e^{z_1}, z_2) \ni D_\beta.$$

Neither of these domains is bounded. Moreover, these domains are *not* smoothly bounded. Each boundary is only Lipschitz, and, in particular, their boundaries are Levi flat.

We notice in passing that the slices of D_β, for each fixed ζ_2, are half planes in the variable ζ_1. Likewise the slices of D'_β, for each fixed ζ_2, are strips in the variable ζ_1.

The geometries of these domains are rather different from that of the smooth worm $\mathcal{W}$, which has smooth boundary, and all boundary points, except those on a singular annulus $(0, e^{i \log |z_2|^2})$ in the boundary, are strictly pseudoconvex. However, our worm domain D_β is actually a model for the smoothly bounded $\mathcal{W}$ (see, for instance, [BAR2]), and it can be expected that phenomena that are true on D_β or D'_β will in fact hold on $\mathcal{W}$ as well. We will say more about this symbiotic relationship below.

6.4 Irregularity of the Bergman Projection

We begin this section by discussing the proof of Barrett's result Theorem 6.1.8 in [BAR2]. Now let us describe these ideas in some detail. We begin with some of Kiselman's main ideas.

Let the Bergman space $\mathcal{H} = A^2$ be the collection of holomorphic functions that are square integrable with respect to Lebesgue volume measure dV on a fixed domain. Following Kiselman [KIS] and Barrett [BAR3], using the rotational invariance in the z_2-variable, we decompose the Bergman space for the domains D_β and D'_β as follows. Using the rotational invariance in z_2 and elementary Fourier series, each $f \in \mathcal{H}$ can be written as

$$f = \sum_{j=-\infty}^{\infty} f_j \,,$$

where each f_j is holomorphic and satisfies $f_j(z_1, e^{i\theta}z_2) = e^{ij\theta} f(z_1, z_2)$ for θ real. In fact, such an f_j must have the form

$$f_j(z_1, z_2) = g_j(z_1, |z_2|)z_2^j \,,$$

where g_j is holomorphic in z_1 and locally constant in z_2.
    Therefore

$$\mathcal{H} = \bigoplus_{j \in \mathbb{Z}} \mathcal{H}^j \,,$$

where

$$\mathcal{H}^j = \{f \in L^2 : f \text{ is holomorphic and } f(w_1, e^{i\theta}w_2) = e^{ij\theta} f(w_1, w_2)\} \,.$$

If K is the Bergman kernel for $\mathcal{H}$ and K_j the Bergman kernel for $\mathcal{H}^j$, then we may write

$$K = \sum_{j=-\infty}^{\infty} K_j \,.$$

Notice that, by the invariance property of $\mathcal{H}^j$, with $z = (z_1, z_2)$ and $w = (w_1, w_2)$, we have that

$$K_j(z, w) = H_j(z_1, w_1)z_2^j \overline{w}_2^j \,.$$

Our job, then, is to calculate each H_j and thereby each K_j. The first step of this calculation is already done in [BAR3]. We outline the calculation here for the sake of completeness.

Proposition 6.4.1. *Let $\beta > \pi/2$. Then*

$$H_j(z_1, w_1) = \frac{1}{2\pi} \int_{-\infty}^{\infty} \frac{e^{i(z_1 - \overline{w}_1)\xi} \xi \left(\xi - \frac{j+1}{2}\right)}{\sinh(\pi\xi) \sinh\left[(2\beta - \pi)\left(\xi - \frac{j+1}{2}\right)\right]} \, d\xi \,. \qquad (6.4.1.1)$$

The articles [KIS] and [BAR2] calculate and analyze only the Bergman kernel for $\mathcal{H}^{-1}$ (i.e., the Hilbert subspace with index $j = -1$). This is attractive to do because certain "resonances" cause cancellations that make the calculations tractable when $j = -1$. One of the main thrusts of the work [KRP1] is to perform the more difficult calculations for all j and then to sum them over j.

We begin by following the calculations in [KIS, BAR3] in order to get our hands on the Bergman kernels of the $\mathcal{H}^j$. Let $f_j \in \mathcal{H}^j$ and fix w_2. Then $f_j(w_1, w_2) = h_j(w_1)w_2^j$ (where we of course take into account the local independence of h_j from w_2). Now, writing $w_1 = x + iy$, $w_2 = re^{i\theta}$, and then making the change of variables $\log r^2 = s$, we have

$$
\|f_j\|_{\mathcal{H}}^2 = \int_{D'_\beta} |h_j(w_1)|^2 |w_2|^{2j}\, dV(w)
$$

$$
= \int_{-\infty}^{\infty} \int_{|y - \log r^2| < \frac{\pi}{2}} 2\pi |h_j(x + iy)|^2 \int_{|\log r^2| < \beta - \frac{\pi}{2}} r^{2j+1}\, dr\, dy\, dx
$$

$$
= \pi \int_{\mathbb{R}} \int_{|y - s| < \frac{\pi}{2}} |h_j(x + iy)|^2 \int_{|s| < \beta - \frac{\pi}{2}} e^{s(j+1)}\, ds\, dy\, dx
$$

$$
= \pi \int_{|y| < \beta, x \in \mathbb{R}} |h_j(x + iy)|^2 \int_{-\infty}^{\infty} e^{(j+1)s} \chi_{\pi/2}(y - s)\chi_{\beta - \pi/2}(s)\, ds\, dx\, dy
$$

$$
= \int_{S_\beta} |h_j(w_1)|^2 \left(\chi_{\pi/2} * \left[e^{(j+1)(\cdot)} \chi_{\beta - \pi/2}(\cdot) \right] \right)(y)\, dx\, dy\,; \qquad (6.4.1.2)
$$

here we have set

$$
S_\beta = \{x + iy \in \mathbb{C} : |y| < \beta\}
$$

and used the notation

$$
\chi_\alpha(y) = \begin{cases} 1 \text{ if } |y| < \alpha\,, \\ 0 \text{ if } |y| \geq \alpha\,. \end{cases}
$$

For $\beta > \frac{\pi}{2}$, we now set

$$
\lambda_j(y) = \left(\chi_{\pi/2} * \left[e^{(j+1)(\cdot)} \chi_{\beta - \pi/2}(\cdot) \right] \right)(y)\,.
$$

So line (6.4.1.2) equals

$$
\int_{S_\beta} |h_j(w_1)|^2 \lambda_j(y)\, dx\, dy\,.
$$

Thus we have shown that, if $f_j \in \mathcal{H}^j$, $f_j = h_j(w_1)w_2^j$, then

$$\|f_j\|_{\mathcal{H}}^2 = \int_{S_\beta} |h_j(w_1)|^2 \lambda_j(y)\, dxdy\,.$$

Now let $\varphi \in A^2(S_\beta, \lambda_j\, dA)$. That is, φ is square integrable on S_β with respect to the measure $\lambda_j\, dA$ (here $dA = dxdy$ is two-dimensional area measure). Note that λ_j depends only on the single variable y. Let $\tilde{\varphi}$ denote the partial Fourier transform of $\varphi(x + iy)$ in the x-variable. Then (by standard Littlewood–Paley theory)

$$\tilde{\varphi}(\xi, y) = \int \varphi(x + iy)e^{-ix\xi}\, dx = e^{-y\xi}\tilde{\varphi}_0(\xi)\,,$$

where $\varphi_0(x) = \varphi(x + i0)$. Therefore denoting by $B_\beta = B_\beta^{(j)}$ the Bergman kernel for the strip S_β with respect to the weight λ_j and writing $\omega = s + it$ and denoting by ξ the variable dual to s, we have

$$\int_{\mathbb{R}} \tilde{\varphi}_0(\xi)e^{i\zeta\xi}\, d\xi = 2\pi\varphi(\zeta) = 2\pi\int_{S_\beta} \varphi(\omega)B_\beta(\zeta, \omega)\lambda_j(\operatorname{Im}\omega)\, dA(\omega)$$

$$= \int_{-\beta}^{\beta}\int_{\mathbb{R}} \tilde{\varphi}(\xi, t)\tilde{B}_\beta(\zeta, (\xi, t))\lambda_j(t)\, d\xi dt$$

$$= \int_{\mathbb{R}} \tilde{B}_\beta(\zeta, (\xi, 0))\int_{-\beta}^{\beta} \tilde{\varphi}_0(\xi)e^{-2\xi t}\lambda_j(t)\, dt\, d\xi\,.$$

Notice that there is a factor of $e^{-\xi t}$ from each of the Fourier transform functions in the integrand.

This gives a formula for $\tilde{B}_\beta$:

$$\tilde{B}_\beta(\zeta, (\xi, 0)) = \frac{e^{i\zeta\xi}}{\int_{-\beta}^{\beta} e^{-2t\xi}\lambda_j(t)\, dt} = \frac{e^{i\zeta\xi}}{\hat{\lambda}_j(-2i\xi)}\,.$$

Amalgamating all our notation, and using the fact that the (Hermitian) diagonal in $\mathbb{C}^2$ is a set of determinacy, we find that

$$B_\beta(z, w) = \frac{1}{2\pi}\int_{\mathbb{R}} \frac{e^{i(z-\bar{w})\xi}}{\hat{\lambda}_j(-2i\xi)}\, d\xi\,.$$

But of course $(\chi_{\pi/2})\hat{\ }(\xi) = (e^{i\xi\pi/2} - e^{-i\xi\pi/2})/\xi$, so that

$$(\chi_{\pi/2})\hat{\ }(-2\xi i) = \frac{1}{\xi}\sinh(\pi\xi)\,.$$

Furthermore,

$$\left(e^{(j+1)s}\chi_{\beta-\frac{\pi}{2}}(s)\right)^{\wedge} = \frac{\sinh\left((2\beta-\pi)\left(\xi-\frac{j+1}{2}\right)\right)}{\xi-\frac{j+1}{2}}.$$

Thus

$$\hat{\lambda}_j(-2i\xi) = \frac{\sinh(\pi\xi)\sinh\left((2\beta-\pi)\left(\xi-\frac{j+1}{2}\right)\right)}{\xi\left(\xi-\frac{j+1}{2}\right)},$$

and

$$\frac{1}{\hat{\lambda}_j(-2i\xi)} = \frac{\xi\left(\xi-\frac{j+1}{2}\right)}{\sinh(\pi\xi)\sinh\left((2\beta-\pi)\left(\xi-\frac{j+1}{2}\right)\right)}.$$

In conclusion,

$$H_j(z_1,w_1) = \frac{1}{2\pi}\int_{-\infty}^{\infty} \frac{\left(e^{i(w_1-\bar{z}_1)\xi}\right)\xi\left(\xi-\frac{j+1}{2}\right)}{\sinh(\pi\xi)\sinh\left((2\beta-\pi)\left(\xi-\frac{j+1}{2}\right)\right)}\,d\xi,$$

thus proving Proposition 6.4.1.   □

At this point we sketch the proof of the main result of Barrett in [BAR4]. Namely, we show that the Bergman projection on the worm does not act continuously on the Sobolev space W^s.

Proof: The proof starts from the observation that the Bergman projection $\mathcal{P}$ on $\mathcal{W}$ preserves each $\mathcal{H}^j$. Therefore in order to show that $\mathcal{P}$ is not continuous on W^s, for some s, it suffices to show that $\mathcal{P}$ fails to be continuous in this topology when restricted to some $\mathcal{H}^j$.

The first step is to calculate the asymptotic expression for the kernel when $j = -1$. Recall that we are working on the non-smooth domain D'_β. Using the method of contour integrals, it is not difficult to obtain that

$$K'_{-1}(z,w) = \left(e^{-\nu_\beta|z_1-\bar{w}_1|} + \mathcal{O}(e^{-\nu|\operatorname{Re}z_1-\operatorname{Re}w_1|})\right)\cdot(z_2\bar{w}_2)^{-1}$$

as $|\operatorname{Re}z_1 - \operatorname{Re}w_1| \longrightarrow +\infty$, uniformly in all closed strips $\{|\operatorname{Im}z_1|, |\operatorname{Im}w_1| \le \lambda\}$, with $\nu > \nu_b$.

By applying the biholomorphic transformation between D_β and D'_β, one obtains an asymptotic expression for the kernel K_{-1} relative to the domain D_β:

$$K_{-1}(\zeta,\omega) = (|\zeta_1||\omega_1|)^{-1}\cdot\left(\frac{|\omega_1|^{\nu_\beta}}{|\zeta_1|^{\nu_\beta}} + \mathcal{O}(|\omega_1|^{\nu_\beta}/|\zeta_1|^{\nu_\beta})^{-\nu}\right)\cdot(\zeta_2\bar{\omega}_2)^{-1},$$

with $v > v_\beta$, as $|\zeta_1| - |\omega_1| \longrightarrow 0^+$. The proof of these two assertions can be found in [BAR4] (or see [CHS]).

The next step is a direct calculation to show that $K_{-1}(\cdot, w) \notin W^s(D_\beta)$ for $s \geq v_\beta$. This assertion is proved by using the characterization of the Sobolev norms for holomorphic functions on a domain Ω: For $-1/2 < t < 1/2$, m a nonnegative integer, the norm

$$\sum_{|\alpha| \leq m} \left\| |\rho|^t \partial_z^\alpha h \right\|_{L^2(\Omega)}$$

is equivalent to the H^{m-t}-norm of the holomorphic function h. The proof of such a characterization can be found in [LIG2].

Next, one notices that the reproducing kernel $K_{-1}(\cdot, w)$ can be written as the projection of a radially symmetric smooth cutoff function χ, translated at w. That is, if we denote by P_{-1} the projection relative to the subspace $\mathcal{H}^{-1}$, then

$$K_{-1}(\cdot, w) = P_{-1}\big(\chi(\cdot - w)\big).$$

Therefore since $K_{-1}(\cdot, w) \notin W^s(D_\beta)$ for $s \geq v_\beta$, then P_{-1}, and therefore P_{D_β} is not continuous on $W^s(D_\beta)$.

The final step of the proof is to transfer this negative result from D_β to $\mathcal{W}$. This is achieved by an exhaustion argument—see [BAR2]. We adapt this kind of argument to obtain a negative result in the L^p-norm for the Bergman projection on the smooth worm of Diederich and Fornæss. We refer the reader to the literature for all the details. $\square$

6.5 Irregularity Properties of the Bergman Kernel

We now examine the boundary asymptotics for the Bergman kernel on the domains D_β and D'_β and determine various irregularity properties of the corresponding Bergman kernels.

Begin with the asymptotic formula in the discussion above of Barrett's result. We point out particularly that there are two kinds of behavior: one kind at the "finite portion of the boundary" and the other one as $|\operatorname{Re} z_1 - \operatorname{Re} w_1| \longrightarrow +\infty$.

These two different behaviors are expressed by the first and second terms in the expansion. For the former type, we notice that the lead terms have expressions in the denominator of products of two terms like

$$(i(z_1 \pm \overline{w}_1) + 2\beta)^2, \quad (z_2 \overline{w}_2 - e^{\pm(\beta - \pi/2)})^2, \quad \text{and} \quad (z_2 \overline{w}_2 - e^{-[i(z_1 - \overline{w}_1) \pm \pi]/2})^2.$$

These singularities are similar to the ones of a Bergman kernel of a domain in $\mathbb{C}^2$ which is essentially a product domain. It is important to observe that the kernel does not become singular only when z, w tend to the same point on the boundary.

For instance, it becomes singular as $(i(z_1 \pm \overline{w}_1) + 2\beta) \longrightarrow 0$, while there is no restriction on the behavior of z_2 and w_2. We will be more detailed below in the case of the domain D_β. For the case of this domain, we finally notice that the main term at infinity, that is, when $|\mathrm{Re}\, z_1 - \mathrm{Re}\, 1| \longrightarrow +\infty$, behaves like $e^{-\nu_\beta |z_1 - \overline{w}_1|} \cdot (z_2 \overline{w}_2)^{-1}$.

Next, we consider the case of D_β. The mapping $(z_1, z_2) \in D'_\beta \mapsto (\zeta_1, \zeta_2) = (e^{z_1}, z_2) \in D_\beta$ sends the point at infinity (in z_1) into the origin (in ζ_1). Keeping into account the Jacobian factor, when $|\zeta_1| - |\omega_1| \longrightarrow 0^+$, the kernel on D_β is asymptotic to

$$\frac{|\omega_1|^{\nu_\beta - 1}}{|\zeta_1|^{\nu_\beta + 1}} \cdot (\zeta_2 \overline{\omega}_2)^{-1} .$$

Recall the inequalities that define D_β:

$$D_\beta = \left\{ \zeta \in \mathbb{C}^2 : \mathrm{Re}\left(\zeta_1 e^{-i \log |\zeta_2|^2}\right) > 0, \; \left| \log |\zeta_2|^2 \right| < \beta - \frac{\pi}{2} \right\}.$$

If we take ζ, $\omega \in D_\beta$ and let ω_1 tend to 0, then clearly $\omega \longrightarrow \partial D_\beta$ and ζ_1, ζ_2, ω_2 are unrestricted. Therefore $K_{D_\beta}(\cdot, \omega) \notin C^\infty(D_\beta)$ for $\omega \in \{(0, \omega_2)\}$, with $|\log |\omega_2|^2| < \beta - \pi/2$.

Notice that this is in contrast, for instance, to the situation on the ball or, more generally, on a strictly pseudoconvex domain. On either of those types of domains Ω, the kernel is known to be smooth on $\overline{\Omega} \times \overline{\Omega} \setminus (\Delta \cap [\partial\Omega \times \partial\Omega])$. See [KER2] and [KRA1, Chap. 1].

By the same token (by almost the same calculation), it is easy to conclude that the Bergman projection on D_β *cannot* map functions in $C^\infty(\overline{D}_\beta)$ to functions in $C^\infty(\overline{D}_\beta)$. This, of course, is the failure of Condition R on these domains.

In Sect. 6.1 we have seen that $P_W : C^\infty(\overline{W}) \not\longrightarrow C^\infty(\overline{W})$, that is, that W does not satisfy Condition R. A philosophically related fact, due to Chen [CHE] and Ligocka [LIG1] independently, is that the Bergman kernel of W cannot lie in $C^\infty(\overline{W} \times \overline{W} \setminus \Delta)$ (where Δ is the boundary diagonal). In fact, in [CHE] it is shown that this phenomenon is a consequence of the presence of a complex variety in the boundary of W.

The proof of the general result of So-Chin Chen follows a classical paradigm for establishing propagation of singularities for the $\overline{\partial}$-Neumann problem and similar phenomena.

Theorem 6.5.1. *Let $\Omega \subseteq \mathbb{C}^n$ be a smoothly bounded, pseudoconvex domain with $n \geq 2$. Assume that there is a complex variety V, of complex dimension at least 1, in $\partial\Omega$. Then*

$$K_\Omega(z, w) \notin C^\infty(\overline{\Omega} \times \overline{\Omega} \setminus \Delta(\partial\Omega)),$$

where $\Delta(\partial\Omega) = \{(z, z) : z \in \partial\Omega\}$.

Proof: Let $p \in V$ be a regular point. Let n_p be the unit outward normal vector at p. Then there are small numbers $\delta, \varepsilon_0 > 0$ such that $w - \varepsilon n_p \in \Omega$ for all $w \in \partial\Omega \cap B(p, \delta)$ and all $0 < \varepsilon < \varepsilon_0$. Let **d** be an analytic disc in $\partial\Omega \cap B(p, \delta) \cap V$. We may assume that this disc is centered at p. In other words, **d** is the image of the unit disc in the plane mapped into $\mathbb{C}^n$ with the origin going to p.

Seeking a contradiction, we assume that $K_\Omega(z, w) \in C^\infty(\overline{\Omega} \times \overline{\Omega} \setminus \Delta(\partial\Omega))$. Then we certainly have

$$\sup_{w \in \partial d} |K_\Omega(p, w)| \le M < +\infty \tag{6.5.1.1}$$

for some positive, finite number M. On the other hand, we know (see [BLP]) that

$$\lim_{\varepsilon \to 0} K_\Omega(p - \varepsilon n_p, p - \varepsilon n_p) = +\infty. \tag{6.5.1.2}$$

By the maximum modulus principle, we then obtain

$$\sup_{w \in \partial d_\varepsilon} |K_\Omega(p - \varepsilon n_p, w)| \ge K_\Omega(p - \varepsilon n_p, p - \varepsilon n_p),$$

where $d_\varepsilon = d - \varepsilon n_p \subseteq \Omega$. We conclude that

$$\sup_{w \in \partial d} |K_\Omega(p, w)| = \lim_{\varepsilon \to 0} \sup_{w \in \partial d_\varepsilon} |K_\Omega(p - \varepsilon n_p, w)| = +\infty.$$

This gives a contradiction, and the result is established. $\square$

6.6 The Kohn Projection Formula

In the 1960s, Kohn produced an elegant formula that relates the Bergman projection to the $\overline{\partial}$-Neumann problem. We refer the reader to [FOK] or [KRA4] for details of this important topic. Here we only briefly review the key concepts.

In studying the $\overline{\partial}$ operator, it is convenient to treat the second order, self-adjoint operator given by

$$\square = \overline{\partial}\overline{\partial}^* + \overline{\partial}^*\overline{\partial}.$$

It is shown that this partial differential operator has a right inverse N, which is known as the $\overline{\partial}$-Neumann operator.

Let $\Omega \subset\subset \mathbb{C}^n$ be a fixed domain on which the equation $\overline{\partial}u = \alpha$ is always solvable when α is a $\overline{\partial}$ closed $(0, 1)$ form (i.e., a domain of holomorphy—in other words, a pseudoconvex domain). Let $P : L^2(\Omega) \to A^2(\Omega)$ be the Bergman projection. If u is any solution to $\overline{\partial}u = \alpha$, then $w = w_\alpha = u - Pu$ is the unique solution that is orthogonal to holomorphic functions. Thus w is well defined, independent of the choice of u. Define the mapping

$$T : \alpha \mapsto w_\alpha.$$

Then, for $f \in L^2(\Omega)$, it holds that

$$Pf = f - T(\overline{\partial} f). \tag{6.6.1}$$

To see this, first notice that $\overline{\partial}[f - T(\overline{\partial} f)] = \overline{\partial} f - \overline{\partial} f = 0$, where all derivatives are interpreted in the weak sense. Thus $f - T(\overline{\partial} f)$ is holomorphic. Also $f - [f - T(\overline{\partial} f)]$ is orthogonal to holomorphic functions by design. This establishes the identity (6.6.1). But we have a more useful way of expressing T : namely, $T = \overline{\partial}^* N$. Thus we have derived the following important result:

$$P = I - \overline{\partial}^* N \overline{\partial}. \tag{6.6.2}$$

This is the Kohn's formula.

Formula (6.6.2) is particularly useful for studying Condition R. For if Ω is a strictly pseudoconvex domain, or a finite type domain in the sense of Kohn–D'Angelo–Catlin, then it is known that N satisfies a regularity estimate. In the strictly pseudoconvex case, the estimate is

$$\|N\alpha\|_{W^{s+1}} \leq C \|\alpha\|_{W^s} .$$

Here W^s is the standard Sobolev space of order s (see [KRA4]). It immediately follows from (6.6.2) that the Bergman projection P maps W^s to W^{s-2}. And that is a form of Condition R (we actually prove something stronger elsewhere in the book). We conclude, then, that a biholomorphic mapping of smoothly bounded, finite type domains extends to a diffeomorphism of the closures.

We can also learn something from applying formula (6.6.2) to the Dirac δ mass at a point w of the domain. For the Bergman kernel $K(z, w) = P(\delta_w)$. Thus according to (6.6.2),

$$K(z, w) = P(\delta_w) = \delta_w - \overline{\partial}^* N \overline{\partial} \delta_w .$$

By the pseudolocality of N on a finite type domain, we may conclude that the kernel is smooth—up to the boundary—away from w. (Here an operator T is said to be *pseudolocal* if $T\varphi$ is smooth wherever φ is smooth. Thus partial differential operators are trivially pseudolocal. What is interesting, and nontrivial, is when an integral operator is pseudolocal.)

6.7   Boundary Behavior of the Bergman Kernel

The earliest work on the boundary behavior of the Bergman kernel was done by S. Bergman himself [BER2]. A more modern approach, based on estimates for the $\overline{\partial}$-Neumann problem, appears in [HOR1]. Hörmander's results later were given some technical refinements in [DIE1, DIE2]. Here we present the statements and proofs of Hörmander's results.

We will see that the main tool used in the proof is a comparison technique. We determine information about the Bergman kernel for the domain Ω under study by comparing that kernel with the kernel of another (nearby) domain for which the kernel is more accessible. This is a very powerful idea and is used pervasively in this field. Even Fefferman's decisive result about the boundary behavior of the Bergman kernel uses a (rather sophisticated) comparison technique.

6.7.1 Hörmander's Result on Boundary Behavior

Theorem 6.7.1. Let $\Omega \subseteq \mathbb{C}^n$ be a bounded domain of holomorphy with defining function given by

$$\rho(z) = \begin{cases} -\text{dist}_{\partial\Omega}(z) & \text{if } z \in \Omega \\ \text{dist}_{\partial\Omega}(z) & \text{if } z \notin \Omega. \end{cases}$$

Assume that the operator $\overline{\partial} : L^2_{(0,0)}(\Omega) \to L^2_{0,1}(\Omega)$ has closed range. Let P be a point of $\partial\Omega$ such that $\partial\Omega$ is C^2 in a neighborhood of P. Also suppose that $\partial\Omega$ is strictly pseudoconvex at P. Let $k(P)$ be the product of the $n-1$ eigenvalues of the Levi form at P. Then

$$\lim_{\Omega \ni z \to P} |\rho(z)|^{n+1} |K_\Omega(z,z)| \to k(P) \cdot \frac{n!}{4\pi^n}.$$

The proof will be broken up into a sequence of lemmas and will occupy most of the rest of this section. At the end we shall comment on the Fefferman's asymptotic expansion and how it trumps the work of Bergman, Hörmander, and Diederich. We note that, by the main theorem of [HOR1], the hypothesis of closedness of the $\overline{\partial}$ operator is automatically fulfilled on a bounded domain of holomorphy.

Lemma 6.7.2. If Ω is as in the theorem and $\Omega' \subseteq \Omega$ is another domain, then

$$|K_{\Omega'}(z,z)| \geq |K_\Omega(z,z)|.$$

Proof: This is obvious from the characterization

$$|K_\Omega(z,z)| = \sup_{\substack{u \in A^2\Omega \\ \|u\|_{A^2}=1}} \frac{|u(z)|^2}{\|u\|^2_{A^2}}. \tag{6.7.2.1}$$

$\square$

Lemma 6.7.3. Let Ω be a bounded, pseudoconvex domain. Let $P \in \partial\Omega$ and suppose that, for some neighborhood U of P, there is a holomorphic function u_0 on $\Omega' \equiv \Omega \cap U$ such that $|u_0| \leq 1$ in Ω' and $|u_0(z)| \to 1$ when $z \to P$. We also suppose that $|u_0(z)|$ has an upper bound less than 1 in $\Omega' \cap {}^cU_0$ for some neighborhood U_0 of P with compact closure contained in U. Then we have

$$\lim_{z \to P} \frac{|K_\Omega(z, z)|}{|K_{\Omega'}(z, z)|} = 1.$$

Remark 6.7.4. Certainly a holomorphic peaking function will suffice for the function u_0 in this lemma. It is known (see [KRA1]) for example, on strictly pseudoconvex domains, that holomorphic peaking functions always exist. $\square$

Proof of Lemma 6.7.3. Let χ be a C^∞ function with compact support in U which is identically equal to 1 on U_0. Assume that $0 \le \chi \le 1$ everywhere. If $u' \in L^2_{(0,0)}(\Omega')$ and is holomorphic there, then (for an integer ν to be specified later) we set

$$u = \chi \cdot u' \cdot u_0^\nu - \nu.$$

Our goal is to choose ν so that $\bar\partial u = 0$ on Ω'. Thus we must solve the equation

$$\bar\partial\nu = (\bar\partial\chi)u'u_0^\nu.$$

Of course the right-hand side is $\bar\partial$-closed and supported in $U \cap \Omega$. We solve this $\bar\partial$ problem on the domain Ω. Hörmander's theorem in [HOR1] tells us that there will be a solution ν satisfying

$$\int_\Omega |\nu|^2 \, dV \le C \int_\Omega |u'u_0^\nu|^2 \, dV = C \int_{\Omega' \cap {}^cU_0} |u'u_0^\nu|^2 \, dV \qquad (6.7.3.1)$$

because $\bar\partial\chi = 0$ on U_0.

If $\epsilon > 0$, then we have

$$\int_{\Omega'} |u - u'u_0^\nu|^2 \, dV \le 2(C + 1) \int_{\Omega' \cap {}^cU_0} |u'u_0^\nu|^2 \, dV \le \epsilon^2 \int_{\Omega'} |u'|^2 \, dV \,,$$

provided that ν is chosen so large that $|u_0|^{2\nu} \le \epsilon^2/[2(C + 1)]$ in $\Omega' \cap {}^cU_0$. From the definition of the kernel function in Ω' and (6.7.3.1), we see now that for $z \in \Omega'$ we have

$$|u(z) - u'(z)u_0(z)^\nu|^2 \le \epsilon^2 |K_{\Omega'}(z, z)| \int_{\Omega'} |u'|^2 \, dV \,.$$

Thus

$$|u(z)| \ge |u'(z)||u_0(z)|^\nu - \epsilon \left(|K_{\Omega'}(z, z)| \int_{\Omega'} |u'|^2 \, dV \right)^{1/2} \,.$$

Since the supremum in (6.7.2.1) is clearly attained, we can, for every $z \in \Omega'$, choose $u' \not\equiv 0$ so that

$$|u'(z)|^2 = |K_{\Omega'}(z,z)| \int_{\Omega'} |u'|^2 \, dV \, .$$

For the corresponding function u, we then obtain the estimate

$$|u(z)|^2 \geq |K_{\Omega'}(z,z)| \, (|u_0(z)|^{\nu} - \epsilon)^2 \int_{\Omega'} |u'|^2 \, dV \, , \qquad (6.7.3.2)$$

as long as $z \in \Omega'$ and $|u_0(z)|^{\nu} > \epsilon$. Now the triangle inequality and (6.7.3.1) tell us that

$$\int_{\Omega} |u|^2 \, dV \leq (1+\epsilon)^2 \int_{\Omega'} |u'|^2 \, dV \, .$$

Together with (6.7.3.2), this last estimate implies that

$$|K_{\Omega}(z,z)| \geq |K_{\Omega'}(z,z)| \cdot (|u_0(z)|^{\nu} - \epsilon)^2 \, (1+\epsilon)^{-2} \quad \text{if } z \in \Omega' \text{ and } |u_0(z)|^{\nu} > \epsilon \, .$$

Thus

$$\liminf_{z \to z_0} \frac{|K_{\Omega}(z,z)|}{K_{\Omega'}(z,z)|} \geq (1-\epsilon)^2 (1+\epsilon)^2$$

and, since $\epsilon > 0$ is arbitrary, this (together with the preceding lemma) proves the current lemma. $\square$

This last lemma will enable us to reduce the proof of our main result to the study of a special domain on which the kernel function is relatively easy to compute.

Lemma 6.7.5. *Let E_0 be the ellipsoid in $\mathbb{C}^n$ defined by*

$$E_0 = \{z \in \mathbb{C}^n : a_1|z_1|^2 + z_2|z_2|^2 + \cdot + a_n|z_n|^2 < a_0\}$$

where $a_0, a_1, a_2, \ldots$ are positive numbers. Then

$$|K_{E_0}(z,z)| = n! \pi^{-n} a_0 \cdot a_1 \cdots \cdots a_n \cdot (a_0 - a_1|z_1|^2 - a_2|z_2|^2 - \cdots - a_n|z_n|^2)^{-n-1} \, .$$

Proof: We may assume that $a_0 = 1$. After a linear change of variables, we may also suppose that $a_1 = a_2 = \cdots a_n = 1$. Because of the unitary invariance of the kernel K, we may let $z = (0, 0, \ldots, 0, \zeta)$. If u is an element of $A^2(E_0)$, then a unitary transformation λ of the variables $z_1, z_2, \ldots, z_{n-1}$ leaves E_0, $u(0, 0, \ldots, 0, \zeta)$, and $\int |u|^2 \, dV$, and $\int |u|^2 \, dV$ invariant. Let $z' = (z_1, z_2, \ldots, z_{n-1})$. If we form

$$u_1(z) = \int u(\lambda z', z_n) \, d\mu \, ,$$

where $d\mu$ is normalized Haar measure on the unitary group, we then obtain a function $u_1 \in A^2(E_0)$ so that

$$u_1(0,0,\ldots,0,\zeta) = u(0,0,\ldots,0,\zeta).$$

By Minkowski's inequality, we now see that

$$\int_{E_0} |u_1|^2 \, dV \le \int_{E_0} |u|^2 \, dV.$$

But u_1 is invariant for unitary transformations of $z_1, z_2, \ldots, z_{n-1}$ and therefore must be a function of z_n only. In determining the supremum in the proof of the first lemma, we may therefore assume that u is a holomorphic function of z_n for $|z_n| < 1$.

Now put

$$u(z) = \sum_0^\infty c_j z_n^j.$$

Since the volume of the unit ball in $\mathbb{R}^{2n-2}$ is $\pi^{n-1}/(n-1)!$, we find that

$$\int_{E_0} |u|^2 \, dV = \frac{\pi^{n-1}}{(n-1)!} \int_0^{2\pi} \int_0^1 |u(re^{i\theta}|^2 r(1-r^2)^{n-1} \, dr d\theta$$

$$= \frac{\pi^{n-1}}{(n-1)!} \sum_0^\infty |c_j|^2 2\pi \int_0^1 r^{2j+1}(1-r^2)^{n-1} \, dr$$

$$= \pi^n \sum_0^\infty \frac{|c_j|^2 j!}{(j+n)!}.$$

By the Cauchy–Schwarz inequality, it follows now that

$$|u(0,0,\ldots,0,\zeta)|^2 \le \pi^{-n} \sum_0^\infty |\zeta|^{2j} \frac{(j+n)!}{j!} \int_{E_0} |u|^2 \, dV,$$

where equality is attained for some u. Since the sum of the series is $n!(1-|\zeta|^2)^{-n-1}$, the lemma is now proved. □

The following variant of the last lemma will be particularly useful in the proof of our main result:

Lemma 6.7.6. *Let a_{jk} ($j,k = 1,\ldots,n$) be a positive definite Hermitian symmetric matrix. Set*

$$F_0 = \left\{ z \in \mathbb{C}^n : \operatorname{Im} z_n > \sum_{j,k=1}^n a_{jk} z_j \bar{z}_k \right\}.$$

Then

$$|K_{F_0}(z,z)| = \sigma n! 4^{-1} \pi^{-n} \left(\operatorname{Im} z_n - \sum_{j,k=1}^{n} a_{jk} z_j \bar{z}_k \right)^{-n-1},$$

where $\sigma = \det(a_{jk})_{j,k=1}^{n-1}$.

Proof: By a unitary transformation of the variables $z_1, z_2, \ldots, z_{n-1}$, we may reduce the matrix $(a_{jk})_{j,k=1}^{n-1}$ to diagonal form; and the statement of the theorem remains invariant. Assuming this reduction to have been made, we can introduce new variables

$$w_j = a_j + \frac{z_n a_{nj}}{a_{jj}}, \quad j = 1, 2 \ldots, n-1$$

and $w_n = z_n$. The determinant of this transformation of variables is equal to 1, so again the statement of the theorem is invariant. So we may assume that the entire matrix (a_{jk}) has diagonal form. If we write

$$\operatorname{Im} z_n - a_{nn}|z_n|^2 = \frac{1}{4a_{nn}} - a_{nn} \left| z_n - \frac{i}{2a_{nn}} \right|^2,$$

then the lemma now follows from Lemma 6.7.3. $\qquad\square$

Proof of Theorem 6.7.1. It is a standard fact (see Lemma 3.3.3 of [HOR1]) that there is a real-valued function $\psi \in C^2$ which is strictly plurisubharmonic in a neighborhood of P so that Ω is defined by the equation $\psi < 0$ and gradψ is the exterior unit normal to $\partial\Omega$ at P. We choose local coordinates at P so that P is the origin and the differentials dz_j are orthonormal at P. Thus we have that the Riemannian element of integration has density 1 with respect to the Lebesgue measure in the coordinate space. We further choose our coordinates so that $\psi(z) + \operatorname{Im} z_n = O(|z|^2)$ at P. By the Taylor's formula, Ω is thus defined in a neighborhood of P by an inequality of the form

$$\operatorname{Im} z_n > \sum_{j,k=1}^{n} \frac{\partial^2 \psi(0)}{\partial z_j \partial \bar{z}_k} z_j \bar{z}_k + \operatorname{Re} A(z) + o(|z|^2),$$

where A is an analytic, homogeneous, second-degree polynomial. If we replace the coordinate z_n by $z_n - iA(z)$, the differential at P is not affected. So we may assume without loss of generality that $A = 0$ from the outset. Put $a_{jk} = \partial^2 \psi(0)/\partial z_j \partial \bar{z}_k$, which is a Hermitian symmetric, positive definite matrix at P.

For any $\epsilon > 0$, we set

$$\Omega_\epsilon = \left\{ z : \operatorname{Im} z_n > \sum_{j,k=1}^{n} a_{jk} z_j \bar{z}_k + \epsilon|a|^2 \right\}.$$

Then

$$\Omega_\epsilon^\delta \equiv \Omega_\epsilon \cap \{z : |z| < \delta\}$$

is contained in Ω if δ is sufficiently small. We now see, using Lemma 6.7.2, that

$$|K_\Omega(z, z)| \le e^\epsilon |K_{\Omega_\epsilon^\delta}(z, z)| .$$

If we let $z \to 0$ so that $\operatorname{Im} z_n / |a|$ has a positive lower bound, then it follows from Lemmas 6.7.3 and 6.7.6 applied to Ω_ϵ^δ and Ω_ϵ that

$$\limsup_{z \to 0} (\operatorname{Im} z_n)^{n+1} |K_\Omega(z, z)| \le e^\epsilon \limsup_{z \to 0} (\operatorname{Im} z_n)^{n+1} |K_{\Omega_\epsilon^\delta}(z, z)|$$

$$= e^\epsilon \limsup_{z \to 0} (\operatorname{Im} z_n)^{n+1} |K_{\Omega_\epsilon}(z, z)|$$

$$= n! \, 4^{-1} \pi^{-n} e^\epsilon \det(a_{jk} + \epsilon \delta_{jk})_{j,k=1}^{n-1} .$$

Since ϵ is arbitrary, this last proves (with the notation from our main theorem) that

$$\limsup_{z \to P} |\rho(z)|^{n+1} |K_\Omega(z, z)| \le \frac{k(P) n!}{4\pi^n} , \qquad (6.7.1.1)$$

assuming that z approaches P inside a small cone in the coordinate space around the normal to $\partial\Omega$ at P. But we can see that in fact the result is valid uniformly in P, so it remains true for arbitrary approach to P.

Let ϵ be positive but smaller than the least eigenvalue of the matrix (a_{jk}). For sufficiently small $\delta > 0$, we have

$$\Omega^\delta = \{z : z \in \Omega, |z| < \delta\} \subseteq \Omega_{-\epsilon} .$$

Hence, Lemma 6.7.3 can be applied with $U = \{z : |z| < \delta\}$ and $u_0(z) = e^{izn}$. From Lemma 6.7.3 and the monotonocity in Lemma 6.5.2, we then find, for δ sufficiently small, that

$$\limsup_{z \to P} (\operatorname{Im} z_n)^{n+1} |K_\Omega(z, z)| = \limsup_{z \to P} (\operatorname{Im} z_n)^{n+1} |K_{\Omega^\delta}(z, z)|$$

$$\ge e^{-\epsilon} \limsup_{z \to P} (\operatorname{Im} z_n)^{n+1} |K_{\Omega_{-\epsilon}}(z, z)|$$

when $z \to P$ and remains in a small cone about the normal to $\partial\Omega$. Arguing exactly as in the proof of (6.5.1.1), we conclude now that

$$\liminf_{z \to P} |\rho(z)|^{n+1} |K_\Omega(z, z)| \ge \frac{k(P) n!}{4\pi^n} .$$

Details are left to the reader. And that completes the proof.   □

6.7.2 The Fefferman's Asymptotic Expansion

The 1974 result of Fefferman [FEF1] subsumes all the earlier work. For Fefferman shows that, near a boundary point of a smooth, strictly pseudoconvex domain,[2] the Bergman kernel for Ω can be written (in suitable local coordinates) as

$$K_\Omega(z, w) = K_B(z, w) + \mathcal{E}(z, w).$$

Here K_B is the Bergman kernel for the unit ball, and $\mathcal{E}$ is an error term which is of measurably lower order than K_B. So we see that this theorem is much more explicit than Theorem 6.5.1 that we proved above, and it is also valid *off the diagonal*.

The Fefferman's argument is quite lengthy and complicated, and we cannot present it here. What we *can* do, however, is to explicate the approximation part of his reasoning. This is a nice piece of logic and shows how to approximate a strictly pseudoconvex point by (the image of) the unit ball.

We shall exploit the following fact:

FACT: Given $p \in \partial\Omega$ strictly pseudoconvex, we can find a region $\tilde{\Omega}$ internally tangent to $\partial\Omega$ to third order at p and an explicit biholomorphic change of coordinates F mapping a neighborhood of p in $\partial\tilde{\Omega}$ to a neighborhood of $F(p)$ in the boundary of the unit ball. Of course we can then pull back the Bergman kernel from the unit ball to obtain an explicit formula for the kernel $K_{\tilde{\Omega}}$ of $\tilde{\Omega}$.

Now we will give a more formal enunciation of this fact and provide a proof.

Proposition 6.7.7. *Let $p \in \partial\Omega$ be a strictly pseudoconvex point. Then there is a neighborhood V of p with the following property: For any point $Q \in V$ there is a biholomorphic mapping ζ_Q sending V to a neighborhood of the origin, sending Q to the point $(\psi(Q), 0, \ldots, 0)$, and sending $\Omega \cap V$ to a region of the form*

$$\operatorname{Re}\zeta_1 = |\zeta'|^2 - p_4(\zeta', \operatorname{Im}\zeta_1) + (\textit{fifth- and higher-order terms in } \zeta', \overline{\zeta}', \operatorname{Im}\zeta_1)$$

with $(\zeta_1, \zeta') \in \zeta_w(V)$. Here $p_4(\zeta', \operatorname{Im}\zeta_1)$ is a real-valued, fourth-order polynomial in $\zeta', \overline{\zeta}', \operatorname{Im}\zeta_1$ satisfying $p_4(\zeta', \operatorname{Im}\zeta_1) \geq C(|\zeta'|^4 + |\operatorname{Im}\zeta_1|^4)$. Furthermore, we can make $\zeta_Q(\cdot)$ depend smoothly on $P \in V$.

*Here we may think (refer to the **FACT** above) of $\{\operatorname{Re}\zeta_1 = |\zeta'|^2\}$ as the boundary of $\tilde{\Omega}$.*

Proof: Let q be the point of $\partial\Omega$ that is closest to Q. Then $Q - q$ is normal to $\partial\Omega$. After a suitable translation and rotation of $\mathbb{C}^n$, we may suppose that $q = 0$ and that

---

[2] At the time that Fefferman wrote his paper, it really was necessary to assume that the entire domain was strictly pseudoconvex in order to get certain global estimates for the $\overline{\partial}$-Neumann problem. However, more recent results of Catlin [CAT1, CAT2] and others show that one need only assume that the boundary is strictly pseudoconvex near the boundary point in question. The more global hypotheses can be something considerably weaker—like finite type.

the tangent plane to $\partial\Omega$ at q is $\{z : \operatorname{Re} z_1 = 0\}$. In particular, in the new coordinate system, we have $Q = (\tau, 0, \ldots, 0)$ with τ real and positive. In the complex part of the tangent space, specified by $\{z : z_1 = 0\}$, make a linear change of coordinates so that the Levi form at Q becomes the identity matrix. As a result, we have a new coordinate system, (z_1, z'), in which $Q = (\tau, 0)$ and $\partial\Omega$ takes the form

$$\operatorname{Re} z_1 = \operatorname{Re}\left\{ a(i\operatorname{Im} z_1)^2 + \sum_{j=2}^{n} b_j \cdot (i\operatorname{Im} z_1) \cdot z_j + \sum_{j,k=2}^{n} c_{jk} z_j z_k \right\}$$

$$+ |z'|^2 + (\text{third- and higher-order terms}).$$

We can make the coefficient a real. Here $|z'|^2$ appears as the Levi form $(z', \bar{z}')$.
  If we set

$$\zeta_1 = z_1 - a z_1^2 - \sum_{j=2}^{n} b_j z_1 z_j - \sum_{j,k=2}^{n} c_{jk} z_j z_k,$$

$$\zeta' = z',$$

then, in the new (ζ_1, ζ') coordinate system, $\partial\Omega$ takes the form

$$\operatorname{Re} \zeta_1 = |\zeta'|^2 + (\text{third and higher terms}), \tag{6.7.7.1}$$

and Q still has the form $(\tau, 0)$ for τ real. If we write out the most general real-valued homogeneous third-degree polynomial in $\operatorname{Im} \zeta_1$, ζ', $\bar{\zeta}'$, then we find that (6.7.7.1) is equivalent to

$$\operatorname{Re} \zeta_1 = |\zeta'|^2 + \operatorname{Re}\left\{ \sum_{j,k,\ell=2}^{n} a_{jk\ell} \zeta_j \zeta_k \zeta_\ell + \sum_{j,k,\ell=2}^{n} b_{jk\ell} \zeta_j \zeta_k \bar{\zeta}_\ell \sum_{j,k=2}^{n} c_{jk}(i\operatorname{Im} \zeta_1)\zeta_j \zeta_k \right.$$

$$\left. + \sum_{j,ik=2}^{n} d_{jk}(i\operatorname{Im} \zeta_1)\zeta_j \bar{\zeta}_k \sum_{j=2}^{n} e_j (i\operatorname{Im} \zeta_1)^2 \zeta_j + f \cdot (i\operatorname{Im} \zeta_1)^3 \right\}, \tag{6.7.7.2}$$

where the as, bs, cs, ds, es, and fs are complex numbers. We may take f to be purely imaginary. In the new coordinates then

$$z_1 = \zeta_1 - f\zeta_1^3 - \sum_{j=2}^{n} e_j \zeta_1^2 \zeta_j - \sum_{j,k=2}^{n} c_{jk} \zeta_1 \zeta_j \zeta_k - \sum_{j,k,\ell=2}^{n} a_{jk\ell} \zeta_j \zeta_k \zeta_\ell,$$

$$z_r = \zeta_r + \frac{1}{2} \sum_{j,k=2}^{n} b_{jkr} \zeta_j \zeta_k + \frac{1}{2} \sum_{j=2}^{n} d_{jr} \zeta_1 \zeta_j,$$

the surface (6.7.7.2) is transformed to $\operatorname{Re} z_1 = |z'|^2 + $ (fourth and higher r-order terms), while Q is mapped to $(\tau + if\tau^3, 0)$.

The rest of the proof depends on the following claim:

Claim: *There is a linear fractional transformation of the Siegel upper half-space* $\{(z_1, z') : \operatorname{Im} z_1 > |z'|^2\}$ *taking 0 to 0 and taking* $(\tau + if\tau^3, 0)$ *to* $(\tau, 0)$. *Moreover, the transformation is biholomorphic in a fixed neighborhood of* $(0, 0)$.

To see this, it is enough to check the case of the upper half plane in

$$\mathbb{C}^1 = \{z = (z_1, z') : z' = 0\}.$$

For, because of the explicit formulas for linear fractional transformations of the Siegel upper half-space, we see that every linear fractional transformation of the upper half plane to itself extends to a linear fractional transformation of the full Siegel domain to itself. Moreover, the property of being biholomorphic in a fixed neighborhood of 0 is preserved. The easy one-dimensional case of the claim is left for the interested reader.

Therefore we can make a holomorphic change of coordinates transforming $\partial\Omega$ to the surface

$$\operatorname{Re} \zeta_1 = |\zeta'|^2 - \text{(fourth-order terms)} + \text{(fifth- and higher-order terms)}. \quad (6.7.7.3)$$

It remains to show that the fourth-order terms can be made positive. However, the reader may easily check that the form (6.7.7.3) of the surface is preserved, while the fourth-order terms are forced to be nonnegative if we simply make the change of coordinates

$$z' = \zeta' + C\zeta_1\zeta'$$
$$z_1 = \zeta_1 - C\zeta_1^4,$$

for $C > 0$ large enough. The reader may check that all the above changes of coordinates depend smoothly on Q. $\qquad\qquad\square$

So we see quite explicitly with this proposition that a strictly pseudoconvex point has defining function which (after suitable modifications of the local coordinates) agrees with the ball up to the fourth order. This result is tricky but elementary. It is noteworthy that it is a decisive improvement over earlier results of this type (see, for instance, [HOR1]). Now we proceed to the guts of the approximation argument.

Fix a domain Ω and a point P in the boundary which is strictly pseudoconvex. This means that the boundary near P is C^2 and that the Levi form at P is positive definite. Of course an obvious implication is that boundary points near to P are also strictly pseudoconvex. Now we consider $A^2(\Omega)$ and the orthogonal space $A^2(\Omega)^\perp$. For a fixed point $w \in \Omega$, the Dirac mass δ_w breaks up uniquely into A^2 and $(A^2)^\perp$ components by the equation

$$\delta_w = K_\Omega(\cdot, w) + \big(\delta_w - K_\Omega(\cdot, w)\big). \quad (6.7.8)$$

We approximate the Bergman kernel K_Ω by producing an explicit $K^0 \in A^2$ and $K^+ \in (A^2)^\perp$ which add up to a small perturbation of δ_w. Let p be the point of $\partial\Omega$ which is nearest to w. We apply Proposition 6.7.7, setting $K^0(\cdot) = K_{\tilde\Omega}(\cdot, w)$, $\tilde{K}^+(\cdot) = \delta_w - K_{\tilde\Omega}(\cdot, w)$ on $\tilde\Omega$. We have $K^0 \in A^2(\tilde\Omega)$, $\tilde{K}^+ \in (A^2_{\tilde\Omega})^\perp$, and $\delta_w = K^0 + \tilde{K}^+$ on $\tilde\Omega$, so that $\delta_w = K^0\chi_{\tilde\Omega} + \tilde{K}^+\chi_{\tilde\Omega}$ on Ω. Clearly, since $\tilde{K}^+ \in (A^2_{\tilde\Omega})^\perp$ and $\tilde\Omega \subseteq \Omega$, we know that $K^+ = \tilde{K}^+\chi_{\tilde\Omega} \in (A^2_\Omega)^\perp$. Furthermore, from the explicit formula for $K_{\tilde\Omega}$, it follows immediately that K^0 continues analytically from $\tilde\Omega$ to all of Ω. (For fixed w in the unit ball, $(1 - z\cdot\overline{w})^{-(n+1)}$ continues analytically beyond the unit sphere.) Thus we may write

$$\delta_w + K^0\chi_{\Omega\setminus\tilde\Omega} = K^0 + K^+ \tag{6.7.9}$$

on Ω, with $K^0 \in A^2_\Omega$ and $K^+ \in (A^2_\Omega)^\perp$ given by explicit formulas in terms of (z, w).

In a sense that can be made precise, the term $K^0\chi_{\Omega\setminus\tilde\Omega}$ on the left-hand side is small, since $\Omega \setminus \tilde\Omega$ is such a thin subset of Ω (recall that $\partial\tilde\Omega$ is highly tangent to $\partial\Omega$). To make this idea precise and quantitative, we associate to each kernel $A(z, w)$ on $\Omega \times \Omega$ the operator $\mathbf{A}f(z) = \int_\Omega A(z, w)f(w)\,dV(w)$ on $L^2(\Omega)A$. Thus (6.7.9) becomes an operator equation

$$\mathbf{I} + \mathcal{E} = \mathbf{K^0} + \mathbf{K^+},$$

where $\mathbf{K^0}$, $\mathbf{K^+}$, $\mathcal{E}$ are given by kernels with explicit formulas and $\mathbf{K^0}f \in A^2(\Omega)$, $\mathbf{K^+}f \in (A^2_\Omega)^\perp$ for all $f \in L^2(\Omega)$. The thinness of $\Omega \setminus \tilde\Omega$ shows that $\mathcal{E}$ has small norm as an operator on $L^2(\Omega)$, so that for $f \in L^2(\Omega)$,

$$
\begin{aligned}
f &= (\mathbf{I} + \mathcal{E})(\mathbf{I} + \mathcal{E})^{-1}f \\
&= (\mathbf{K^0} - \mathbf{K^0}\mathcal{E} + \mathbf{K^0}\mathcal{E}^2 - \cdots)f + (\mathbf{K^+} - \mathbf{K^+}\mathcal{E} + \cdots)f \\
&\equiv F + G
\end{aligned}
$$

with $F \in A^2_\Omega$ and $G \in (A^2_\Omega)^\perp$. Comparing this result with (6.7.8), we see that the Bergman kernel for Ω must be given by the operator equation

$$\mathbf{P}_\Omega = \mathbf{K^0} - \mathbf{K^0}\mathcal{E} + \mathbf{K^0}\mathcal{E}^2 - \cdots, \tag{6.7.10}$$

the series converging in the norm topology of operators on $L^2(\Omega)$. Note that $\mathbf{P}_\Omega$ is the Bergman projection on Ω. We can obtain an asymptotic expansion for the Bergman kernel itself by applying both sides of this last operator equation to the Dirac delta mass. It is actually rather difficult to evaluate the right-hand sides of (6.7.10). Fefferman needs to develop an entire calculus of integral operators in order to do it. We cannot provide the details here but instead refer the reader to [FEF1, Part I]. We note that a consequence of these calculations (and this should be compared to the result of Hörmander discussed in the earlier part of this section) is that

$$K_\Omega(z, z) = \Phi(z)\delta_\Omega^{-(n+1)}(z) + \tilde{\Phi}(z) \log \delta_\Omega(z)$$

for z near $\partial\Omega$. Here $\Phi, \tilde{\Phi} \in C^\infty(\Omega)$, $\delta_\Omega(z)$ is the distance of z to $\partial\Omega$ and $\Phi(z) \neq 0$ for $z \in \partial\Omega$. This is in fact the corollary on page 45 of [FEF1].

It is worth noting here that Boutet de Monvel and Sjöstrand [BOS] show (Corollary 1.7) that, for the Szegő kernel,

$$K_\Omega(z, w) = \Phi(z, w)\delta_\Omega^{-(n+1)}(z, w) + \tilde{\Phi}(z, w) \log \delta_\Omega(z, w)$$

for some $\Phi, \tilde{\Phi} \in C^\infty(\overline{\Omega} \times \overline{\Omega})$. They are then able to relate their formula to Fefferman's.

A word of explanation is needed for our last statement. Where did the logarithmic term come from? This was one of the dramatic results of the Fefferman's work— that the Bergman kernel of a strictly pseudoconvex domain can (at least in principle) contain a logarithmic term. And in fact Fefferman provides in his paper an explicit example of domain for which the logarithmic term actually occurs (see Example 6.7.11 below).

It is a noted conjecture of Ramadanov (see [RAM2]) that a strongly pseudoconvex domain with no logarithmic term for the Bergman kernel asymptotic expansion at any boundary point must be spherical. Dan Burns (unpublished) in fact proved such a result. Boutet de Monvel [BOU] in dimension two and Robin Graham [GRA3] in general gave rigorous proofs of the result. See also the work of Hirachi [HIR1]. There are unbounded domains and also roughly bounded domains on which the analogue of this result for the Szegő is known to fail—see [HIR2].

The logarithmic term arises in the Fefferman's calculations because he is analyzing certain integral expressions—which are in effect negative powers of a nonisotropic distance function—using integration by parts. When the power of the nonisotropic distance is -1, then the next integration gives rise to a log term.

One of the nice features of the work of Boutet de Monvel and Sjöstrand is that they derive their asymptotic expansion for the Szegő kernel in a rather natural fashion from the explicit formula on the ball using the theory of Fourier integral operators (see [HOR5]). From this study they obtained the formula (see Theorem 1.5 in [BOS])

$$K(z, \zeta) = \int_0^\infty e^{i\delta_\Omega(z, w)} b(z, w, t) \, dt ,$$

where $b \in C^\infty(\overline{\Omega}, \overline{\Omega}, \mathbb{R}^+)$.

We conclude this section with a presentation of the Fefferman's example of a domain whose Bergman kernel has a logarithmic term. We will exploit this form of the asymptotic expansion:

$$K_\Omega(z, w) = C |\text{grad}\delta_\Omega(w)|^2 \det\mathcal{L}(w) X^{-(n+1)}(z, w) + \tilde{K}(z, w) ,$$

where $\tilde{K}$ is an admissible kernel of weight $\geq n - 1$. Here $\mathcal{L}$ is the Levi form. Also "weight" is a concept that Fefferman introduces as part of his integral calculus. We can safely take it for granted.

Example 6.7.11. Let Ω be the connected component of

$$\{(z_1, z_2) \in \mathbb{C}^2 : |z_1|^2 + |z_2|^2 - c|z_2|^8 < 1\}$$

containing the origin, where c is small and positive. Note that this Ω is highly tangential to the unit ball B at the point $(1, 0)$. Also $B \subseteq \Omega$. Now set

$$\psi(w) = 1 - |w_1|^2 - |w_2|^2 + c|w_3|^8$$

and $w^0 = (\tau, 0)$ with $0 < \tau < 1$. Now the above formulation of the asymptotic expansion for the Bergman kernel tells us that

$$K_\Omega(z, w^0)) = \varphi_0(w^0)(1 - \tau z_1)^{-3} + \sum_{j=1}^{M} \varphi_j(z, w^0)(1 - \tau z_1)^{-m_j}$$
$$+ \tilde{\varphi}(z, w^0) \log(1 - \tau z_1) + O(1), \qquad (6.7.11.1)$$

with $\varphi_0(w^0) \neq 0$ and weight$(\varphi_j) - m_j \geq -5/2$. We shall show that $\lim_{t \to 1^-} \tilde{\varphi}(w^0, w^0) \neq 0$. To do so, we apply the reproducing property of K_Ω to the anti-holomorphic function $F(w) = K_B(w^0, w) = c_1(1 - \tau \bar{w}_1)^{-3}$ to obtain

$$c_1(1 - \tau^2)^{-3} = F(w^0)$$

$$= \int_\Omega F(z) K_\Omega(z, w^0) \, dV(z)$$

$$= \int_\Omega c_1(1 - \tau \bar{z}_1)^{-3} K_\Omega(z, w^0) \, dV(z)$$

$$= \int_B c_1(1 - \tau \bar{z}_1)^{-3} K_\Omega(z, w^0) \, dV(z)$$

$$+ \int_{\Omega \setminus B} c_1(1 - \tau \bar{z}_1)^{-3} K_\Omega(z, w^0) \, dV(z)$$

$$= K_\Omega(w^0, w^0) + \int_{\Omega \setminus B} c_1(1 - \tau \bar{z}_1)^{-3} K_\Omega(z, w^0) \, dV(z) \quad (6.7.11.2)$$

(since $c_1(1 - \tau \bar{z}_1)^{-3} = K_B(w^0, z)$ and $K_\Omega(\cdot, w^0)$ is holomorphic on $B \subseteq \Omega$). Substituting (6.7.11.1) into (6.7.11.2), we find that

$$c_1(1-\tau^2)^{-3} \quad \varphi_0(w^0)(1-\tau^2)^{-3} - \sum_{j=1}^{M} \varphi_j(w^0, w^0)(1-\tau^2)^{-m_j}$$

$$-\tilde{\varphi}(w^0, w^0)\log(1-\tau^2) + O(1)$$

$$= c_1\varphi_0(w^0)\int_{\Omega\setminus B} |1-\tau z_1|^{-6}\,dV(z)$$

$$+c_1\left\{\int_{\Omega\setminus B}(1-\tau\bar{z}_1)^{-3}\left[\sum_{j=1}^{M}\varphi_j(z, w^0)\cdot(1-\tau z_1)^{-m_j}\right.\right.$$

$$\left.\left. +\tilde{\varphi}(z, w^0)\log(1-\tau z_1) + O(1)\right]dV(z)\right\}$$

$$= c_1\varphi_0(w^0)\int_{\Omega\setminus B} |1-\tau z_1|^{-6}\,dV(z) + O(1)$$

$$\sim \log(1-\tau^2) \quad \text{as } \tau \to 1^-.$$

We see immediately that $\lim_{\tau\to 1^-} \tilde{\varphi}(w^0, w^0)$ could not be zero. Therefore the term $\tilde{\varphi}(z)$ is really present in $K_\Omega(z, z)$.

We conclude this section by noting that Fefferman developed his ideas further by formulating a program for studying the geometry and analysis of strictly pseudoconvex domains. The main idea is to consider the Bergman and Szegő kernels as analogues of the heat kernel of a Riemannian manifold. In Riemannian geometry, the coefficients of the asymptotic expansion of the heat kernel can be expressed in terms of the curvature of the metric. Integrating the coefficients, one may obtain index theorems in various settings. See [BEG, HIR3] for some of the details.

We also mention that, in his famous problem list, Yau [YAU] raises the question of classifying pseudoconvex domains whose Bergman metrics are Kähler–Einstein. Cheng [CHENG] conjectured that, if the Bergman metric of a strictly pseudoconvex domain is Kähler–Einstein, then the domain is biholomorphic to the ball. This conjecture was proved by Fu–Wong [FUW] in the case of a simply connected, strictly pseudoconvex domain with smooth boundary.

6.8 The Bergman Kernel for a Sobolev Space

We may define the Bergman kernel for the Sobolev space W^1 and it appears to be (up to a bounded error term)

$$\frac{1}{\pi}\log(1-z\bar{\zeta}).$$

Specifically, set $\varphi_j(\zeta) = \zeta^j$. We calculate that

$$\iint\limits_{D} |\varphi_j(\zeta)|^2 \, dA = \iint\limits_{D} |\zeta^j|^2 \, dA = \frac{\pi}{j+1}$$

and

$$\iint\limits_{D} |\varphi_j'(\zeta)|^2 \, dA = \iint\limits_{D} |j\zeta^{j-1}|^2 \, dA = j\pi \,.$$

Thus

$$\|\varphi_j\|_{W^1} = \sqrt{\pi} \cdot \sqrt{\frac{j^2 + j + 1}{j+1}} \,.$$

Thus the full Bergman kernel for W^1 is given by

$$\sum_{j=0}^{\infty} \frac{1}{\pi} \cdot \frac{j+1}{j^2+j+1} \cdot z^j \overline{\zeta}^j = \frac{1}{\pi} + \sum_{j=1}^{\infty} \frac{1}{\pi} \cdot \frac{j+1}{j^2+j+1} \cdot z^j \overline{\zeta}^j = \frac{1}{\pi} + \sum_{j=1}^{\infty} \frac{1}{\pi} \cdot \frac{1}{j} \cdot z^j \overline{\zeta}^j + \mathcal{E} \,,$$

where $\mathcal{E}$ is an error term which is bounded and has one bounded derivative. So $\mathcal{E}$ is negligible from the point of view of determining where the kernel has singularities (i.e., where it blows up).

We look at

$$\frac{1}{\pi} + \frac{1}{\pi} \sum_{j=1}^{\infty} \frac{1}{j} \alpha^j = \frac{1}{\pi} + \frac{1}{\pi} \sum_{j=1}^{\infty} \int \alpha^{j-1}$$

$$= \frac{1}{\pi} + \frac{1}{\pi} \int \sum_{j=1}^{\infty} \alpha^{j-1}$$

$$= \frac{1}{\pi} + \frac{1}{\pi} \int \frac{1}{\alpha} \sum_{j=1}^{\infty} \alpha^j$$

$$= \frac{1}{\pi} + \frac{1}{\pi} \int \frac{1}{\alpha} \left[\sum_{j=0}^{\infty} \alpha^j - 1 \right]$$

$$= \frac{1}{\pi} + \frac{1}{\pi} \int \frac{1}{\alpha} \left[\frac{1}{1-\alpha} - 1 \right]$$

$$= \frac{1}{\pi} + \frac{1}{\pi} \int \frac{1}{1-\alpha}$$

$$= \frac{1}{\pi} - \frac{1}{\pi} \log(1 - \alpha) \,.$$

Thus the Bergman kernel for the order 1 Sobolev space is given by

$$K(z, \zeta) = \frac{1}{\pi} - \frac{1}{\pi} \log(1 - z\bar{\zeta}).$$

Also the kernel for the space generated just by the monomials with even index seems to be given by (up to a bounded error term)

$$\frac{1}{\pi} \left(\log(z\bar{\zeta}) + \frac{1}{2} \log(1 - z\bar{\zeta}) - \frac{1}{2} \log(1 + z\bar{\zeta}) \right).$$

To see this, we look at

$$\sum_{j=0}^{\infty} \frac{1}{\pi} \cdot \frac{2j + 1}{(2j)^2 + 2j + 1} z^{2j} \bar{\zeta}^{2j} = \frac{1}{\pi} + \frac{1}{\pi} \sum_{j=1}^{\infty} \frac{1}{2j} z^{2j} \bar{\zeta}^{2j} + \mathcal{F}.$$

Here, as in the first calculation, $\mathcal{F}$ is a bounded term with one bounded derivative. So it is negligible from the point of view of our calculation.

Thus we wish to calculate

$$\frac{1}{\pi} + \frac{1}{\pi} \sum_{j=1}^{\infty} \frac{1}{2j} \alpha^{2j} = \frac{1}{\pi} + \frac{1}{\pi} \sum_{j=1}^{\infty} \int \alpha^{2j-1}$$

$$= \frac{1}{\pi} + \frac{1}{\pi} \int \frac{1}{\alpha} \sum_{j=1}^{\infty} \alpha^{2j}$$

$$= \frac{1}{\pi} + \frac{1}{\pi} \int \frac{1}{\alpha} \left[\sum_{j=0}^{\infty} \alpha^{2j} - 1 \right]$$

$$= \frac{1}{\pi} + \frac{1}{\pi} \int \frac{1}{\alpha} \left[\frac{1}{1 - \alpha^2} - 1 \right]$$

$$= \frac{1}{\pi} + \frac{1}{\pi} \int \frac{\alpha}{1 - \alpha^2}$$

$$= \frac{1}{\pi} - \frac{1}{2\pi} \log(1 - \alpha^2).$$

In conclusion, the Bergman kernel for the order 1 Sobolev space using only the basis elements with even index is

$$K'(z, \zeta) = \frac{1}{\pi} - \frac{1}{2\pi} \log(1 - z \cdot \bar{\zeta}) - \frac{1}{2\pi} \log(1 + z \cdot \bar{\zeta}).$$

In short, there are singularities as z and ζ tend to the *same* disc boundary point and also as z and ζ tend to antipodal disc boundary points.

6.9   Regularity of the Dirichlet Problem on a Smoothly Bounded Domain and Conformal Mapping

We begin by giving a precise definition of a domain "with smooth boundary."

Definition 6.9.1. Let $U \subseteq \mathbb{C}$ be a bounded domain. We say that U has *smooth boundary* if the boundary consists of finitely many curves and each of these is locally the graph of a C^∞ function.

In practice it is more convenient to have a different definition of domain with a smooth boundary. A function ρ is called a *defining function* for U if ρ is defined in a neighborhood W of ∂U, $\nabla \rho \neq 0$ on ∂U, and $W \cap U = \{z \in W : \rho(z) < 0\}$. Now we say that U has smooth (or C^k) boundary if U has a defining function ρ that is smooth (or C^k). Yet a third definition of smooth boundary is that it consists of finitely many curves γ_j, each of which is the trace of a a smooth curve $\mathbf{r}(t)$ with nonvanishing gradient. We invite the reader to verify that these three definitions are equivalent.

Our motivating question for the present section is as follows:

Let $\Omega \subseteq \mathbb{C}$ be a bounded domain with smooth boundary. Assume that $f \in \Lambda_\alpha(\partial\Omega)$. If $u \in C(\overline{\Omega})$ satisfies (i) u is harmonic on Ω and (ii) $u|_{\partial\Omega} = f$, then does it follow that $u \in \Lambda_\alpha(\overline{\Omega})$?

Here Λ_α is the usual Lipschitz space (see [KRA12]). Here is a scheme for answering this question:

Step 1: Suppose at first that U is bounded and simply connected.
Step 2: By the Riemann mapping theorem, there is a conformal mapping ϕ : $U \to D$. Here D is the unit disc. We would like to reduce our problem to the Dirichlet problem on D for the data $f \circ \phi^{-1}$.

In order to carry out this program, we need to know that ϕ extends smoothly to the boundary. It is a classical result of Carathéodory [CAR] that, if a simply connected domain U has boundary consisting of a Jordan curve, then any conformal map of the domain to the disc extends univalently and bicontinuously to the boundary. It is less well known that Painlevé, in his thesis [PAI], proved that when U has smooth boundary, then the conformal mapping extends smoothly to the boundary. In fact Painlevé's result long precedes that of Carathéodory.

We shall present here a modern approach to smoothness to the boundary for conformal mappings. These ideas come from [KER1]. See also [BEK] for a self-contained approach to these matters. Our purpose here is to tie the smoothness-to-the-boundary issue for mappings directly to the regularity theory of the Dirichlet problem for the Laplacian.

Let W be a collared neighborhood of ∂U. Set $\partial U' = \partial W \cap U$ and let $\partial D' = \phi(\partial U')$. Define B to be the region bounded by ∂D and $\partial D'$. We solve the Dirichlet problem on B with boundary data

$$f(\zeta) = \begin{cases} 1 \text{ if } \zeta \in \partial D \\ 0 \text{ if } \zeta \subset \partial D' \end{cases}$$

Call the solution u.

Consider $v \equiv u \circ \phi : U \to \mathbb{R}$. Then of course v is still harmonic. By the Carathéodory's theorem, v extends to $\partial\Omega, \partial\Omega'$, and

$$v = \begin{cases} 1 \text{ if } \zeta \in \partial U \\ 0 \text{ if } \zeta \in \partial U' \end{cases}.$$

Suppose that we knew that solutions of the Dirichlet problem on a smoothly bounded domain with C^∞ data are in fact C^∞ on the closure of the domain. Then, if we consider a first-order derivative $\mathcal{D}$ of v, we obtain

$$|\mathcal{D}v| = |\mathcal{D}(u \circ \phi)| = |\nabla u|\,|\nabla\phi| \le C.$$

It follows that

$$|\nabla\phi| \le \frac{C}{|\nabla u|}. \tag{6.9.2}$$

This will prove to be a useful estimate once we take advantage of the following:

Lemma 6.9.3 (Hopf). *Let $\Omega \subset\subset \mathbb{R}^N$ have C^2 boundary. Let $u \in C(\overline{\Omega})$ with u harmonic and nonconstant on Ω. Let $P \in \overline{\Omega}$ and assume that u takes a local minimum at P. Then*

$$\frac{\partial u}{\partial v}(P) < 0.$$

Proof: Suppose without loss of generality that $u > 0$ on Ω near P and that $u(P) = 0$. Let B_R be a ball that is internally tangent to $\overline{\Omega}$ at P. We may assume that the center of this ball is at the origin and that P has coordinates $(R, 0, \ldots, 0)$. Then, by Harnack's inequality (see [GRK12]), we have for $0 < r < R$ that

$$u(r, 0, \ldots, 0) \ge c \cdot \frac{R^2 - r^2}{R^2 + r^2}$$

hence

$$\frac{u(r, 0, \ldots, 0) - u(R, 0, \ldots, 0)}{r - R} \le -c' < 0.$$

Therefore

$$\frac{\partial u}{\partial v}(P) \le -c' < 0.$$

This is the desired result. $\square$

Now let us return to the u from the Dirichlet problem that we considered prior to line (6.9.2). Hopf's lemma tells us that $|\nabla u| \geq c' > 0$ near ∂D. Thus from Corollary 6.9.2, we conclude that

$$|\nabla \phi| \leq C. \tag{6.9.4}$$

Thus we have bounds on the first derivatives of ϕ.

To control the second derivatives, we calculate that

$$C \geq |\nabla^2 v| = |\nabla(\nabla v)| = |\nabla(\nabla(u \circ \phi))|$$
$$= |\nabla(\nabla u(\phi) \cdot \nabla \phi)| = |(\nabla^2 u \cdot [\nabla \phi]^2) + (\nabla u \cdot \nabla^2 \phi)|.$$

Here the reader should think of ∇ as representing a generic first derivative and ∇^2 a generic second derivative. We conclude that

$$|\nabla u| \, |\nabla^2 \phi| \leq C + |\nabla^2 u| \, |(\nabla \phi)^2| \leq C'.$$

Hence, (again using Hopf's lemma)

$$|\nabla^2 \phi| \leq \frac{C}{|\nabla u|} \leq C''.$$

In the same fashion, we may prove that $|\nabla^k \phi| \leq C_k$, any $k \in \{1, 2, \dots\}$. This means (use the fundamental theorem of calculus) that $\phi \in C^\infty(\overline{\Omega})$.

We have arrived at the following situation: Smoothness to the boundary of conformal maps implies regularity of the Dirichlet problem on a smoothly bounded domain. Conversely, regularity of the Dirichlet problem can be used, together with Hopf's lemma, to prove the smoothness to the boundary of conformal mappings. We must find a way out of this impasse.

Our solution to the problem posed in the last paragraph will be to study the Dirichlet problem for a more general class of operators that is invariant under *smooth* changes of coordinates. We will study these operators by (1) localizing the problem and (2) mapping the smooth domain under a diffeomorphism to an upper half-space. It will turn out that *elliptic operators* are invariant under these operations. One can then (we shall not actually do this) use the calculus of pseudodifferential operators to prove local boundary regularity for elliptic operators.

There is an important point implicit in our discussion that deserves to be brought into the foreground. The Laplacian is invariant under conformal transformations (exercise). This observation was useful in setting up the discussion in the present section. But it turned out to be a point of view that is too narrow: We found ourselves in a situation of circular reasoning. We shall thus expand to a wider universe in which our operators are invariant under *diffeomorphisms*. This type of invariance will give us more flexibility and more power. See [KRA4, Chap. 3] for background to this discussion.

Let us conclude this section by exploring how the Laplacian behaves under a diffeomorphic change of coordinates. For simplicity we restrict attention to $\mathbb{R}^2$ with coordinates (x, y). Let

$$\phi(x, y) = (\phi_1(x, y), \phi_2(x, y)) \equiv (x', y')$$

be a diffeomorphism of $\mathbb{R}^2$. Let

$$\Delta = \frac{\partial^2}{\partial x^2} + \frac{\partial^2}{\partial y^2}.$$

In (x', y') coordinates, the operator Δ becomes

$$\phi_*(\Delta) = |\nabla \phi_1|^2 \frac{\partial^2}{\partial x'^2} + |\nabla \phi_2|^2 \frac{\partial^2}{\partial y'^2}$$

$$+ 2 \left[\frac{\partial x'}{\partial x} \frac{\partial y'}{\partial x} + \frac{\partial x'}{\partial y} \frac{\partial y'}{\partial y} \right] \frac{\partial^2}{\partial x \partial y} + \text{(first-order terms)}.$$

In an effort to see what the new operator has in common with the old one, we introduce the notation

$$D = \sum a_\alpha \frac{\partial}{\partial x^\alpha},$$

where

$$\frac{\partial}{\partial x^\alpha} = \frac{\partial}{\partial x_1^{\alpha_1}} \frac{\partial}{\partial x_2^{\alpha_2}} \cdots \frac{\partial}{\partial x_n^{\alpha_n}}$$

is a differential monomial. Its "symbol" is defined to be

$$\sigma(D) = \sum a_\alpha(x) \xi^\alpha , \quad \xi^\alpha = \xi_1^{\alpha_1} \xi_2^{\alpha_2} \cdots \xi_n^{\alpha_n}.$$

The symbol of the Laplacian $\Delta = \frac{\partial^2}{\partial x^2} + \frac{\partial^2}{\partial y^2}$ is

$$\sigma(\Delta) = \xi_1^2 + \xi_2^2.$$

Now associate to $\sigma(\Delta)$ a matrix $A_\Delta = (a_{ij})_{1 \le i,j \le 2}$, where $a_{ij} = a_{ij}(x)$ is the coefficient of $\xi_i \xi_j$ in the symbol. Thus

$$A_\Delta = \begin{pmatrix} 1 & 0 \\ 0 & 1 \end{pmatrix}.$$

The symbol of the transformed Laplacian (in the new coordinates) is

$$\sigma(\phi_*(\Delta)) = |\nabla\phi_1|^2\xi_1^2 + |\nabla\phi_2|^2\xi_2^2$$

$$+ 2\left[\frac{\partial x'}{\partial x}\frac{\partial y'}{\partial y} + \frac{\partial x'}{\partial y}\frac{\partial y'}{\partial y}\right]\xi_1\xi_2 + (\text{lower-order terms}).$$

Then

$$A_{\sigma(\phi^*(\Delta))} = \begin{pmatrix} |\nabla\phi_1|^2 & \left[\frac{\partial x'}{\partial x}\frac{\partial y'}{\partial x} + \frac{\partial x'}{\partial y}\frac{\partial y'}{\partial y}\right] \\ \left[\frac{\partial x'}{\partial x}\frac{\partial y'}{\partial x} + \frac{\partial x'}{\partial y}\frac{\partial y'}{\partial y}\right] & |\nabla\phi_2|^2 \end{pmatrix}.$$

The matrix $A_{\sigma(\phi^*(\Delta))}$ is positive definite provided that the change of coordinates ϕ is a diffeomorphism (i.e., has nondegenerate Jacobian). It is this positive definiteness property of the symbol that is crucial to the success of the theory of pseudodifferential operators (see [KRA4, Chap. 3]). For our study of the boundary regularity of conformal mappings, the transformation properties of the Laplacian under holomorphic mappings were sufficient.

6.10 Existence of Certain Smooth Plurisubharmonic Defining Functions for Strictly Pseudoconvex Domains and Applications

6.10.1 Introduction

In classical analysis, an important theorem of Painlevé [PAI] and Kellogg [KEL] states that any conformal mapping between two smoothly bounded domains in the complex plane $\mathbb{C}$ can be extended to be a diffeomorphism on the closures of the domains. This theorem was generalized by Fefferman [FEF1, Part I] in 1974 to strictly pseudoconvex domains in $\mathbb{C}^n$. Fefferman's original proof of this theorem is very technical, relying as it does on deep work on the boundary asymptotics of the Bergman kernel and on the regularity of $\bar{\partial}$-Neumann operator that is due to Folland and Kohn [FOK]. Bell–Ligocka [BELL], and later Bell [BEL1], gave a simpler proof which deals with more general domains, including pseudoconvex domains of finite type, by using regularity of the Bergman projection and the $\bar{\partial}$-Neumann operator as studied by Folland and Kohn [FOK], Catlin [CAT2], Boas–Straube [BOS1], and others.

We know from [KER3] that Painlevé and Kellogg's theorem can be proved by using the regularity of the Dirichlet problem for the Laplacian in a smoothly bounded planar domain, where the property of the Laplacian being conformally invariant plays an important role in the proof. (See [BEK] for another point of view, and also the discussion in the last section.) The natural generalization of the Laplacian in one complex variable to several complex variables, with these considerations in mind, is the complex Monge–Ampère equation. In [KER3], Kerzman observed that the proof of the Fefferman mapping theorem would follow from the C^∞ global regularity of the Dirichlet problem of a degenerate complex Monge–Ampère equation. However, counterexamples in Bedford–Fornæss [BEF2] as well as in Gamelin–Sibony [GAS] show that, in general, the degenerate Dirichlet problem for the complex Monge–Ampère equation does not have C^2 boundary regularity. Thus Kerzman's idea does not work in the sense of its original formulation.

The main purpose of the present section is to construct a plurisubharmonic defining function ρ for a smoothly bounded strictly pseudoconvex domain in $\mathbb{C}^n$ with det $H(\rho)$ vanishing to higher order near the boundary, where $H(\rho)$ denotes the complex hessian of ρ. In other words, we shall prove the following theorem:

Theorem 6.10.1. *Let Ω be a bounded strictly pseudoconvex domain in $\mathbb{C}^n$ with C^∞ boundary $\delta\Omega$. For any $0 < \epsilon << 1$ and any positive integer q, there is a plurisubharmonic defining function $\rho_q \in C^\infty(\overline{\Omega})$ for Ω so that*

$$\det H(\rho_q)(z) \leq C \operatorname{dist}(z, \delta\Omega)^q, \quad z \in \Omega. \qquad (6.10.1.1)$$

Combining Theorem 6.10.1 with a result of Caffarelli et al. [CKNS] on the Dirichlet problem for the complex Monge–Ampère equation, we provide a new proof for the following result, which was proved in [BEC, DIF4]:

Corollary 6.10.2. *Let $\Omega_j, j = 1, 2$ be two bounded strictly pseudoconvex domains in $\mathbb{C}^n$ with C^∞ boundary. Let $\varphi : \Omega_1 \to \Omega_2$ be a proper holomorphic mapping. Then for any $\epsilon > 0$ we have*

(i) φ can be extended as a $\operatorname{Lip}_1(\overline{\Omega}_1)$ mapping.
(ii) $\det(\varphi') \in \operatorname{Lip}_{1/2}(\overline{\Omega}_1)$.
(iii) There is a C^2 defining function ρ for Ω_2 so that, for any $k = 1, \ldots, n$, we have
  $\sum_{j=1}^{n} \left[\frac{\delta\rho}{\delta w_j} \circ \varphi \right] \cdot \frac{\delta\varphi_j}{\delta z_k} \in \operatorname{Lip}_1(\overline{\Omega}_1).$

As a corollary of (ii) and of a theorem in [FOR, NWY, PIH, WEB2], we may give a new proof of the Fefferman mapping theorem.

Our presentation is organized as follows. In the next section, Theorem 6.10.1 is proved. In the following section, some applications are given.

6.11 Proof of Theorem 6.10.1

We now supply the proof of the first theorem.

Proof of Theorem 6.10.1. Let $\delta(z)$ denote the (signed) distance function from z to $\partial\Omega$ (positive inside and negative outside). Since Ω is a bounded domain in $\mathbb{C}^n$ with C^∞ boundary $\partial\Omega$, then $\delta(z) \in C^\infty(\overline{\Omega})$ (after modification of the distance function on a compact set in the interior). By a rotation we see that, for any fixed point $z^0 \in \Omega$ near the boundary, we may assume that the z_n direction is the normal direction $\left(\frac{\partial \delta}{\partial \bar{z}_1}, \ldots, \frac{\partial \delta}{\partial \bar{z}_n} \right)$ at the point z^0. Let

$$H(\delta)_{n-1}(z) = \left[\frac{\partial^2 \delta}{\partial z_\ell \partial \bar{z}_q}(z) \right]_{1 \leq \ell, q \leq n-1}$$

Since Ω is strictly pseudoconvex, there is an $\epsilon > 0$ so that

$$-H(\delta)_{n-1}(z^0) \geq \epsilon I_{n-1}$$

for all $z \in \bar{\Omega}$ with $\delta(z) \leq \epsilon$. We may assume that $H(\delta)_{n-1}$ is diagonalized at z^0. Note that

$$H(-\delta)(z) = \begin{bmatrix} -H(\delta)_{n-1}(z) & 0 \\ 0 & \frac{-\partial^2 \delta}{\partial z_n \partial \bar{z}_n} \end{bmatrix}.$$

If it happens that $\sum_{i,j} \frac{\partial \delta}{\partial z_i} \frac{\partial \delta}{\partial \bar{z}_j} \frac{\partial^2 \delta}{\partial z_i \partial \bar{z}_j} = 0$, then $\rho_q(z) = -\delta(z) + \delta(z)^{2+q}$ is the desired defining function. (In fact it is these terms that distinguish the study of the real Hessian from the more subtle study of the complex Hessian. In particular, we know that $\partial^2 \delta / \partial x_n^2$ equals 0, but the term $\partial^2 \delta / \partial z_n \partial \bar{z}_n$ may not be zero. Therefore estimate (6.10.1.1) is much easier to check for the determinant of the real Hessian of δ; matters are much trickier for the complex Hessian.)

Now we let

$$\rho(z) = -\delta(z)$$

and

$$r^{[m]}(z) = \rho(z) + \sum_{k=2}^{m} a_k(z) \delta(z)^k \quad z \in \Omega.$$

We will prove inductively that

$$\det(H(r^{[m]})(z)) = b_{m-1}(z) \cdot \rho^{m-1},$$

where b_{m-1} is some smooth function. In particular, $\det(H(r^{[m]})$ vanishes to order $m - 1$ at the boundary.

Now

$$\partial_{ij} r^{[m]}(z) = \partial_{ij} \rho(z) + \sum_{k=2}^{m} \left[\partial_{ij} a_k \, \rho(z)^2 + k(k-1) a_k \partial_i \rho \partial_{\bar{j}} \rho) \right] \rho^{k-2}$$

$$+ \sum_{k=2}^{m} k \left[\partial_i a_k \partial_{\bar{j}} \rho + \partial_i \rho \partial_{\bar{j}} a_k + a_k \rho_{ij} \right] \rho^{k-1}.$$

Let $H(r^{[m]})_{n-1}(z) = [\frac{\delta^2 r^{[m]}}{\delta z_i \delta \bar{z}_j}]_{(n-1)\times(n-1)}$, and let $B(z)^* = [r_{n\bar{1}}, \ldots, r_{n\overline{n-1}}]$ be a row vector. Then

$$H(r^{[m]})(z) = \begin{bmatrix} H(r^{[m]})_{n-1}(z) & B(z) \\ B(z)^* & r_{n\bar{n}}^{[m]}(z) \end{bmatrix}.$$

Then

$$\det(H(r^{[m]})(z)) = \det(H(r^{[m]})_{n-1}(z))[r_{n\bar{n}}^{[m]} - \{B^* H(r^{[m]})_{n-1}(z)\}^{-1} B(z)].$$

We know that

$$H(r^{[m]})_{n-1}(z) \geq \epsilon I_{n-1}$$

for all $z \in \Omega_\epsilon$, where ϵ is a positive number depending only on Ω and $\|a_j\|_{C^2(\overline{\Omega})}$. At first, we let

$$(d\rho)(z) = \sum_{i,j} \rho_i \rho_{\bar{j}} \rho_{i\bar{j}};$$

then we define

$$a_2(z) = -\frac{(d\rho)(z)}{2((d\rho)(z) + |\partial \rho|^4)}.$$

Thus we have

$$r_{n\bar{n}}(z) = \rho_{n\bar{n}} + \sum_{k=2}^{m} [\partial_{n\bar{n}} a_k \rho^k + k(k-1)a_k |\partial \rho|^2 \rho^{k-2}]$$

$$+ \sum_{k=2}^{m} k[\partial_n a_k \partial_{\bar{n}} \rho + \partial_n \rho \partial_{\bar{n}} a_k + a_k \rho_{n\bar{n}}] \rho^{k-1}$$

$$= \sum_{k=2}^{m} [\partial_{n\bar{n}} a_k \rho^k + k(k+1)a_{k+1} |\partial \rho|^2 \rho^{k-1}]$$

$$+ \sum_{k=2}^{m} k[\partial_n a_k \partial_{\bar{n}} \rho + \partial_n \rho \partial_{\bar{n}} a_k + a_k \rho_{n\bar{n}}] \rho^{k-1}$$

$$= O(\rho).$$

Therefore

$$\det(H(r^{[2]})(z)) = b_1(z)\rho(z).$$

Assume that we have constructed $r^{[m]} = \rho(z) + \sum_{k=2}^{m} a_k \rho(z)^k$ such that

$$\det(H(r^{[m]})(z)) = b_{m-1}(z)\rho^{m-1}, \quad z \in \Omega.$$

We consider

$$r^{[m+1]} = r^{[m]} + a_{m+1}\rho(z)^{m+1}.$$

Since

$$H(r^{[m+1]})(z) = H(r^{[m]}) + \rho(z)^{m+1}H(a_{m+1})(z) + (m+1)a_{m+1}\rho(z)^m H(\rho)$$
$$+ (m+1)ma_{m+1}\rho^{m-1}\partial\rho(z) \otimes \overline{\partial\rho}$$
$$+ (m+1)\rho(z)^m[\partial\rho \otimes \overline{\partial a_{m+1}} + \partial a_{m+1} \otimes \overline{\partial\rho}],$$

it is easy to see that

$$\det(H(r^{[m+1]})(z)) = b_m(z)\rho(z)^m + \det[H(r^{[m]}) + (m+1)ma_{m+1}\rho^{m-1})].$$

By a rotation, we may let z_n be the complex normal direction of $\partial\Omega_{\partial(z)}$ at z. Thus

$$\det\left(H(r^{[m]})(z) + (m+1)m\rho(z)^{m-1}\partial\rho(z) \otimes \overline{\partial\rho}\right) = \det M,$$

where

$$M = \begin{bmatrix} H(r^{[m]})_{n-1}(z) & B(m)(z) \\ B(m)(z)^* & r^{[m]}_{n\bar{n}}(z) + (m+1)m\rho^{m-1}\rho_n\rho_{\bar{n}} \end{bmatrix}.$$

But this

$$= \det(H(r^{[m]})_{n-1}(z)[r^{[m]}_{n\bar{n}} + B(m)^* H(r^{[m]})_{n-1}(z)^{-1}B(m)(z)$$
$$+ (m+1)m\rho^{m-1}\rho_n\rho_{\bar{n}}]$$
$$= \det(H(r^{[m]})(z)) + \det(H(r^{[m]})_{n-1}(z))(m+1)m\rho_n\rho_{\bar{n}}\rho(z)^{m-1}$$
$$= b_{m-1}(z)\rho(z)^{m-1} + \det(H(r^{[m]})_{n-1}(z))(m+1)m\rho_n\rho_{\bar{n}}\rho(z)^{m-1}.$$

We need to choose a_{m+1} such that

$$b_{m-1}(z) + (m+1)m \det(H(r^{[m]})_{n-1}(z))\rho_n\rho_{\bar{n}}(z) = 0.$$

From this it will follow that

$$\det(H(r^{[m+1]})(z)) = b_m(z)\rho(z)^m.$$

From the construction, we know that $H(r^{[m]})_{n-1}(z)$ is positive definite with least positive eigenvalue ϵ_m for all $z \in \Omega \setminus \Omega_{\epsilon_m}$. Thus if c_m is large enough so that $|\nabla \delta|^2 c_m \geq 2|b_m(z)|$, then, by choosing ϵ_m small enough, we will easily see that the function

$$\rho_m(z) \equiv r^{[m+1]}(z) + c_m \delta(z)^{m+2} \qquad (6.10.1.2)$$

is strictly plurisubharmonic in $\Omega \setminus \Omega_{\epsilon_m}$ and (6.10.1) holds on $\Omega \setminus \Omega_{\epsilon_m}$. Here $\Omega_t = \{z \in \Omega : \delta(z) < t\}$. Then we use arguments in [CKNS,LI] to extend ρ_m to be defined on Ω and strictly plurisubharmonic on Ω. The proof of the theorem is complete. $\square$

6.12   Application of the Complex Monge–Ampère Equation

In this section, we shall prove Corollary 6.10.2. Let us recall a theorem of Caffarelli et al. [CKNS].

Theorem 6.12.1. *Let Ω be a bounded, strictly pseudoconvex domain in $\mathbb{C}^n$ with C^4 boundary $\partial \Omega$. Let f be a nonnegative function on $\overline{\Omega}$ such that $f(z)^{1/n} \in C^{1,1}(\overline{\Omega})$. Let $H(u)$ denote the complex Hessian matrix of the function u. Then there is a unique plurisubharmonic function $v \in C^{1,1}(\overline{\Omega})$ satisfying*

$$\det H(v)(z) = f(z) \quad \text{for } z \in \Omega;$$

$$v = 0 \qquad \text{for } z \in \delta\Omega . \qquad (6.12.1.1)$$

Proof of Corollary 6.10.2. Let $\varphi : \Omega_1 \to \Omega_2$ be a proper holomorphic mapping, and let ρ_{4n} be the plurisubharmonic defining function for Ω_2, with $m = 4n$, that we constructed in Theorem 6.10.1. We let

$$r(z) = \rho_{4n}(\varphi(z)), \quad z \in \Omega_1.$$

Then we have

$$\det H(r)(z) = \det(H(\rho_{4n})(\varphi(z)))| \det \varphi'(z)|^2 = f(z), \quad z \in \Omega_1.$$

Since $\det(H(\rho_{4n}(\varphi(z))) \leq O(\rho(\varphi(z))^{4n} \leq C\,\mathrm{dist}(z, \partial\Omega_1)^4 n$ and $|\det \varphi'(z)|^2 \leq C\,\mathrm{dist}(z, \partial\Omega)^{-2n}$. We have $f(z)^{1/n} \in C^{1,1}(\overline{\Omega}_1)$ and $f \geq 0$. Theorem 6.12.1 implies that $r \in C^{1,1}(\overline{\Omega}_1) \cap C^2(\Omega_1)$. So

$$\|r\|_{C^{1,1}(\overline{\Omega}_1)} \leq C.$$

It is obvious that $\exp(\rho_{4n})$ is strictly plurisubharmonic in Ω_2 and that there is a constant $\epsilon > 0$ so that

$$H(\exp(\rho_{4n})(w) \geq \epsilon I_n, \quad \text{for all} \ \ w \in \Omega_2.$$

Thus

$$\frac{\partial^2 \exp(r(z))}{\partial z_\ell \partial \overline{z}_\ell} = \sum_{p,\ell=1}^{n} \frac{\partial^2 \exp(\rho_{4n})}{\partial w_p \partial \overline{w}_\ell}(\varphi(z)) \frac{\partial \varphi_p}{\partial z_\ell} \frac{\overline{\partial \varphi_\ell}}{\partial z_\ell}(z)$$

$$\geq \epsilon \sum_{p=1}^{n} \left| \frac{\partial \varphi_p}{\partial z_\ell}(z) \right|^2.$$

This shows that

$$\|\varphi\|^2_{\mathrm{Lip}(\overline{\Omega}_1)} \leq C \lambda_0^{-1} \|\exp(r)\|_{C^{1,1}(\overline{\Omega}_1)}.$$

Thus we have $\det(\varphi'(z)) \in L^\infty(\Omega_1)$.

If we apply $\frac{\partial^2}{\partial z_\ell \partial z_m}$ to $r(z)$ and use the above result, then we have

$$\left| \sum_{p=1}^{n} \frac{\partial \rho_{4n}}{\partial w_p}(\varphi(z)) \frac{\partial^2 \varphi_p}{\partial z_\ell \partial z_m} \right| \leq C. \tag{6.10.2.1}$$

Let $z^0 \in \Omega_1$ be sufficiently near to $\partial \Omega_1$. Without loss of generality, by applying a rotation, we may assume that $z_1^0, \ldots, z_{n-1}^0$ are complex tangential at z^0 and also that at the point $\varphi(z_0)$ the directions $w_1, \ldots, w_{n-1}$ are complex tangential. Thus

$$\frac{\partial}{\partial z_p} \log(\det(\varphi'(z_0)))(z)$$

$$= \sum_{\ell,m=1}^{n} \varphi^{\ell m} \frac{\partial^2 \varphi_m}{\partial z_\ell \partial z_p}(z)$$

$$= \sum_{\ell < n} \varphi^{\ell m} \frac{\partial^2 \varphi_m}{\partial z_\ell \partial z_p}(z) + \sum_{m < n} \varphi^{nm} \frac{\partial^2 \varphi_m}{\partial z_n \partial z_p}(z) + \varphi^{nn} \frac{\partial^2 \varphi_n}{\partial z_n \partial z_p}(z),$$

where $(\varphi^{\ell m})$ is the inverse matrix of $\varphi'(z)$. Since $\varphi \in \mathrm{Lip}_1(\overline{\Omega}_1)$, we have

$$\left| \frac{\partial^2 \varphi_m}{\partial z_\ell \partial z_p}(z_0) \right| \leq C \partial_1(z)^{-1/2}$$

for all $1 \leq \ell \leq n - 1$ and $1 \leq m, p \leq n$. By (6.10.2.1), we have

$$\left| \frac{\partial^2 \varphi_n}{\partial z_n \partial z_p}(z_0) \right| \leq C.$$

Now we consider the terms with $1 \leq m \leq n - 1$. Since $(0, \ldots, 0, \varphi_n)$ is normal at $\varphi(z_0)$ and $z_1, \ldots, z_{n-1}$ are complex tangential to $\delta\Omega_1$ at z_0, we have

$$\left|\frac{\partial\varphi_n}{\partial z_\ell}\right| \leq C\delta_2(\varphi(z_0))^{1/2} \leq C\delta_1(z_0)^{1/2}.$$

Since $\det(\varphi'(z))$ is bounded and

$$|\det(\varphi(z))\varphi^{\ell m}(z)| \leq C,$$

we see that

$$|\det(\varphi'(z)\varphi^{nm}(z_0)\frac{\partial^2\varphi_m}{\partial z_n \partial z_n}(z_0)|$$

$$\leq C\partial_1(z_0)^{-1}\left|\det(\frac{\partial\varphi_p}{\partial z_\ell})_{1 \leq \ell \leq n-1, p \neq m})\right|$$

$$\leq C\partial_1(z)^{-1/2}.$$

Combining all the estimates, we have proved that

$$|\nabla \det(\varphi'(z))| = |\det(\varphi'(z))\nabla \det(\varphi'(z)\log\det(\varphi'(z))| \leq C\delta(z)^{-1/2},$$

for all $z \in \Omega_1$. Hence, $\det(\varphi'(z)) \in \mathrm{Lip}_{1/2}(\overline{\Omega}_1)$. The proof of Corollary 6.10.2 is complete.   □

Note: Combining Corollary 6.10.2 and results in [FOR, NWY, PIH, WEB1, WEB2], we obtain a new proof of the Fefferman's mapping theorem in [FEF1]. Fefferman's theorem is treated in some detail in Sect. 3.3, 6.7.

6.13   An Example of David Barrett

As discussed in Sects. 2.1 and 6.1, Bell's Condition R is a matter of some interest. While there was originally hope that Condition R would hold on any smoothly bounded, pseudoconvex domain, we now know (see [CHR1]) that Condition R fails on the worm domain of Diederich and Fornæss. Before Christ's result, David Barrett [BAR1] exhibited a smoothly bounded, *non-pseudoconvex* domain on which Condition R fails. We treat that example now.

We begin by defining three real-valued functions, r_1, r_2, and c, on the interval $[1, 6]$. Each of these is smooth on $(1, 6)$. They are given by

Fig. 6.2 The function
$y = r_1(x)$

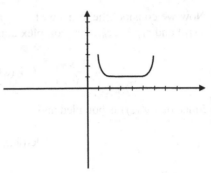

Fig. 6.3 The function
$y = r_2(x)$

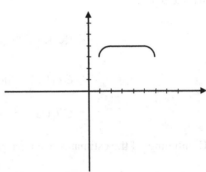

$$r_1(x) = \begin{cases} 3 - \sqrt{x-1} & \text{for} & x \geq 1 \text{ near } 1 \\ \text{decreasing} & \text{for} & 1 \leq x \leq 2 \\ 1 & \text{for} & 2 \leq x \leq 5 \\ \text{increasing} & \text{for} & 5 \leq x \leq 6 \\ 3 - \sqrt{6-x} & \text{for} & x \leq 6 \text{ near } 6. \end{cases}$$

The function is exhibited in Fig. 6.2.
   Second,

$$r_2(x) = \begin{cases} 3 + \sqrt{x-1} & \text{for} & x \geq 1 \text{ near } 1 \\ \text{increasing} & \text{for} & 1 \leq x \leq 2 \\ 4 & \text{for} & 2 \leq x \leq 5 \\ \text{decreasing} & \text{for} & 5 \leq x \leq 6 \\ 3 + \sqrt{6-x} & \text{for} & x \leq 6 \text{ near } 6. \end{cases}$$

The function is exhibited in Fig. 6.3.

Fig. 6.4 The function
$y = c(x)$

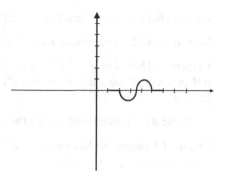

Finally, for a positive integer k,

$$c(x) = \begin{cases} 0 & \text{for} & 1 \leq x \leq 2 \\ \text{decreasing} & \text{for} & 2 \leq x \leq 3 \\ (x-3)^{2k} - 1 & \text{for} & 3 - \epsilon < x < 3 + \epsilon, \epsilon > 0 \text{ small} \\ \text{increasing} & \text{for} & 3 \leq x \leq 4 \\ -(x-4)^{2k} + 1 & \text{for} & 4 - \epsilon < x < 4 + \epsilon, \epsilon > 0 \text{ small} \\ \text{decreasing} & \text{for} & 4 \leq x \leq 5 \\ 0 & \text{for} & 5 \leq x \leq 6. \end{cases}$$

The function is exhibited in Fig. 6.4.
   Now define

$$\Omega\{(z, w) \in \mathbb{C}^2 : 1 < |w| < 6, |z| < r_2(|w|), |z - c(|w|)| > r_1(|w|)\}.$$

Let

$$\Omega_w = \{z \in \mathbb{C} : (z, w) \in \Omega\}$$

be the cross section of Ω at w.
   We note the following:

(a) If $2 \leq |w| \leq 5$, then Ω_w consists of the disc $|z| < 4$ minus a unit disc of varying center.
(b) If $1 < |w| \leq 2$ or $5 \leq |w| < 6$, then the slices Ω_w are annuli collapsing towards the circle $|z| = 3$ at the limiting values $|w| = 1$ and $|w| = 6$.
(c) The domain Ω is smooth near the limiting values $|w| = 1$ and $|w| = 6$ since it is defined there by the inequalities $(|z| - 3)^2 < |w| - 1$ and $(|z| - 3)^2 < 6 - |w|$, respectively.
(d) The union of the Ω_w for $1 \leq |w| \leq 6$ is the punctured disc $0 < |z| < 4$.

We now have three key lemmas. In what follows, we let P denote the Bergman projection on Ω and $\mathcal{O}(\Omega)$ the space of all holomorphic functions on Ω. Further we let $C_c^\infty(\Omega)$ be the C^∞ functions with compact support in Ω.

Lemma 6.13.1. *The space $P(C_c^\infty(\Omega))$ is dense in $L^2(\Omega) \cap \mathcal{O}(\Omega)$.*

Lemma 6.13.2. *The function $1/z$ lies in $L^2(\Omega) \cap \mathcal{O}(\Omega)$.*

Lemma 6.13.3. *Let $p \geq 2 + 1/k$. Any function $g \in L^p(\Omega) \cap \mathcal{O}(\Omega)$ which is independent of the variable w extends to a holomorphic function of z on the disc $|z| < 4$.*

These three lemmas will imply the main result. We now prove the lemmas.

Proof of Lemma 6.13.1. Let $h \in L^2(\Omega) \cap \mathcal{O}(\Omega)$ be orthogonal to $P(C_c^\infty(\Omega))$. Then, for every $\eta \in C_c^\infty(\Omega)$,

$$\int_\Omega \eta \cdot |h|^2 \, dV(z, w) = \int_\Omega (\eta h) \cdot \overline{h} \, dV(z, w) = \int_\Omega P(\eta h) \cdot \overline{h} \, dV(z, w) = 0.$$

If η is taken to be nonnegative, then we must conclude that $h = 0$ on the support of η. Since the choice of η is otherwise arbitrary, we see that $h \equiv 0$. That concludes the proof. □

Proof of Lemma 6.13.2. We calculate that

$$\|1/z\|_{L^2(\Omega)}^2 = \int_\Omega 1/|z|^2 \, dV(z, w)$$

$$= 2\pi \int_1^6 |w| \, d|w| \int_{\Omega_{|w|}} 1/|z|^2 \, dA(z). \tag{6.13.2.1}$$

Now define

$$I(t) = t \int_{D_t} 1/|z|^2 \, dA(z)$$

for t real. Clearly, in view of (6.13.2.1), our job is to show that $\int_1^6 I(t) \, dt < \infty$.

It is plain that, for t away from 3 to 4, $I(t)$ is well behaved (indeed bounded). So we must examine the behavior for t near 3 and for t near 4. For t near 3,

$$\Omega_t \subseteq \{z \in \mathbb{C} : (t - 3)^{2k} < |z| < 4\}.$$

Hence,

$$I(t) \leq t \int_{(t-3)^{2k} < |z| < 4} 1/|z|^2 \, dA(z)$$

$$= 2\pi t \log[4/(t - 3)^{2k}].$$

Certainly the logarithm function is integrable. So $I(t)$ is integrable near 3. A similar argument can be applied to analyze the situation for t near 4. Thus $1/|z|$ is in L^2 and the result is proved. $\square$

Proof of Lemma 6.13.3. This is the trickiest of the three lemmas.

First observe that if $p = \infty$, then the result follows from the Riemann removable singularity theorem. So assume that $p < \infty$, and let $h \in L^p(\Omega) \cap \mathcal{O}(\Omega)$ be independent of w. Write

$$h(z, w) = h(z) = \sum_{j=-\infty}^{\infty} a_j z^j \, .$$

Certainly this series converges for $0 < |z| < 4$. We calculate that

$$\|h\|_{L^p(\Omega)}^p = 2\pi \int_1^6 |w| \, d|w| \int_{\Omega_{|w|}} |h(z)|^p \, dA(z)$$

$$\geq 2\pi \int_{-\epsilon}^{\epsilon} dt \int_{\Omega_{3+t} \cup \Omega_{4+t}} |h(z)|^p \, dA(z) \, .$$

For small t, we see that

$$\Omega_{3+t} \cup \Omega_{4+t} = \{z \in \mathbb{C} : |z| < 4\} \setminus \{z \in \mathbb{C} : |z-1+t^{2k}| < 1, |z+1-t^{2k}| < 1\}$$

$$\supseteq \{z \in \mathbb{C} : 2|t|^{2k} < |z| < 4\} \, .$$

As a result,

$$\|h\|_{L^p(\Omega)}^p \geq 4\pi \int_0^{\epsilon} dt \int_{2t^k < |z| < 4} |h(z)|^p \, dA(z) \, . \qquad (6.13.3.1)$$

Since $h \in L^2(\Omega) \cap \mathcal{O}(\Omega)$, we may use line (6.13.3.1) to calculate that

$$\infty > 4\pi \int_0^{\epsilon} dt \int_{2t^k < |z| < 4} \left| \sum_j a_j z^j \right|^2 dA(z)$$

$$= 4\pi \sum_j |a_j|^2 \int_0^{\epsilon} dt \int_{2t^k < |z| < 4} |z|^{2j} \, dA(z)$$

$$= 8\pi^2 \left\{ |a_{-1}|^2 \int_0^{\epsilon} \log(2t^{-k}) \, dt + \sum_{j \neq -1} |a_j|^2 \frac{4^{j+1}}{2j+2} \int_0^{\epsilon} (4^{j+1} - t^{2(j+1)k}) \, dt \right\} \, .$$

We see then that $a_j = 0$ when $2(j+1)k \leq -1$, in other words when $j \leq -2$. Thus h may be written as $h(z) = b/z + g(z)$, where g is holomorphic in the disc $|z| < 4$.

Certainly $g \in L^p(\Omega)$ (because $1/z \notin L^2$). But line (6.13.3.1) now tells us that

$$\|1/z\|_{L^p(\Omega)} \geq 4\pi \int_0^\epsilon dt \int_{2t^k < |z| < 4} |z|^{-p} \, dA(z)$$

$$= 8\pi^2 \cdot \frac{2^{2-p}}{2-p} \int_0^\epsilon \left(2^{2-p} - t^{(2-p)k}\right) dt$$

$$= +\infty$$

if $(2 - p)k \leq -1$, that is, if $p \geq 2 + 1/k$. Thus, the only way that h can be in $L^p(\Omega)$ is if $b = 0$. That proves the lemma. $\square$

Theorem 6.13.4. *Let $p \geq 2 + 1/k$. Then the space $P(C_c^\infty(\Omega))$ is not a subset of $L^p(\Omega)$.*

Proof: Lemma 6.13.1 tells us that it suffices to show that $L^p(\Omega) \cap \mathcal{O}(\Omega)$ is not dense in $L^2(\Omega) \cap \mathcal{O}(\Omega)$ when $p \geq 2 + 1/k$. By Lemma 6.13.2, it thus suffices to show that the function $1/z$ cannot be approximated in the L^2 topology by functions in $L^p(\Omega) \cap \mathcal{O}(\Omega)$.

Seeking a contradiction, let us then suppose that we have functions $\{h_j\}_{j=1}^\infty$ in $L^p(\Omega) \cap \mathcal{O}(\Omega)$ with $h_j \to 1/z$ in $L^2(\Omega)$. Set

$$g_j(z, w) = \frac{1}{2\pi} \int_0^{2\pi} h_j(z, e^{i\theta} w) \, d\theta \,.$$

Then $g_j \in L^p(\Omega) \cap \mathcal{O}(\Omega)$ and $g_j \to 1/z$ in $L^2(\Omega)$. Moreover, each g_j is constant on circles of the form $\{(z, e^{i\theta} w) : 0 \leq \theta \leq 2\pi\}$; thus each g_j is locally independent of w.

We claim that actually each g_j is *globally* independent of w. We already know that g_j is independent of w in a neighborhood of the product set

$$\{(z, w) \in \mathbb{C}^2 : 1 < |w| < 6, |z| = 3\} \subseteq \Omega \,.$$

Thus for any two points w_1, w_2 in the annulus $1 < |w| < 6$, the functions $g_j(\cdot, w_1)$ and $g_j(\cdot, w_2)$ have as common domain of definition the connected open set $\Omega_{w_1} \cap \Omega_{w_2}$. Since the two functions agree near the circle $|z| = 3$, we have that $g_j(z, w_1) = g_j(z, w_2)$ whenever (z, w_1) and (z, w_2) are in Ω, as we wished to show.

As a consequence, Lemma 6.13.3 tells us that the g_j extend to holomorphic functions of z on the disc $|z| < 4$, so that the residue

$$\frac{1}{2\pi i} \int_{|z|=3} g_j(z) \, dz$$

must vanish. But $g_j \to 1/z$ uniformly on compact subsets of Ω, so that

$$\frac{1}{2\pi i} \int_{|z|=3} g_j(z)\, dz \;\to\; \frac{1}{2\pi i} \int_{|z|=3} \frac{1}{z}\, dz = 1 \,.$$

This contradiction proves the theorem. □

This example of Barrett has the advantage of being elementary and accessible. But it must be stressed that the domain he produces is *not* pseudoconvex. As previously noted, M. Christ has given us the more dramatic and important example in that he shows that the Diederich–Fornæss worm domain does not satisfy Condition R (see [CHS] for information about the worm). Y.-T. Siu [SIU] has offered another proof of Christ's result. Neither proof is easy.

Here we shall offer a few words about why Christ's proof works. We begin with a quick review of the properties of the worm. Refer to Sects. 6.1 and 6.2 for basic properties of worm domains.

Now we turn to Christ's theorem. In order to discuss the failure of Condition R on the Diederich–Fornæss worm domain, we recall the basic facts about the $\bar{\partial}$-Neumann problem.

Let $\Omega \subset \mathbb{C}^n$ be a bounded domain with smooth boundary and let ρ be a smooth defining function for Ω. The $\bar{\partial}$-Neumann problem on Ω is a boundary value problem for the elliptic partial differential operator

$$\Box = \bar{\partial}\bar{\partial}^* + \bar{\partial}^*\bar{\partial} \,.$$

Here $\bar{\partial}^*$ denotes the L^2-Hilbert space adjoint of the (unbounded) operators $\bar{\partial}$. Of course $\Box$ acts componentwise, just as a multiple of the Laplacian.

In order to apply $\Box$ to a form or current u, one needs to require that $u, \bar{\partial}u \in \mathrm{dom}(\bar{\partial}^*)$. These conditions translate into two differential equations on the boundary for u and they are called the $\bar{\partial}$-*Neumann boundary conditions* [FOK] or [TRE]. These equations are

$$u \lrcorner \bar{\partial}\rho = 0, \quad \text{and} \quad \bar{\partial}u \lrcorner \bar{\partial}\rho = 0, \quad \text{on } \partial\Omega \,. \tag{6.13.5}$$

Thus the equation $\Box u = f$ becomes a boundary value problem.

$$\Box u = f \quad \text{on } \Omega$$
$$u \lrcorner \bar{\partial}\rho = 0, \, \bar{\partial}u \lrcorner \bar{\partial}\rho = 0 \quad \text{on } \partial\Omega \,. \tag{6.13.6}$$

This is an equation defined on *forms*. The significant problem is for $(0, 1)$-forms, and we restrict to this case in the present discussion.

It follows from Hörmanders' original article on the solution of the $\bar{\partial}$-equation [HOR1] that the $\bar{\partial}$-Neumann problem is always solvable on a bounded pseudoconvex domain Ω in $\mathbb{C}^n$ with smooth boundary for any data $f \in L^2(\Omega)$. We denote by N—the Neumann operator—such a solution operator. Moreover, N turns out to be continuous in the L^2-topology:

$$\|Nu\|_{L^2} \leq c\|u\|_{L^2} .$$

An important formula of Kohn (see Sect. 6.7) says that

$$P = I - \overline{\partial}^* N \overline{\partial} .$$

The proof of this is a formal calculation—see [KRA4]. Important work by Boas and Straube [BOS2] essentially established that the Neumann operator N has a certain regularity (i.e., it maps some Sobolev space W^s to itself, for instance) if and only if P will have the same regularity property. In particular, if N is continuous on a Sobolev space W^s for some $s > 0$ (of $(0, 1)$-forms), then the Bergman projection P is continuous on the same Sobolev space W^s (of functions).

Such regularity is well known to hold on strictly pseudoconvex domains [FOK, KRA4]. In addition, Catlin proved a similar regularity result on finite type domains (see [CAT1, CAT2, KRA1]).

Michael Christ's milestone result [CHR1] has proved to be of central importance for the field. It demonstrates concretely the seminal role of the worm and points to future directions for research. Certainly the research program being described here, including the calculations in [KRP1], is inspired by Christ's work.

Christ's work is primarily concerned with global regularity or global hypoellipticity. A partial differential operator L is said to be *globally hypoelliptic* if whenever $Lu = f$ and f is globally C^∞, then u is globally C^∞. We measure regularity, here and in what follows, using the standard Sobolev spaces W^s, $0 < s < \infty$ (see [KRA4, HOR2]).

Christ's proof of the failure of global hypoellipticity is a highly complex and recondite calculation with pseudodifferential operators. We cannot replicate it here. But the ideas are so important that we feel it worthwhile to outline his argument. We owe a debt to the elegant and informative article [CHR2] for these ideas.

As a boundary value problem for an elliptic operator, the $\overline{\partial}$-Neumann problem may be treated by Caldéron's method of reduction to a pseudodifferential equation on $\partial\Omega$. The sources [HOR3, TRE] give full explanations of the classical ideas about this reduction. In the more modern reference [CNS], Chang et al. elaborate the specific application of these ideas to the $\overline{\partial}$-Neumann problem in $\mathbb{C}^2$. (Thus in the remaining part of this discussion, Ω will denote a smoothly bounded pseudoconvex domain in $\mathbb{C}^2$.) The upshot is that one reduces the solution of the equation $\Box u = f$ to the solution of an equation $\Box^+ v = g$ on the boundary. Here, u and f are $(0, 1)$-forms, while v and g are sections of a certain complex line bundle on $\partial\Omega$. (The fact that this bundle is one dimensional is a consequence of the inclusion $\Omega \subset \mathbb{C}^2$.)

To be more explicit, the solution u of (6.13.6) can be written as $u = \mathcal{G}f + \mathcal{R}v$, where $\mathcal{G}$ is *Green's operator* and $\mathcal{R}$ is the *Poisson operator*[3] for the operator $\square$ and v is chosen in such a way as to satisfy the boundary conditions. In fact,

$$\square\left(\mathcal{G}f + \mathcal{R}v\right) = f + 0 = f \qquad \text{on } \Omega$$

$$\left(\mathcal{G}f + \mathcal{R}v\right)\lrcorner\bar{\partial}\rho = v\lrcorner\bar{\partial}\rho = 0 \qquad \text{on } \partial\Omega$$

$$\bar{\partial}\left(\mathcal{G}f + \mathcal{R}v\right)\lrcorner\bar{\partial}\rho = \bar{\partial}\mathcal{G}\lfloor\bar{\partial}\rho + \bar{\partial}v\lrcorner\bar{\partial}\rho = 0 \qquad \text{on } \partial\Omega .$$

The section v has two components, but one of these vanishes because of the first $\bar{\partial}$-Neumann boundary condition. The second $\bar{\partial}$-Neumann boundary condition may be written as an equation $\square^{+}v = g$ on $\partial\Omega$, where $\square^{+}$ is a pseudodifferential operator of order 1. Also we note that $g = (\bar{\partial}\mathcal{G}f\lrcorner\bar{\partial}\rho)$ restricted to $\partial\Omega$.

Christ's argument begins with a real-variable model for the $\bar{\partial}$-Neumann problem that meshes well with the geometry of the boundary of the worm domain $\mathcal{W}$.

Let M be the two-torus $\mathbb{T}^2$ and let X, Y two smooth real vector fields on M. Fix a coordinate patch V_0 in M and suppose that V_0 has been identified with $\{(x, t) \in (-2, 2) \times (-2\delta, -2\delta)\} \subset \mathbb{R}^2$. Let $J = [-1, 1] \times \{0\} \subset V_0$.

Call a piecewise smooth path γ on M *admissible* if every tangent to γ is in the span of X, Y. Assume that:

1. The vector fields $X, Y, [X, Y]$ span the tangent space to M at every point of $M \setminus J$.
2. In V_0, $X \equiv \partial_x$ and $Y \equiv b(x, t)\partial_t$.
3. For all $|x| \le 1$ and $|t| \le \delta$, we have that $b(x, t) = \alpha(x)t + \mathcal{O}(t^2)$, where $\alpha(x)$ is nowhere vanishing.

It follows then that every pair $x, y \in M$ is connected by an admissible path.

Theorem 6.13.7. *With X, Y, M as above, let L be any partial differential operator on M of the form $L = -x^2 - Y^2 + a$, where $a \in C^\infty(M)$ and*

$$\|u\|^2 \le C\langle Lu, u\rangle \tag{6.13.7.1}$$

for all $u \in C^2(M)$. Then L is not globally regular in C^∞.

We note that our hypotheses, particularly inequality (6.13.7.1), imply that L has a well-defined inverse L^{-1} which is a bounded linear operator on $L^2(M)$.

The following theorem gives a more complete, and quantitative, version of this result:

---

[3]Thus $\mathcal{G}$ is the solution operator for the elliptic boundary value problem $\square\,(\mathcal{G}f) = f$ on Ω and $\mathcal{G}f = 0$ on $\partial\Omega$, while $\mathcal{R}$ is the solution operator for the elliptic boundary value problem $\square\,(\mathcal{R}v) = 0$ on Ω and $\mathcal{R}v = v$ on $\partial\Omega$.

Theorem 6.13.8. *Let X, Y, M, L be as above. Then L has the following global properties:*

(a) *There is a positive number s_0 such that, for every $0 < s < s_0$, L^{-1} preserves $W^s(M)$.*
(b) *For each $s > s_0$, L^{-1} fails to map $C^\infty(M)$ to $W^s(M)$.*
(c) *There is a sequence of values $s < r$ tending to infinity such that if $u \in W^s(M)$ satisfies $Lu \in H^r(M)$ then $u \in H^r$;*
(d) *There are arbitrarily large values of s with a constant $C = C_s$ such that if $u \in W^s(M)$ is such that $Lu \in W^s(M)$, then*

$$\|u\|_{W^s} \le C \|Lu\|_{W^s} . \qquad (6.13.8.1)$$

(e) *For each value of s as in part (d), $\{ f \in W^s(M) : L^{-1} f \in W^s(M)\}$ is a closed subspace of W^s with finite codimension.*

The proof of Theorem 6.13.8.1 breaks into two parts. The first part consists of proving the a priori inequality (17). The second part, following ideas of Barrett in [BAR2], shows that, for any $s \ge s_0$, the operator L cannot be exactly regular on $W^s(M)$. We refer the reader to [CHR1] for the details. Section 8 of [CHR2] also provides a nice outline of the analysis.

The next step is to reduce the analysis of the worm domain, as defined in our Sects. 2.1 and 6.1, to the study of the manifold M as above. With this idea in mind, we set $\overline{L} = \overline{\partial}_b$ and L its complex conjugate. The characteristic variety[4] of $\overline{L}$ is a real line bundle Σ that splits smoothly as two rays: $\Sigma = \Sigma^+ \cup \Sigma^-$.

The principal symbol of $\square^+$ vanishes only on Σ^+ that is half the characteristic variety. We may compose $\square^+$ with an elliptic pseudodifferential operator of order $+1$ to change $\square^+$ to the form

$$\mathcal{L} = \overline{L}L + B_1\overline{L} + B_2L + B_3 \qquad (6.13.9)$$

microlocally in a conical neighborhood of Σ^+, where each B_j is a pseudodifferential operator with order not exceeding 0. Since $\square^+$ is elliptic on the complement of Σ^+, our analysis may thus be microlocalized to a small conical neighborhood of Σ^+.

For a worm domain $\mathcal{W}$, there is circular symmetry in the second variable. This induces a natural action on functions and on forms (as indicated in Sect. 6.1). As indicated earlier, the Hilbert space of square-integrable $(0, k)$-forms has the orthogonal decomposition $\oplus_j \mathcal{H}_k^j$. The Bergman projection and the Neumann operator preserve $\mathcal{H}_0^j$ and $\mathcal{H}_1^j$. We now have the following key result:

---

[4]The characteristic variety of a pseudodifferential operator is the conic subset of the cotangent bundle on which its principal symbol vanishes.

Proposition 6.13.10. *Let* $\mathcal{W}$ *be the worm. Then there is a discrete subset* $S \subset \mathbb{R}^+$ *such that, for each* $s \notin S$ *and each* $j \in \mathbb{Z}$, *there is a constant* $C = C(s, j) < \infty$ *such that, for each* $(0, 1)$ *form* $u \in \mathcal{H}_1^j \cap C^\infty(\overline{\mathcal{W}})$ *such that* $N u \in C^\infty$, *it holds that*

$$\|Nu\|_{W^s(\mathcal{W})} \leq C \cdot \|u\|_{W^s(\mathcal{W})}.$$

The operators $\mathcal{L}$, $\overline{L}$, L, B_j in (6.13.9) may be constructed so as to commute with the circle action in the second variable; hence, they will preserve each $\mathcal{H}^j$. In summary, for each j, the action of $\mathcal{L}$ on $\mathcal{H}^j(\partial\mathcal{W})$ may be identified with the action of an operator $\mathcal{L}_j$ on $L^2(\partial\mathcal{W}/S^1)$.

Of course $\partial\mathcal{W}$ is three dimensional; hence, $\partial\mathcal{W}/S^1$ is a real two-dimensional manifold. It is convenient to take coordinates (x, θ, t) on $\partial\mathcal{W}$ so that

$$z_2 = \exp(x + i\theta) \quad \text{and} \quad z_1 = \exp(i2x)(e^{it} - 1);$$

here $|\log|z_2|^2| \leq r$ and $\mathcal{L}_j$ takes the form $\overline{L}L + B_1\overline{L} + B_2 L + B_3$ (just as in (6.13.9)!). In this last formula, $\overline{L}$ is a complex vector field which has the form $\overline{L} = \partial_x + it\alpha(t)\partial_t$, where $|x| \leq r/2$, $\alpha(0) \neq 0$, and each B_j is a classical pseudodifferential operator of order not exceeding 0—depending on j in a nonuniform manner.

We set $J = \{(x, t) : |x| \leq r/2, t = 0\}$ and write $\overline{L} = X + iY$; then the vector fields X, Y, $[X, Y]$ span the tangent space to $\partial\mathcal{W}/S^1$ at each point of the complement of J and are tangent to J at every point of J. We conclude that the operator $\mathcal{L}_j$ on $\partial\mathcal{W}/S^1$ is quite similar to the two-dimensional model that we discussed above.

There are two complications which we must note (and which are not entirely trivial): (1) There are pseudodifferential factors, and the reduction of the $\bar{\partial}$-Neumann problem to $\mathcal{L}$, and thereafter to $\mathcal{L}_j$, requires only a microlocal a priori estimate for $\mathcal{L}_j$ in a conic subset of phase space; (2) the lower-order terms $B_1\overline{L}$, $B_2 L$, B_3 are *not* negligible; indeed, they determine the values of the exceptional Sobolev exponents, but the analysis can be carried out for these terms as well.

It should be noted that a special feature of the worm is that the rotational symmetry in z_2 makes possible (as we have noted) a reduction to a two-dimensional analysis, and this in turn produces a certain convenient ellipticity. There is no uniformity of estimates with respect to j, but the analysis can be performed for each fixed j.

6.14 The Bergman Kernel as a Hilbert Integral

Perhaps the most important and ubiquitous integral operator in all of analysis is the *Hilbert transform*:

$$f \longmapsto \int_{\mathbb{R}} \frac{f(t)}{x - t} \, dt \, . \tag{6.14.1}$$

This integral is interesting because it does not converge in the sense of the Lebesgue integral. That is to say, the singularity of the function $1/(x-t)$ is non-integrable. So the integral (6.14.1) must be interpreted in the sense of the *Cauchy principal value*. It is the most classical example of a *singular integral*. Essential to the understanding of this integral is that the function $1/(x - t)$ has certain cancellation built into it. In simplest terms, $1/t$ is odd, so it integrates to 0 on "spheres" (in dimension one a sphere is just a pair of points centered at the origin) centered at 0. Because of this cancellation property, the convolution with the kernel $1/t$ induces a distribution. Thus the integral turns out to be tractable.

The concept of singular integral, at least in its classical form, really only makes sense on a space with a homogeneous structure. Such a structure arises naturally from a (Euclidean) group acting on the space. For $\mathbb{R}$ or $\mathbb{R}^N$, the natural group to consider is the group of translations (although rotations and dilations also play a distinctive role). The translations give rise to the notion of convolution, and the Hilbert transform (6.14.1) is a convolution operator.

The Hilbert transform is important in classical analysis because it governs the norm convergence (and, in a more subtle form, also the pointwise convergence) of Fourier series. It also arises in the study of the regularity of important partial differential operators. It is a fundamental result of M. Riesz that the Hilbert transform is bounded on L^p for $1 < p < \infty$. It is unbounded on L^1 and L^∞.

The Bergman kernel does *not* fit into the context just described. This kernel is defined on a *domain* Ω. There certainly is no translation structure and no concept of convolution. A better model for the Bergman kernel is the classical *Hilbert integral*. In its most basic form, the Hilbert integral is the operator given by

$$f \longmapsto Hf \equiv \int_0^\infty \frac{f(t)}{x + t} \, dt \, .$$

Here we think of f as a function with domain $[0, \infty)$ and the Hilbert integral Hf also has domain $[0, \infty)$.

Certainly, on the half line, the kernel $1/(x + t)$ has no cancellation built into it. Nonetheless, H is bounded on L^p for $1 < p < \infty$. Our treatment of the Hilbert integral derives from [PHS1], [PHS2].

Proposition 6.14.2. *Let* $1 < p < \infty$. *We have the estimate*

$$\int_0^\infty |Hf(x)|^p \, dx \le C_p \int_0^\infty |f(x)| \, dx \, .$$

Proof: Let $\epsilon > 0$ and consider the integral

$$\int_0^\infty \frac{\tau^{-\epsilon}}{x + \tau} \, d\tau \, . \tag{6.14.2.1}$$

The substitution $\tau = x\mu$ gives that

$$(6.14.2.1) = x^{-\epsilon} \cdot \int_0^\infty \frac{\mu^{-\epsilon}}{1+\mu}\, d\mu .$$

If $0 < \epsilon < 1$ then the new integrand is clearly integrable at the origin. Also it is of size $\mu^{-1-\epsilon}$ at infinity, so it is integrable there. Thus

$$(6.14.2.1) = \gamma_\epsilon \cdot x^{-\epsilon} .$$

Here γ_ϵ is a positive constant whose value can be determined, but is of no intrinsic interest.

Now we write the Hilbert integral as

$$Hf(x) = \int_0^\infty \left(\frac{t^{\epsilon/q} f(t)}{(x+t)^{1/p}} \right) \cdot \left(\frac{t^{-\epsilon/q}}{(x+t)^{1/q}} \right) dt .$$

Here p and q are conjugate Hölder's exponents: $1/p + 1/q = 1$ with $1 < p, q < \infty$. Now apply Hölder's inequality with exponent p on the first integral and exponent q on the second integral. The result is

$$|Hf(x)| \leq \left(\int_0^\infty \frac{t^{\epsilon p/q} |f(t)|^p}{x+t}\, dt \right)^{1/q} \cdot \left(\int_0^\infty \frac{t^{-\epsilon}}{x+t}\, dt \right)^{1/q} .$$

So we see that

$$|Hf(x)|^p \leq \gamma_\epsilon^{p/q} \cdot x^{-\epsilon p/q} \cdot \left(\int_0^\infty \frac{t^{\epsilon p/q} |f(t)|^p}{x+t}\, dt \right) .$$

We integrate both sides with respect to x on the interval $[0, \infty)$ and apply Fubini. Of course we must use (6.14.2.1) again (with ϵ replaced by $\epsilon p/q$) to see that

$$\int_0^\infty |Hf(x)|^p\, dx \leq C_p \int_0^\infty |f(x)|\, dx .$$

Note that, by the second application of (6.14.2.1), the term $t^{\epsilon p/q}$ cancels out. This all works provide $0 < \epsilon < 1$ and $0 < \epsilon p/q < 1$. That completes the proof.   □

Next we turn to a higher-dimensional version of the Hilbert integral:

Proposition 6.14.3. *Let $L(x,t)$ be a nonnegative kernel for $(x,t) \in \mathbb{R}^N \times \mathbb{R}_+$ which satisfies:*

1. $L(\lambda x, \lambda t) = \lambda^{-(N+1)} \cdot L(x,t)$ *for all $\lambda > 0$, all $(x,t) \in \mathbb{R}^N \times \mathbb{R}_+$.*
2. $\int_{\mathbb{R}^N} L(x,1)\, dx = C < \infty$.

Assume that the kernel $K(x, y, t, u)$ *for* $(x, y, t, u) \in \mathbb{R}^N \times \mathbb{R}^N \times \mathbb{R}_+ \times \mathbb{R}_+$ *satisfies*

$$|K(x, y, t, u)| \le L(x - y, t + u).$$

Define

$$H f(x, t) = \int_{\mathbb{R}^N} \int_0^\infty K(x, y, t, u) f(y, u) \, dy \, du, \quad (x, t) \in \mathbb{R}^N \times \mathbb{R}_+.$$

Then

$$\int_{\mathbb{R}^N} \int_0^\infty |H(f)(x, t)|^p \, dx \, dt \le \tilde{C}_p \int_{\mathbb{R}^N} \int_0^\infty |f(y, u)|^p \, dy \, du \qquad (6.14.3.1)$$

whenever $1 < p < \infty$ *and* $f \in L^p(\mathbb{R}_+^{N+1})$.

Proof: It is enough to prove (6.14.3.1) with K replaced by $L(x - y, t + u)$ and we do so. By homogeneity and change of variables in the integral,

$$\int_{\mathbb{R}^N} L(x, t + u) \, dx = (t + u)^{-(N+1)} \int_{\mathbb{R}^N} L(\frac{x}{t + u}, 1) \, dx$$

$$= (t + u)^{-1} \int_{\mathbb{R}^N} L(x, 1) \, dx$$

$$= C \cdot (t + u)^{-1}.$$

Thus our estimate of line (6.14.3.1) gives

$$\int_{\mathbb{R}^N} \int_0^\infty L(x, t + u) u^{-\epsilon} \, du = C \cdot \gamma_\epsilon \cdot t^{-\epsilon}. \qquad (6.14.3.2)$$

Now, just as in the proof of Proposition 6.14.2, we see that

$$|H(f)(x, t)| = \left| \int_{\mathbb{R}^N} \int_0^\infty L(x - y, t + u) f(y, u) \, dy \, du \right|$$

$$= \left| \int_{\mathbb{R}^N} \int_0^\infty \left[u^{\epsilon/q} \cdot f(y, u) \cdot (L(x - y, t + u))^{1/p} \right] \right.$$

$$\left. \times \left[u^{-\epsilon/q} \cdot (L(x - y, t + u))^{1/q} \right] \, dy \, du \right|$$

$$\le \int_{\mathbb{R}^N} \int_0^\infty u^{\epsilon p/q} |f(y, u)|^p L(x - y, t + u) \, dy \, du^{1/p}$$

$$\times \int_{\mathbb{R}^N} \int_0^\infty u^{-\epsilon} L(x - y, t + u) \, dy \, du^{1/q}$$

$$\le (C \gamma_\epsilon t^{-\epsilon})^{1/q} \cdot \int_{\mathbb{R}^N} \int_0^\infty u^{\epsilon p/q} |f(y, u)|^p L(x - y, t + u) \, dy \, du^{1/p}.$$

As a result,

$$|Hf(x,t)|^p \le (C\gamma_\epsilon t^{-\epsilon})^{p/q} \cdot \int_{\mathbb{R}^N} \int_0^\infty u^{\epsilon p/q} |f(y,u)|^p \cdot L(x-y,t+u)\,dy\,du\,.$$

Now we integrate in x and t, apply Fubini, and utilize (6.14.3.2). Then, just as in the proof of Proposition 6.14.2, we get the desired estimate. □

Now let us split $\mathbb{R}^N = \mathbb{R}^p \times \mathbb{R}^q$, where $x = (x^p, x^q)$, $x \in \mathbb{R}^N$, $x^p \in \mathbb{R}^p$, and $x^q \in \mathbb{R}^q$. Of course $N = p + q$. For $s > 0$ we consider the nonisotropic dilations

$$(x^p, x^q) \longmapsto (\lambda x^p, \lambda^s x^q)\,, \quad \lambda > 0\,.$$

We assume that $\mathbb{R}^N$ is equipped with a group structure so that Lebesgue measure $dx = dx^p dx^q$ is the bi-invariant Haar measure. (An example of such a group structure is ordinary translation. Another is the Heisenberg group structure, which we shall consider below.) We denote the group operation by $\cdot$ and the group inverse of g by g^{-1}. Now a similar proof to that of the last proposition gives the following generalization:

Proposition 6.14.4. *Let $L(x,t)$ be a nonnegative kernel satisfying:*

1. $L(\lambda x^p, \lambda^s x^q, \lambda^s t) = \lambda^{-(p+(q+1)s)} L(x^p, x^q, t)$.
2. $\int_{\mathbb{R}^N} L(x,1)\,dx = C < \infty$.
3. $|K(x,y,t,u)| \le L(y^{-1} \cdot x, t+u)$.

Then the operator

$$H(f)(x,t) = \int_{\mathbb{R}^N} \int_0^\infty K(x,y,t,u) f(y,u)\,dy\,du$$

is bounded on L^p.

The papers [PHS1] and [PHS2] apply these results on the Hilbert integral to derive boundedness theorems for the Bergman integral. The methods are rather detailed and technical, and we cannot treat them here.

Exercises

1. Show that the smooth worm domain may be exhausted by biholomorphic copies of the non-smooth worm. [**Hint:** See [BAR2] for more on this idea.]
2. Let $h^2(\Omega)$ be the real-valued harmonic functions on a domain Ω in $\mathbb{R}^N$ which satisfy a natural square integrability property. Show that h^2 is a Hilbert space with reproducing kernel in the sense of Aronszajn. How is the corresponding kernel related to the Poisson kernel?

3. Use the Fefferman asymptotic expansion for the Bergman kernel to actually calculate an asymptotic formula for the holomorphic sectional curvature of the Bergman metric on a strictly pseudoconvex domain. See [KLE] for details.

4. Show that the complex Monge–Ampère operator is invariant under biholomorphic mappings.

5. Calculate explicitly the Bergman kernel for the Sobolev space W^1 on the unit disc in the plane.

6. Calculate explicitly the Bergman kernel for the Sobolev space W^1 on the unit ball in complex space.

7. The Kohn projection formula easily shows that Condition R holds on any smoothly bounded domain in the plane. Explain.

8. Consider the domains

$$\Omega_m = \{z \in \mathbb{C} : \operatorname{Re} z > 0, |\operatorname{Im} z| < (\operatorname{Re} z)^m\}$$

for m a positive, even integer. What can you say about Condition R on such a domain?

9. The original Diederich–Fornæss worm domain is strictly pseudoconvex except on a boundary set of $(2n - 1)$-dimensional measure zero. Explain.

10. Consider the linear space $C^\infty(\overline{\Omega})$ for a smoothly bounded domain Ω in $\mathbb{R}^N$.

11. Is there any smoothly bounded domain in the complex plane on which Condition R fails?

12. Show that the Diederich–Fornæss worm domain is topologically trivial.

13. Show that the Diederich–Fornæss worm domain is not biholomorphic to the unit ball.

Chapter 7
Curvature of the Bergman Metric

We begin with a little introductory material on the scaling method. Then we use these ideas to discuss Klembeck's theorem about the boundary asymptotics of the curvature of the Bergman metric on a strictly pseudoconvex domain.

7.1 What is the Scaling Method?

See [GKK] for the details behind the discussion here. The presentation here owes much to that reference. If a bounded domain Ω in $\mathbb{C}^n$ has a noncompact automorphism group, then all the orbits of the automorphism group are noncompact. Thus each orbit must "go out to the boundary" of the domain Ω, since orbits are closed in Ω. Boundary orbit accumulation points are pseudoconvex (see [GRK10]). Under some reasonable hypothesis on the domain as a whole, e.g., that it is a domain of holomorphy, one expects in general terms that localization properties of the $\overline{\partial}$-operator would imply that the geometric essentials of the domain would be localized at boundary points. What does this mean? If a sequence of automorphisms $\varphi_j \in \mathrm{Aut}(\Omega)$, $j = 1, 2, \ldots$, has, for some $X \in \Omega$, $\lim \varphi_j(X) = P \in \partial\Omega$, then the structure of Ω as a whole should be controlled by the nature of $\partial\Omega$ near P.

It has been conjectured by Greene and Krantz that, when such a P is a C^∞ boundary point, it must be of "finite type" in the sense of D'Angelo (see [KRA1, Sect. 11.5] [DAN1, DAN4]). In this case, rather precise information on $\overline{\partial}$-localization is also available (cf., [CAT2] and [CAT3]).

It has turned out that, for many purposes, the study of these matters has best been achieved by way of a renormalized normal families process rather than by looking at $\overline{\partial}$ results as such (although it should be stressed that future work may give greater emphasis to $\overline{\partial}$ methods). The collection of techniques and results of this sort has become known as the *scaling method*.[1] This chapter is devoted

---

[1] In some geometric contexts this technique is also known as the "method of flattening." We thank M. Gromov for this comment.

S.G. Krantz, *Geometric Analysis of the Bergman Kernel and Metric*,
Graduate Texts in Mathematics 268, DOI 10.1007/978-1-4614-7924-6_7,
© Springer Science+Business Media New York 2013

to exploring this method and its results. In particular, we shall describe a new result, the asymptotic constancy of holomorphic sectional curvature for C^2 strictly pseudoconvex domains. This result improves the result of Klembeck [KLE], which used the Fefferman expansion and hence required C^∞ boundary. See [GKK] for the origin of these ideas, and of this approach. This asymptotic constancy yields important consequences for automorphism groups.

A good source for basic ideas about the scaling method, together with some elegant applications, is [PIN].

In the Appendix to this chapter, we describe a one-dimensional scaling technique. There is nothing new here, but it offers a microcosm of what this chapter is about. In the next section we offer the fully developed version of scaling in higher dimensions.

7.2   Higher Dimensional Scaling

7.2.1   Nonisotropic Scaling

We now continue our discussion in complex dimension 2. Considerations in dimension n are analogous but a bit more tedious. We demonstrate the scaling of the complex two-dimensional ball

$$B^2 = \{(z_1, z_2) \in \mathbb{C}^2 \mid |z_1|^2 + |z_2|^2 < 1\}$$

at the boundary point $(1, 0)$.

Denote by a_j a sequence of real numbers satisfying

$$0 < a_j < a_{j+1} < 1 \ \forall j = 1, 2, \ldots$$

and

$$\lim_{j \to \infty} a_j = 1,$$

and let $q_j = (a_j, 0)$ for each $j = 1, 2, \ldots$. Then consider the translation

$$T(z_1, z_2) = (z_1 - 1, z_2).$$

The domain $T(B^2)$ is now defined by the inequality

$$|\zeta_1 + 1|^2 + |\zeta_2|^2 < 1$$

or, equivalently, by

$$2 \operatorname{Re} \zeta_1 < -|\zeta_1|^2 - |\zeta_2|^2.$$

Notice that the mapping

$$\varphi_j(z_1, z_2) = \left(\frac{z_1 + a_j}{1 + a_j z_1}, \frac{\sqrt{1 - |a_j|^2}}{1 + a_j z_1} z_2 \right)$$

is an automorphism of B^2 satisfying $\varphi_j(0) = q_j$ for every j. Finally consider

$$L_j(z_1, z_2) = \left(\frac{z_1}{\lambda_j}, \frac{z_2}{\sqrt{\lambda_j}} \right)$$

where $\lambda_j = 1 - a_j$ for each j. Imitating the one-dimensional case (see the Appendix), we consider the scaling sequence

$$\psi_j(z_1, z_2) = L_j \circ T \circ \varphi_j(z_1, z_2).$$

Notice here that L_j is a dilation but, unlike the one-dimensional case, it is nonisotropic in the sense that the eigenvalues are not uniformly comparable.

We now compute the limit map $\hat{\psi}(z_1, z_2) \equiv \lim_{j \to \infty} \psi_j(z_1, z_2)$ and the set $\hat{\psi}(B^2)$. A direct computation yields the following:

$$\psi_j(z_1, z_2) = \left(\frac{1}{\lambda_j} \left(\frac{z_1 + a_j}{1 + a_j z_1} - 1 \right), \frac{1}{\sqrt{\lambda_j}} \cdot \frac{\sqrt{1 - a_j^2}}{1 + a_j z_1} \cdot z_2 \right)$$

$$= \left(\frac{z_1 - 1}{1 + a_j z_1}, \frac{\sqrt{1 + a_j}\, z_2}{1 + a_j z_1} \right).$$

Therefore we see immediately that

$$\hat{\psi}(z_1, z_2) = \left(\frac{z_1 - 1}{z_1 + 1}, \frac{\sqrt{2}\, z_2}{z_1 + 1} \right)$$

and that ψ_j converges to $\hat{\psi}$ uniformly on compact subsets of B^2.

Observe that the map $\hat{\psi} : B^2 \to \mathbb{C}^2$ is an injective holomorphic mapping and that its image coincides with the Siegel half-space

$$\mathcal{U} = \{(z_1, z_2) \in \mathbb{C}^2 \mid 2\, \mathrm{Re}\, z_1 < -|z_2|^2\}.$$

Therefore $\hat{\psi} : B^2 \to \mathcal{U}$ is in fact a biholomorphic mapping.

Observe also that one can see the convergence of the sets $\psi_j(B^2)$ here. A direct argument yields

$$\psi_j(B^2) = L_j \circ \alpha_j \circ \varphi_j(B^2)$$
$$= L_j \circ \alpha_j(B^2)$$
$$= L_j \left(\{ z \in \mathbb{C}^2 \mid |z_1 + 1|^2 < 1 - |z_2|^2 \} \right)$$
$$= L_j \left(\{ z \in \mathbb{C}^2 \mid 2 \operatorname{Re} z_1 < -|z_1|^2 - |z_2|^2 \} \right)$$
$$= \{ z \in \mathbb{C}^2 \mid 2 \operatorname{Re} \lambda_j z_1 < -\lambda_j{}^2 |z_1|^2 - \lambda_j |z_2|^2 \}$$
$$= \{ z \in \mathbb{C}^2 \mid 2 \operatorname{Re} z_1 < -\lambda_j |z_1|^2 - |z_2|^2 \}.$$

Since $\lambda_j \searrow 0$, it follows immediately that

$$\psi_j(B^2) \subset \sigma_{j+1}(B^2) \quad \forall j = 1, 2, \ldots$$

and

$$\bigcup_{j=1}^{\infty} \psi_j(B^2) = \mathcal{U}.$$

In this sense, it seems sensible to say that $\hat{\psi}(B^2)$ is in fact the limit domain of the sequence of domain $\psi_j(B^2)$.

This simple example already illustrates an important aspect of the scaling technique in complex dimension two (as well as in higher complex dimensions).

In view of the discussion of the one-dimensional scaling (see the Appendix), the following theorem may be duly noted (see also the discussion in Sect. 3.3):

Theorem 7.2.1 (Wong 1977, Rosay 1979). *Let Ω be a bounded domain in $\mathbb{C}^n$ with a boundary point $P \in \partial\Omega$ satisfying the following:*

(i) $\partial\Omega$ is C^2 smooth and strictly pseudoconvex near P.
(ii) There exists a sequence $\varphi_j \in \operatorname{Aut}\Omega$ and an interior point $X \in \Omega$ such that $\lim_{j \to \infty} \varphi_j(X) = P$.

Then the domain Ω is biholomorphic to the unit ball in $\mathbb{C}^n$.

We shall present a proof of this result, which illustrates the scaling method in detail, in subsequent sections. First we shall present a detailed exposition of the background theory starting with the notion of *normal convergence of domains*.

7.2.2 Normal Convergence of Sets

We first describe the concept of normal convergence of domains.

Definition 7.2.2. Let Ω_j be domains in $\mathbb{C}^n$ for each $j = 1, 2, \ldots$. The sequence Ω_j is said to *converge normally* to a domain $\hat{\Omega}$ if the following two conditions hold:

(i) If a compact set K is contained in the interior (i.e., the largest open subset) of $\bigcap_{j>m} \Omega_j$ for some positive integer m, then $K \subset \hat{\Omega}$.

(ii) If a compact subset K' lies in $\hat{\Omega}$, then there exists a constant $m > 0$ such that $K' \subset \bigcap_{j>m} \Omega_j$.

The reason for introducing such ideas of convergence of sets is because these are what are used for the scaling method and normal families with source and target domains varying (compare with the idea of convergence in the Hausdorff metric—see [FED]).

Proposition 7.2.3. *If Ω_j is a sequence of domains in $\mathbb{C}^n$ that converges normally to the domain $\hat{\Omega}$, then:*

(1) *If a sequence of holomorphic mappings $\varphi_j : \Omega_j \to \Omega'$ from Ω_j to another domain Ω' converges uniformly on compact subsets of $\hat{\Omega}$, then its limit is a holomorphic mapping from $\hat{\Omega}$ into the closure of the domain Ω'.*

(2) *If a sequence of holomorphic mappings $g_j : \Omega' \to \Omega_j$ converges uniformly on compact subsets of G, then its limit is a holomorphic mapping from the domain Ω' into the closure of $\hat{\Omega}$.*

7.2.3 Localization

Local Holomorphic Peak Functions

Definition 7.2.4. Let Ω be a domain in $\mathbb{C}^n$. A boundary point $p \in \partial\Omega$ is said to admit a *holomorphic peak function* if there exists a continuous function $h : \overline{\Omega} \to \mathbb{C}$ that satisfies the following properties:

(i) h is holomorphic on Ω.

(ii) $h(p) = 1$.

(iii) $|h(z)| < 1$ for every $z \in \overline{\Omega} \setminus \{p\}$.

Such a function h is called a *holomorphic peak function* for Ω at p. And p is called a *peak point*.

Furthermore, we say that a boundary point p of Ω admits a *local holomorphic peak function* if there exists an open neighborhood U of p such that there exists a holomorphic peak function for $\Omega \cap U$ at p (see [GAM, p. 52 ff.]).

Proposition 7.2.5. *Let Ω be a bounded domain in $\mathbb{C}^n$ with a C^2 smooth, strictly pseudoconvex boundary point p. Let B^n be the unit open ball in $\mathbb{C}^n$. Let η be a positive real number satisfying $0 < \eta < 1$. Then, for every $\epsilon > 0$, there exists $\delta > 0$ such that*

$$|f(z) - p| < \epsilon, \quad \forall z \text{ with } |z| < \eta,$$

for every holomorphic mapping $f : B^n \to \Omega$ with $|f(0) - p| < \delta$.

Proof. Assume to the contrary that there exist holomorphic mappings $\varphi_j : B^n \to \Omega$ satisfying the following two conditions:

(a) $\lim_{j\to\infty} \varphi_j(0) = p$.
(b) $\exists \epsilon > 0$ for which there exists a sequence $z_j \in B^n$ such that $|z_j| < \eta$ and $|\varphi_j(z_j) - p| \geq \epsilon$ for every $j = 1, 2, \ldots$.

Let U be an open neighborhood of p such that there exists a local holomorphic peak function $h : \overline{\Omega} \cap U \to \mathbb{C}$ at p. (Here we use the fact that a strictly pseudoconvex boundary point always admits a local holomorphic peak function—see [KRA1, Chap. 5].)

Since Ω is bounded, Montel's theorem yields that φ_j admits a subsequence that converges uniformly on compact subsets. By an abuse of notation, we denote the subsequence by the same notation φ_j and then the subsequential limit mapping by $F : B^n \to \overline{\Omega}$.

Take an open neighborhood V of 0 sufficiently small so that it satisfies the properties:

(1) $\overline{V} \subset B^n$.
(2) There exists $N > 0$ such that $\varphi_j(\overline{V}) \subset U \cap \Omega$ for every $j > N$.

Consider the sequence of mappings $h \circ \varphi_j|_V : V \to D$, where D is the open unit disc in $\mathbb{C}$. Apply Montel's theorem again to this sequence. Choosing a subsequence from φ_j again, we may assume that $h \circ \varphi_j|_V$ converges uniformly on compact subsets of V to a holomorphic map $G : V \to \overline{D}$. Since $G(0) = 1$ and $|G(\zeta)| < 1$ for every $\zeta \in V$, the maximum principle implies that $G(\zeta) \equiv 1$ for every $\zeta \in V$.

By the properties of the local holomorphic peak function h at p, this implies that $F(\zeta) = p$ for every $\zeta \in V$. Since V is open, and since F is holomorphic, it follows that $F(z) = p$ for every $z \in B^n$. Since the convergence of φ_j to F is uniform on compact subsets, it is impossible to have z_j with $|z_j| \leq \eta$ such that $\varphi_j(z_j)$ stays away from p for every j. This contradiction completes the proof. $\square$

Plurisubharmonic Peak Functions

There is an effective method of localization in a more general setting [SIB]). A main point of this method is that it avoids Montel's theorem altogether. Thus for instance, the assumption that Ω is bounded is no longer needed:

Definition 7.2.6. Let Ω be a domain in $\mathbb{C}^n$ and let p be a boundary point. If there exists a continuous function $h : \overline{\Omega} \to \mathbb{R}$ satisfying:

(i) h is plurisubharmonic on Ω, and
(ii) $h(p) = 0$ and $h(z) < 0$ for every $z \in \overline{\Omega} \setminus \{0\}$,

then we call h a *plurisubharmonic peak function at p for Ω*. In such a case, p is called a *plurisubharmonic peak point for Ω*.

Likewise, a boundary point p of the domain Ω is called a *local plurisubharmonic peak point* if there exists an open neighborhood of p in $\mathbb{C}^n$ such that p is a plurisubharmonic peak point for $\Omega \cap U$.

We present first the following lower-bound estimate for the Kobayashi metric near a local plurisubharmonic peak boundary point. See [SIB].

Proposition 7.2.7. *Let Ω be a bounded domain in $\mathbb{C}^n$ with a boundary point $p \in \partial\Omega$ which admits a local plurisubharmonic peak function for Ω. Then, for every open neighborhood U of p in $\mathbb{C}^n$, there exists an open neighborhood V with $p \in V \subset U$ such that the inequality*

$$k_\Omega(z, \xi) \geq \frac{1}{2}\, k_{\Omega \cap U}(z, \xi), \quad \forall (z, \xi) \in (\Omega \cap V) \times \mathbb{C}^n,$$

where k_Ω denotes the infinitesimal Kobayashi pseudometric of a domain Ω.

Proof. Denote by D_r the open disc in $\mathbb{C}$ of radius r centered at the origin. For the open unit disc, write $D = D_1$.

By the definition of the Kobayashi metric, it suffices to prove the following statement:

(7.2.7.1)   *It is possible to choose V so that the following holds: Given $(z, \xi) \in (\Omega \cap V) \times \mathbb{C}^n$, every holomorphic mapping $f : D \to \Omega$ from the unit disc D into Ω satisfying $f(0) = z$, $df|_0(\lambda) = \xi$ for some $\lambda > 0$ enjoys the property that $f(D_{1/2}) \subset U$.*

Replacing U by a smaller neighborhood of p if necessary, let $\psi_1 : U \cap \overline{\Omega}$ be a local plurisubharmonic peak function at p. Choose an open neighborhood U_1 of p inside U and a constant $c_1 > 0$ such that

$$\sup\{\psi_1(z) \mid z \in \overline{\Omega} \cap \partial U_1\} = -c_1.$$

Choose a neighborhood V_1 of p inside U_1 such that

$$V_1 = \{z \in \Omega \cap U_1 \mid \psi_1(z) > -\frac{c_1}{2}\}.$$

Then we can extend ψ_1 to a new function $\psi_2 : \overline{\Omega} \to \mathbb{R}$ by

$$\psi_2(z) = \begin{cases} \psi_1(z) & \text{if } z \in \overline{\Omega} \cap \overline{V_1} \\ \max\{\psi_1(z), -3c_1/2\} & \text{if } z \in \overline{\Omega} \cap (U_1 \setminus \overline{V_1}) \\ -3c_1/2 & \text{if } z \in \overline{\Omega} \setminus U_1. \end{cases}$$

Notice that ψ_2 is a global plurisubharmonic peak function for Ω at p.

Towards the proof of (7.2.7.1), there is no harm in assuming (by a simple dilation) that the analytic disc f is holomorphic in a neighborhood of the closed unit disc $\overline{D}$.

Let $a > 0$ be such that $\psi_2 \circ f(0) > -a$. Consider

$$E_a = \{\theta \in [0, 2\pi] \mid \psi_2 \circ f(e^{i\theta}) \geq -2a\}.$$

By the sub-mean-value inequality, we see that

$$-a < \psi_2 \circ f(0)$$

$$\leq \frac{1}{2\pi} \int_{[0, 2\pi]} \psi_2 \circ f(e^{i\theta}) \, d\theta$$

$$\leq \frac{1}{2\pi} \int_{[0, 2\pi] \setminus E_a} (-2a) \, d\theta$$

$$\leq -\frac{a}{\pi}(2\pi - |E_a|),$$

where $|S|$ denotes the Lebesgue measure of the set S. Hence, we see that

$$|E_a| > \pi.$$

Now consider a plurisubharmonic function at p given by

$$\upsilon_\epsilon(z) = \epsilon \log \|z - p\|,$$

where ϵ is a certain positive constant to be chosen shortly.

Let

$$\inf\{\psi_1(z) + \upsilon_\epsilon(z) \mid z \in \overline{\Omega} \cap \partial V_1\} = -c_2,$$

and

$$\sup\{\psi_1(z) + \upsilon_\epsilon(z) \mid z \in \overline{\Omega} \cap \partial U_1\} = -c_3.$$

Choose $\epsilon > 0$ so that

$$-c_3 < -c_2 < 0.$$

Extend $\psi_1 + \upsilon_\epsilon$ to the plurisubharmonic function $\Upsilon : \overline{\Omega} \to \mathbb{R}$ defined by

$$\Upsilon(z) = \begin{cases} \psi_1(z) + \upsilon_\epsilon(z) & \text{if } z \in \overline{\Omega} \cap \overline{V_1} \\ \max\{\psi_1(z) + \upsilon_\epsilon(z), -\dfrac{c_2 + c_3}{2}\} & \text{if } z \in \overline{\Omega} \cap (\overline{U_1} \setminus \overline{V_1}) \\ -\dfrac{c_2 + c_3}{2} & \text{if } z \in \overline{\Omega} \setminus \overline{U_1}. \end{cases}$$

Observe that $\Upsilon^{-1}(-\infty) = \{p\}$.

For each $\zeta \in D_{1/2}$, apply the Poisson integral formula to obtain

$$\Upsilon \circ f(\zeta) \leq \frac{1}{10\pi} \int_0^{2\pi} \Upsilon \circ f(e^{i\theta}) \, d\theta.$$

We now focus upon the peak function ψ_2 and the companion function Υ. Since the sets

$$G_k = \{z \in \overline{\Omega} \mid \psi_2(z) \geq -1/k\}$$

for $k = 1, 2, \ldots$ form a neighborhood basis for p in $\overline{\Omega}$, we see that for each $L > 0$ there exists $a > 0$ with a arbitrarily small such that

$$\{z \in \overline{\Omega} \mid \psi_2(z) \geq -2a\} \subset \{z \in \overline{\Omega} \mid \Upsilon(z) < -L\}.$$

Then we present

Claim. If a holomorphic function $f : \overline{D} \to \Omega$ satisfies $\psi_2 \circ f(0) > -a$, then $\Upsilon \circ f(\zeta) \leq -L/10$ for every $\zeta \in D_{1/2}$.

The proof is immediate; simply check for each $\zeta \in D_{1/2}$ that

$$\Upsilon \circ f(\zeta) \leq \frac{1}{10\pi} \int_0^{2\pi} \Upsilon \circ f(e^{i\theta}) \, d\theta$$

$$\leq \frac{1}{10\pi} \int_{E_a} (-L) \, d\theta + \frac{1}{10\pi} \int_{[0,2\pi] \setminus E_a} 0 \, d\theta$$

$$= -\frac{L}{10}.$$

Finally we are ready to finish the proof. Observe that the sets

$$U_k = \{z \in \overline{\Omega} \mid \Upsilon(z) < -\frac{k}{10}\}$$

for $k = 10, 11, \ldots$ also form a neighborhood basis for p in $\overline{\Omega}$. By Claim 7.2.3 above, for each k we may choose $a_k > 0$ such that

(1) $\Upsilon(z) > -k$ whenever $\psi_2(z) > -2a_k$.
(2) $a_{10} > a_{11} > \ldots \to 0$.

Consequently, if we choose $V_k = \{z \in \overline{\Omega} \mid \psi_2 > -a_k\}$ for each k, then it follows immediately that

$$f(0) \in V_k \Rightarrow f(D_{1/2}) \subset U_k$$

for every $k = 10, 11, \ldots\ldots$. That is the proof. $\square$

Proposition 7.2.8. *Let Ω be a bounded domain in $\mathbb{C}^n$ with a boundary point $p \in \partial\Omega$ which admits a local holomorphic peak function for Ω. Let K be a compact subset of Ω and let $q \in \Omega$. Then, for every open neighborhood U of p in $\mathbb{C}^n$, there exists an open set V with $p \in V \subset U$ such that $f(K) \subset U$ whenever $f : \Omega \to \Omega$ is a holomorphic mapping satisfying $f(q) \in V$.*

Proof. Note first that a local holomorphic peak function h at p generates the local plurisubharmonic peak function $\log|h|$ at p.

Since the Kobayashi pseudo-distance $d_M : M \times M \to \mathbb{R}$ is continuous for any complex manifold M, we may select $R > 0$ such that the Kobayashi distance ball

$$B_\Omega^K(q, R) = \{z \in \Omega \mid d_\Omega(z, q) < R\}$$

contains K.

Then use the local holomorphic peak function $h : U \cap \overline{\Omega} \to D$ at p. The distance-decreasing property implies that

$$\lim_{\Omega \cap U \ni p_j \to p} d_{\Omega \cap U}(z, p_j) \geq \lim_{\Omega \ni p_j \to p} d_D(h(z), h(p_j)) = \infty.$$

Moreover, the local holomorphic peak function and the distance-decreasing property guarantee the existence of an open set U' with $p \in U' \subset \overline{U'} \subset U$ and an open neighborhood V with $p \in V \in U'$ such that

$$d_{\Omega \cap U}(z, w) > 3R$$

for every $z \in V$ and every $w \in \partial U' \cap \Omega$.

Now let $\zeta \in \Omega \setminus U$. Then, by the definition of the Kobayashi metric, there exists a piecewise smooth "almost-the-shortest" connector $\gamma : [0, 1] \to \Omega$ with $\gamma(0) = z, \gamma(1) = \zeta$ induced from the holomorphic chain in the definition of the Kobayashi metric such that

$$L_\Omega^K(\gamma) - \frac{R}{2} < d_\Omega(z, \zeta) < L_\Omega^K(\gamma),$$

where $L_\Omega^K(\gamma)$ denotes the length of γ measured by the Kobayashi metric of Ω. Since $\gamma([0, 1])$ has to cross $\partial U' \cap \Omega$, we let $t \in (0, 1)$ such that $\gamma([0, t)) \subset U' \cap \Omega$ and $\gamma(t) \in \partial U' \cap \Omega$. Then it follows that

$$L_\Omega^K(\gamma) > L_\Omega^K(\gamma|_{[0,t]}) > \frac{1}{2} L_{\Omega \cap U}^K(\gamma|_{[0,t]}) > \frac{1}{2} d_{\Omega \cap U}(z, \gamma(t)) > \frac{3}{2} R.$$

This therefore implies that

$$d_\Omega(z, \zeta) > \frac{3}{2}R - \frac{R}{2} = R.$$

In particular, $B_\Omega^K(z, R) \subset U$ whenever $z \in V$.

Since f given in the hypothesis is a holomorphic mapping, the distance-decreasing property of the Kobayashi distance yields that

$$f(K) \subset f(B_\Omega^K(q, R)) \subset B_\Omega^K(f(q), R) \subset U.$$

Since $f(q) \in V$, we see that $B_\Omega^K(f(q), R) \subset U$. This is what we wanted to establish. $\qquad \square$

7.3 Klembeck's Theorem with Stability in the C^2 Topology

7.3.1 The Main Goal

We now describe Klembeck's theorem with stability in the C^2 topology.

We shall not prove anything in the ensuing discussion. Details may be found in the book [GKK]. Our goal here is to introduce the reader to a circle of ideas.

The precise target should be described first. Denote by $\mathcal{D}_n$ the collection of all bounded domains in $\mathbb{C}^n$ with C^2 smooth, strictly pseudoconvex boundary. We impose the C^2 topology on $\mathcal{D}_n$ by invoking the C^2 topology on defining functions.

Denote by $S_\Omega(p; \xi)$ the holomorphic sectional curvature at p in the holomorphic two-plane direction ξ of the Bergman metric of the domain Ω.

Theorem 7.3.1. *Let $\hat{\Omega}$ be a bounded strictly pseudoconvex domain with C^2 boundary in $\mathbb{C}^n$. Then, for every $\epsilon > 0$, there exist $\delta > 0$ and an open neighborhood $\mathcal{U}$ of $\hat{\Omega}$ in $\mathcal{D}_n$ such that, whenever $\Omega \in \mathcal{U}$,*

$$\sup\left\{\left|S_\Omega(p; \xi) - \left(-\frac{4}{n+1}\right)\right| : \Omega \in \mathcal{U}, \xi \in \mathbb{C}^n \setminus \{0\}\right\} < \epsilon$$

whenever $p \in \Omega$ satisfies dis $(p, \partial\Omega) < \delta$.

The following gives in effect the localization of Bergman metric holomorphic sectional curvature:

Theorem 7.3.2. *There exists an open neighborhood U of the origin in $\mathbb{C}^n$ such that*

$$\lim_{\nu \to \infty} \sup_{\xi \in \mathbb{C}^n, |\xi|=1} \left|\frac{2 - S_{\Omega_\nu \cap U}(p_\nu; \xi)}{2 - S_{\Omega_\nu}(p_\nu; \xi)} - 1\right| = 0.$$

The conclusion of this statement implies: as soon as $\lim_{\nu \to \infty} S_{\Omega_\nu \cap U}(p_\nu; \xi)$ exists, it will coincide with $\lim_{\nu \to \infty} S_{\Omega_\nu}(p_\nu; \xi)$.

We now demonstrate how the problem on boundary asymptotic behavior of the Bergman curvature (generally considered difficult) can be converted to the problem on the stability of the Bergman kernel function in the interior under perturbation of the boundaries (which is generally easier). This is done by the scaling method, and this conversion is the important, second component of the proof.

Theorem 7.3.3. *Let the sequence $\{(p_\nu; \xi_\nu) \in \Omega_\nu \times (\mathbb{C}^n \setminus \{0\})\}$ be chosen as above. Let B^n denote the open unit ball in $\mathbb{C}^n$. Then there exists a sequence of injective holomorphic mappings $\sigma_\nu : \Omega_\nu \cap U \to \mathbb{C}^n$ satisfying the following properties:*

(i) $\sigma_\nu(p_\nu) = 0$ (the origin of $\mathbb{C}^n$).
(ii) For every $r > 0$, there exists $N > 0$ such that

$$(1 - r)B^n \subset \sigma_\nu(\Omega_\nu \cap U) \subset (1 + r)B^n$$

for every $\nu > N$.

The third component is the following theorem of Ramadanov [RAM1] (see also Sect. 1.12). We treated this result in detail in Sect. 1.12.

Theorem 7.3.4. *Let D be a bounded domain in $\mathbb{C}^n$ containing the origin 0. Let D_ν denote a sequence of bounded domains in $\mathbb{C}^n$ that satisfies the following convergence condition:*

Given $\epsilon > 0$, there exists $N > 0$ such that

$$(1 - \epsilon)D \subset D_\nu \subset (1 + \epsilon)D$$

for every $\nu > N$.

Then, for every compact subset F of D, the sequence of Bergman kernel functions K_{D_ν} of D_ν converges uniformly to the Bergman kernel function K_D of D on $F \times F$.

7.3.2 The Bergman Metric near Strictly Pseudoconvex Boundary Points

As an application of the ideas introduced by far, we will now deduce the boundary behavior estimate for the Bergman metric of a bounded, C^2-smooth strictly pseudoconvex domain, establishing the completeness of the Bergman metric there.

Let Ω be a bounded domain in $\mathbb{C}^n$ with C^2 smooth, strictly pseudoconvex boundary. Let $\nu \in \mathbb{C} \setminus \{0\}$. Then, for any $p \in \Omega$, the Bergman metric length $\|v\|_{\Omega, p}$ at p has the following representation by minimum integrals:

$$\|v\|_{\Omega,p}^2 = \frac{1}{I_0^\Omega(p)I_1^\Omega(p;v)}.$$

Note that this follows from the exposition in Sect. 3.1, where the Bergman's special orthonormal system for $A^2(\Omega)$ was introduced.

Thus the usual localization arguments imply the following:

For any $\hat{p} \in \partial\Omega$, any open neighborhood U of $\hat{p}$ in $\mathbb{C}^n$, and any positive constant $C > 1$, there exists an open set V satisfying $\hat{p} \in V \subset\subset U$ such that

$$\frac{1}{C}\|v\|_{\Omega\cap U,p}^2 \le \|v\|_{\Omega,p}^2 \le C\|v\|_{\Omega\cap U,p}^2$$

for any $p \in V$ and any $v \in \mathbb{C}^n$.

This, together with the scaling method arguments, implies immediately the following:

Let p be as above. Then let $\tilde{p} \in \partial\Omega$ be the closest point to p. (Such a $\tilde{p}$ is uniquely determined if V is chosen sufficiently small.) Write $v \in \mathbb{C}^n$ as

$$v = v' + v''$$

so that v' is complex tangent to $\partial\Omega$ at $\tilde{p}$ whereas v'' is complex normal. Then there exists a constant $C' > 0$ such that

$$\|v\|_{\Omega,p}^2 \ge C'\left(\frac{\|v'\|^2}{\|p - \tilde{p}\|} + \frac{\|v''\|^2}{\|p - \tilde{p}\|^2}\right),$$

where the norm $\|\cdot\|$ in the right-hand side is the Euclidean norm, which is in fact the Bergman metric of the unit ball in $\mathbb{C}^n$ at the origin up to a constant multiple.

Notice that this in particular implies the completeness of the Bergman metric of the bounded strictly pseudoconvex domains, which was used in the exposition of Chap. 3.

Exercises

1. Use the scaling method to construct a proof of the Riemann mapping theorem in one complex variable.
2. Use the scaling method to show that the Poisson kernel of a smoothly bounded domain in the complex plane will satisfy

$$c\frac{\delta_\Omega(x)}{|x - y|^2} \le P_\Omega(x, y) \le C\frac{\delta_\Omega(x)}{|x - y|^2}$$

for positive constants c and C. Here $\delta_\Omega(x)$ is the distance of $x \in \Omega$ to the boundary.

3. Let $\Omega \subseteq \mathbb{C}^2$ have defining function

$$\rho(z_1, z_2) = -1 + |z_1|^2 + |z_2|^{2m} + \text{(terms that are order } 2m + 2 \text{ or higher)}$$

   for m an integer greater than 1. Perform scaling near the boundary point $(1, 0)$. What does the limit domain look like?

4. Perform scaling on the bidisc near the boundary point $(1, 0)$. What does the limit domain look like? Do you have trouble with the convergence of the scaling process?

5. Use the Fefferman's asymptotic expansion to calculate an asymptotic formula for the Bergman metric near a strictly pseudoconvex boundary point.

6. Generalize Exercise 2 to higher dimensions.

7. Why don't we do scaling in the real variable context? What essential feature would be missing?

8. Calculate the curvature of the Bergman metric on the bidisc.

9. Can you use scaling to get information about the Bergman metric near the boundary point $(1, 0)$ of the domain

$$E = \{(z_1, z_2) : |z_1|^2 + |z_2|^4 < 1\}?$$

10. What can you say about the curvature of the Bergman metric near the boundary point $(1, 0)$ of the domain

$$E = \{(z_1, z_2) : |z_1|^2 + |z_2|^4 < 1\}?$$

APPENDIX: Scaling in Dimension One

The Scaling of the Unit Disc

Let D be the open unit disc in the complex plane $\mathbb{C}$. Choose a sequence a_j in D satisfying the conditions

$$0 < a_j < a_{j+1} < \cdots < 1, \; \forall j = 1, 2, \ldots,$$

and

$$\lim_{j \to \infty} a_j = 1.$$

Consider the sequence of dilations

$$L_j(z) = \frac{1}{1 - a_j}(z - 1).$$

Let us write $\lambda_j = 1 - a_j$. Then one sees immediately that

$$L_j(D) = \{\zeta \in \mathbb{C} \mid (1 + \lambda_j \zeta)(1 + \lambda_j \bar{\zeta}) < 1\}$$
$$= \{\zeta \in \mathbb{C} \mid 2\,\mathrm{Re}\,\zeta < -\lambda_j |\zeta|^2\}.$$

It follows that the sequence of sets $L_j(D)$ *converges* to the left half plane $H = \{\zeta \in \mathbb{C} \mid \mathrm{Re}\,\zeta < 0\}$ in the sense that

$$L_j(D) \subset L_{j+1}(D), \; \forall j = 1, 2, \ldots,$$

and

$$\bigcup_{j=1}^{\infty} L_j(D) = H.$$

[Compare the concept of convergence in the Hausdorff metric on sets—see [FED].]

Now we combine this simple observation with the fact that there exists the sequence of maps

$$\varphi_j(z) = \frac{z + a_j}{1 + a_j z}$$

that are automorphisms of D satisfying $\varphi_j(0) = a_j$. Consider the sequence of composite maps

$$\sigma_j \equiv L_j \circ \varphi_j : D \to \mathbb{C}.$$

A direct computation yields that

$$L_j \circ \varphi_j(z) = \frac{1}{1-a_j}\left(\frac{z+a_j}{1+a_jz} - 1\right)$$

$$= \frac{z-1}{1+a_jz}.$$

Hence, in fact we see that the sequence of holomorphic mappings $L_j \circ \varphi_j$ converges uniformly on compact subsets of D to the mapping

$$\hat{\sigma}(z) = \frac{z-1}{z+1}$$

that is a biholomorphic mapping from the open unit disc D onto the left half plane H. (We have in effect discovered here a means to see the Cayley map by way of scaling.)

The point is that we have exploited the automorphism of the disc to see that the disc is conformally equivalent to a certain canonical domain—namely, the half plane. This result is neither surprising nor insightful. But it is a toy version of the main results that we shall present below.

A Generalization

We now expand the simple observations of the preceding subsection to yield the statement and the proof of the following one-dimensional version of the Wong–Rosay theorem:

Proposition: *Let Ω be a domain in the complex plane $\mathbb{C}$ admitting a boundary point p such that*

(i) *There exists an open neighborhood U of p in $\mathbb{C}$ such that $U \cap \partial\Omega$ is a C^1 curve.*
(ii) *There exists a sequence φ_j of automorphisms of Ω and a point $q \in \Omega$ such that*

$$\lim_{j\to\infty} \varphi_j(q) = p.$$

Then Ω is biholomorphic to the open unit disc.

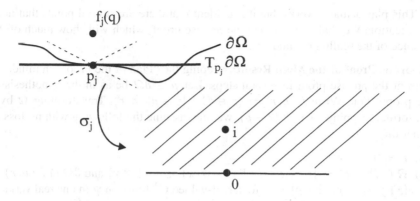

Fig. 7.1 The scaling process

See [KRA13] for this theorem. We use this simple result to illustrate the technique of scaling.

In order to be consistent with the remainder of this chapter, we change a bit the notation for the orbit accumulation point and the point whose orbit we are calculating. This will all make sense in context.

Sketch of the Proof: Let $q_j = \varphi_j(q)$ for each j. Choose the closest point in the boundary to q_j and call it p_j. If the closest boundary point p_j to q_j is not unique, then make a choice. As j tends to infinity, p_j converges to p because q_j converges to p. Then we select θ_j and apply the map $\rho_j(z) \equiv e^{i\theta_j}(z - p_j)$ so that

$$\rho_j(p_j) = 0 \quad \text{and} \quad \rho_j(q_j) > 0$$

for each j. Now consider the sequence of mappings

$$\psi_j(z) = \frac{1}{\rho_j(q_j)} \left(\rho_j \circ \varphi_j(z) \right).$$

Notice that $\psi_j(\Omega) = \dfrac{1}{\rho(q_j)} \rho_j(\Omega)$ for each j. Thus we expect that $\psi_j(\Omega)$ is almost the right half plane as j becomes very large. At least every $\psi_j(\Omega)$ is contained in $\mathbb{C} \setminus \ell$ for some line segment ℓ of positive length and for every j. (Note that ℓ can be chosen independently of j.) Therefore one can select a subsequence from $\{\psi_j\}$ that converges uniformly on compact subsets of Ω. Let $\hat{\psi}$ be the limit mapping. Then we expect $\hat{\psi} : \Omega \to \mathbb{C}$ to be an injective holomorphic mapping, and furthermore, $\hat{\psi}(\Omega)$ is equal to the right half plane. Thus we hope to conclude that Ω is biholomorphic to the right half plane, which in turn is biholomorphic to the open unit disc. See Fig. 7.1.

This plan actually works, but it is evident that there are several points that need clarification. We shall now present the precise proof, which will show much of the essence of the scaling method.

Rigorous Proof of the Main Result: Keeping the "Plan of the Proof" in mind, we present the precise proof in several steps. Let $p \in \partial\Omega$ be as in the hypothesis of the proposition. Write $D(p,r) = \{z \in \mathbb{C} \mid |z - p| < r\}$. Transforming Ω by a conformal mapping $z \mapsto e^{i\alpha}(z - p)$, we may assume the following with no loss of generality:

(a) $p = 0$
(b) $\Omega \cap D(p,r) = \{z = x + iy \mid y > \psi(x), |z - p| < r\}$ and $\partial\Omega \cap D(p,r) = \{z \mid y = \psi(x), |z - p| < r\}$ for a real-valued C^1 function ψ in one real variable satisfying $\psi(0) = 0$ and $\psi'(0) = 0$.

Step 1. THE SCALING MAP. Notice that the sequence $\varphi_j(q)$ now converges to 0 as $j \to \infty$. For each j, we choose a point $p_j \in \partial\Omega$ that is the closest to $\varphi_j(q)$. Since p_j also converges to 0, replacing φ_j by a subsequence if necessary, we may assume that every $p_j \in D(p, r/4)$. Now, for each j, set

$$\alpha_j(z) = i \, \frac{|\varphi_j(q) - p_j|}{\varphi_j(q) - p_j} \, (z - p_j).$$

Notice that $\varphi_j(q) - p_j$ is a positive scalar multiple of the inward unit normal vector to $\partial\Omega$ at p_j. Thus $\dfrac{\varphi_j(q) - p_j}{|\varphi_j(q) - p_j|}$ converges to the inward unit normal vector to $\partial\Omega$ at 0. This implies that α_j in fact converges to the identity map. Consequently, there exist positive constants r_1, r_2 independent of j such that, for each j, there exists a C^1 function $\psi_j(x)$ defined for $|x| < r_1$ satisfying

$$\alpha_j(z) \cap ([-r_1, r_1] \times [-r_2, r_2]) = \{x + iy \mid |x| < r_1, |y| < r_2, y > \psi_j(x)\}.$$

Furthermore, for each $\epsilon > 0$, there exists $\delta > 0$ such that

$$\psi_j(x) < \epsilon|x| \text{ whenever } |x| < \delta$$

regardless of j.

Next, let $\lambda_j = |\varphi_j(q) - p_j|$ for each j. Consider the dilation map

$$L_j(z) = \frac{z}{\lambda_j}.$$

Then the sequence of holomorphic mappings we want to construct is given by

$$\psi_j \equiv L_j \circ \alpha_j \circ \varphi_j : \Omega \to \mathbb{C}.$$

Before starting the next step, we make a few remarks. The automorphism φ_j preserves the domain Ω but moves q to $\varphi_j(q)$ so that $\varphi_j(q)$ converges to the origin—recall that we made changes so that p became the origin at the beginning of the proof. Then the affine map α_j adjusts Ω so that the direction vector $\dfrac{\varphi_j(q) - p_j}{|\varphi_j(q) - p_j|}$ is transformed to a purely imaginary number. The final component L_j in the construction simply magnifies the domain $\alpha_j(\Omega)$, while the map L_j itself diverges.

Step 2. CONVERGENCE OF THE ψ_j. We shall actually choose a subsequence from $\{\psi_j\}$ that converges uniformly on compact subsets of Ω. Observe first that

$$\psi_j(\Omega) = L_j \circ \alpha_j \circ \varphi_j(\Omega) = L_j \circ \alpha_j(\Omega)$$

since $\varphi_j(\Omega) = \Omega$. Choosing a subsequence of ψ_j, we may assume that $\lambda_j < 1$ for every j. Then, since L_j is a simple dilation by a positive number, and since $\alpha_j(\Omega)$ will miss a line segment

$$E = \{-iy \mid 0 \le y \le b\}$$

for some constant b independent of j, we see immediately that

$$\psi_j(\Omega) \subset \mathbb{C} \setminus E$$

for every $j = 1, 2, \ldots$. Therefore Montel's theorem implies that every subsequence of $\{\psi_j\}$ admits a subsequence, which we again (by an abuse of notation) denote by ψ_j, that converges uniformly on compact subsets of Ω. Denote by $\hat{\psi}$ the limit of the sequence ψ_j.

Step 3. ANALYSIS OF $\hat{\psi}(\Omega)$. We want to establish that

$$\hat{\psi}(\Omega) = \mathcal{U},$$

where $\mathcal{U} \equiv \{z \in \mathbb{C} \mid \operatorname{Im} z > 0\}$.

Let ϵ be a positive real number and let K an arbitrary compact subset of Ω. We will show that $\hat{\psi}(K) \subset C_\epsilon$, where $C_\epsilon \equiv \{z \in \mathbb{C} : -\epsilon < \arg z < \pi + \epsilon\}$.

Choose $R > 0$ such that $\hat{\psi}(K)$ is contained in the disc $D(0, R)$ of radius R centered at 0.

The sequence $\varphi_j : \Omega \to \Omega$ is a normal family since $\mathbb{C} \setminus \Omega$ contains a line segment with positive length. Every subsequence of φ_j contains a subsequence that converges uniformly on compact subsets, since $\varphi_j(q)$ converges to p. Let $g : \Omega \to \overline{\Omega}$ be a subsequential limit map. Then $g(q) = p$. Recall that $p \in \partial\Omega$. Hence, the open mapping theorem yields that $g(z) = p$ for every $z \in \Omega$. Thus the sequence

φ_j itself converges uniformly on compact subsets to the constant map with value p. Therefore we may choose $N > 0$ such that $\varphi_j(K)$ is contained in a sufficiently small neighborhood of the origin for every $j > N$, and hence $\alpha_j \circ \varphi_j(K) \subset C_\epsilon$ for every $j > N$. Then it follows immediately that $\psi_j(K) \subset C_\epsilon$ for every $j > N$ and consequently that

$$\hat{\psi}(K) \subset C_\epsilon.$$

Since K is an arbitrary compact subset of Ω, it follows that $\hat{\psi}(\Omega) \subset \overline{\mathcal{U}}$. We also have $\hat{\psi}(q) = i$, since $\psi_j(q) = L_j \circ \alpha_j \circ \varphi_j(q) = i$ for every $j = 1, 2, \ldots$. Therefore $\hat{\psi}(\Omega) \subset \mathcal{U}$.

Step 4. CONVERGENCE OF ψ_j^{-1}. Let $\tilde{K}$ be an arbitrary compact subset of the upper half plane $\mathcal{U}$. Then choose $\epsilon > 0$ so that $\tilde{K} \subset C_\epsilon$. Choose then $r > 0$ such that

$$D(0, r) \cap C_\epsilon \subset \Omega \cap D(0, r).$$

Shrinking $r > 0$ if necessary, since α_j converges to the identity map uniformly on compact subsets of $\mathbb{C}$, there exists $N > 0$ such that

$$D(0, r) \cap C_\epsilon \subset \alpha_j(\Omega) \cap D(0, r)$$

for every $j > N$. Hence, we see that ψ_j^{-1} maps K into Ω. Since $\Omega \subset \mathbb{C} \setminus E$ as observed before, we may again choose a subsequence of ψ_j, which we again denote by ψ_j, so that ψ_j^{-1} converges to a holomorphic map, say $\tau : \mathcal{U} \to \overline{\Omega}$. Since τ is holomorphic and $\tau(i) = q$, we see that τ maps the upper half plane $\mathcal{U}$ into Ω.

Step 5. SYNTHESIS. We are ready to complete the proof. By the Cauchy estimates, the derivatives $d\psi_j$ of ψ_j as well as the derivatives $d[\psi_j^{-1}]$ both converge. Therefore $d\hat{\psi}(q) \cdot d\hat{\tau}(i) = 1$. This means that $\hat{\psi} \circ \tau : \mathcal{U} \to \mathcal{U}$ is a holomorphic mapping satisfying $\hat{\psi} \circ \tau(i) = i$ and $(\hat{\psi} \circ \tau)'(i) = 1$. Then, by the Schwarz's lemma, one concludes that $\hat{\psi} \circ \tau = $ id, where id is the identity mapping. Likewise, the same reasoning applied to $\tau \circ \hat{\psi} : \Omega \to \Omega$ implies that $\tau \circ \hat{\psi} = $ id. So $\hat{\psi} : \Omega \to \mathcal{U}$ is a biholomorphic mapping. $\qquad\square$

Remark: The sequence of mappings ψ_j constructed above is often called a *scaling sequence*. It is constructed from a composition of

(1) The automorphisms carrying one fixed interior point successively to a boundary point
(2) Certain affine adjustments
(3) The stretching dilation map

The proof given above is a good example of the scaling technique. The main thrust of the method is that the image of the limit mapping is determined solely by the affine adjustments and the dilations, while the scaling sequence converges to a conformal mapping. $\qquad\square$

Remark: As observed earlier, the main result here can be proved in a much simpler way. Namely, one may conclude immediately from the argument on the shrinking of $\varphi_j(K)$ into a simply connected subset of Ω that Ω must be simply connected. Then the conclusion follows by the Riemann mapping theorem. But we are trying to skirt around the Riemann mapping theorem. The goal of this argument is to provide a basis for the scaling method which can be applied to the higher-dimensional cases.   □

Chapter 8
Concluding Remarks

We have endeavored in this book to give the reader a look at the ever-evolving Bergman theory. Some of the results here are 90 years old, and others were proved quite recently.

The Bergman's ideas have proved to be remarkably robust and fruitful. They continue to yield new techniques and new paths for research. They have played a key role in the development of complex geometry and of partial differential equations in one and several complex variables, in complex function theory, and in extremal problems. The biholomorphic invariance of the Bergman kernel and metric has proved to be particularly important.

We have with pleasure presented the ideas connected with the Bergman representative coordinates. This is a much underappreciated aspect of the Bergman theory, and one that deserves further development. The proof of the Lu Qi-Keng's theorem serves to illustrate what a powerful idea it is. It also played a decisive role in the original proof of the Greene–Krantz semicontinuity theorem.

Fefferman's work on biholomorphic mappings of strictly pseudoconvex domains gives yet another illustration of the centrality and power of the Stefan Bergman's ideas. The Fefferman's asymptotic expansion has proved to be one of the central ideas in the modern function theory of several complex variables.

The work of Greene and Krantz has also served to illustrate the geometric force of the Bergman kernel and Bergman metric. Krantz and Li, and Kim and Krantz, have developed these ideas even further.

We look forward to many years of future activity in the Bergman geometric theory and the Bergman function theory.

S.G. Krantz, *Geometric Analysis of the Bergman Kernel and Metric*, Graduate Texts in Mathematics 268, DOI 10.1007/978-1-4614-7924-6_8, © Springer Science+Business Media New York 2013

Table of Notation

Notation	Section	Definition
D	1.1	The unit disc in the plane
Ω	1.1	A domain
dV	1.1	The volume element
$H^\infty(\Omega)$	1.1	Bounded holomorphic functions on Ω
$A(\Omega)$	1.1	Functions holomorphic on Ω, continuous on the closure
$A^2(\Omega)$	1.1	Square integrable holomorphic functions on Ω (the Bergman space)
C^k	1.1	The k-times continuously differentiable functions
$\| \ \|_{A^2(\Omega)}$	1.1	The Bergman norm
$K(z,\zeta) = K_\Omega(z,\zeta)$	1.1	The Bergman kernel
$\{\phi_j\}$	1.1	A complete orthonormal basis
P	1.1	The Bergman projection
$J_{\mathbb{C}}(f)$	1.1	The complex Jacobian matrix of f
$J_{\mathbb{R}}(f)$	1.1	The real Jacobian matrix of f
$g_{i,j}(z)$	1.1	The Bergman metric
$|\xi|_{B,z}$	1.1	The Bergman metric
$\ell(\gamma)$	1.1	The Bergman length of γ
$d_\Omega(z,w)$	1.1	The Bergman distance of z to w
$\Gamma(z)$	1.1	Euler's gamma function
$\Gamma(\zeta,z)$	1.1	Fundamental solution for the Laplacian
dA	1.1	The area element
$G(\zeta,z)$	1.1	The Green's function
$S(z,\zeta)$	1.2	The Szegő kernel
$\eta(z)$	1.2	The Leray form
$\bar{\partial}$	1.2	The Cauchy–Riemann operator
$S(z,\zeta)$	1.2	The Szegő projection
$\mathcal{P}(z,\zeta)$	1.2	The Poisson–Szegő kernel
δ_z	1.2	The Dirac delta mass
N	1.2	The $\bar{\partial}$-Neumann operator

S.G. Krantz, *Geometric Analysis of the Bergman Kernel and Metric*,
Graduate Texts in Mathematics 268, DOI 10.1007/978-1-4614-7924-6,
© Springer Science+Business Media New York 2013

Notation	Section	Definition
$H^2(\Omega)$	1.2	The square-integrable Hardy space
$\mathcal{H}$	1.3	A Hilbert space with reproducing kernel
$K(x, y)$	1.3	A reproducing kernel
$\mathbf{h}^2(D)$	1.6	The square-integrable harmonic functions
$\mathbb{Z}^+$	1.7	The nonnegative integers
A	1.8	An annulus
$d\sigma$	1.9	Area measure on the boundary of the ball
W^s	1.11	The Sobolev space
$\mathcal{U}_\epsilon^k(\Omega_0)$	2.1	Domains neighboring Ω_0
$\sim$	2.1	Is biholomorphic to
$\mathrm{Aut}(\Omega)$	2.1	The automorphism group of Ω
$\bar\partial^*$	2.1	The adjoint of $\bar\partial$
Condition R	2.1	A regularity condition for the Bergman projection
$\bigwedge^{0,j}$	2.1	Differential forms
$WH^j(\Omega)$	2.1	$W^j(\Omega) \cap$ (holomorphic functions)
$WH^\infty(\Omega)$	2.1	$W^\infty \cap$ (holomorphic functions)
$W_0^j(\Omega)$	2.1	The W^j closure of $C_c^\infty(\Omega)$
$\partial/\partial\nu_P$	2.1	The outward normal derivative at P
b_j^i	3.1	Bergman representative coordinates
$\bar\partial_b$	3.3	The boundary Cauchy–Riemann operator
$\mathcal{B}(z, \zeta)$	3.3	The Poisson–Bergman kernel
$\Lambda f(w, x)$	3.3	The Berezin transform
$\beta_2(z, r)$	3.3	Nonisotropic ball
$\mathcal{M}f(z)$	3.3	Maximal operator
$\mathcal{A}_\alpha(z)$	3.3	Admissible approach region
$\mathcal{L}$	3.5	Laplace–Beltrami operator
$\rho(z, \zeta)$	3.6	A nonisotropic metric on B
$\mathcal{P}_k$	4.1	All homogeneous polynomials of degree k
$\mathcal{A}_k$	4.1	Kernel of the Laplacian in $\mathcal{P}_k$
$\mathcal{B}_k$	4.1	Image of the Laplacian in $\mathcal{P}_k$
$\mathcal{H}_k$	4.1	The spherical harmonics
$Z_{x'}^{(k)}$	4.2	Zonal harmonic
$P(x, t')$	4.2	The Poisson kernel
$P_k^\lambda(t)$	4.2	Gegenbauer polynomial
$\mathcal{H}^{p,q}$	4.3	Harmonic polynomials of bidegree (p, q)
$Q(f, g)$	4.4	Cauchy–Riemann inner product
$[L, M]$	5.3	First-order commutator
$\nu(\phi)$	5.3	The order of vanishing of ϕ
$\mathcal{H}$	5.7	Hausdorff distance on domains
$A^2(M)$	5.9	The Bergman space on a manifold
ds^2	5.9	The Bergman metric on a manifold
$\mathcal{W}$	6.1	The Diederich–Fornæss worm domain
η	6.1	Function used to construct the worm
$\mathcal{A}$	6.2	Singular annulus in the worm
$B_\Omega^K(q, r)$	7.2	Kobayashi distance ball
$S_\Omega(p; \xi)$	7.3	Holomorphic sectional curvature

Bibliography

[ADA]   R. Adams, *Sobolev Spaces*, Academic Press, 1975.

[AFR]   P. Ahern, M. Flores, and W. Rudin, An invariant volume-mean-value property, *Jour. Functional Analysis* 11(1993), 380–397.

[AHL]   L. Ahlfors, *Complex Analysis*, 3rd ed., McGraw-Hill, New York, 1979.

[ARO]   N. Aronszajn, Theory of reproducing kernels, *Trans. Am. Math. Soc.* 68(1950), 337–404.

[BEG]   T. N. Bailey, M. G. Eastwood, and C. R. Graham, Invariant theory for conformal and CR geometry, *Annals of Math.* 139(1994), 491–552.

[BAR1]   D. Barrett, Irregularity of the Bergman projection on a smooth bounded domain in $\mathbb{C}^2$, *Annals of Math.* 119(1984), 431–436.

[BAR2]   D. Barrett, The behavior of the Bergman projection on the Diederich–Fornaess worm, *Acta Math.*, 168(1992), 1–10.

[BAR3]   D. Barrett, Regularity of the Bergman projection and local geometry of domains, *Duke Math. Jour.* 53(1986), 333–343.

[BAR4]   D. Barrett, Behavior of the Bergman projection on the Diederich–Fornæss worm, *Acta Math.* 168(1992), 1–10.

[BEDF]   E. Bedford and P. Federbush, Pluriharmonic boundary values, *Tohoku Math. Jour.* 26(1974), 505–511.

[BEF1]   E. Bedford and J. E. Fornæss, A construction of peak functions on weakly pseudoconvex domains, *Ann. Math.* 107(1978), 555–568.

[BEF2]   E. Bedford and J. E. Fornæss, Counterexamples to regularity for the complex Monge–Ampère equation, *Invent. Math.* 50 (1978/79), 129–134.

[BEL1]   S. Bell, Biholomorphic mappings and the $\bar{\partial}$ problem, *Ann. Math.*, 114(1981), 103–113.

[BEL2]   S. Bell, Local boundary behavior of proper holomorphic mappings, *Proc. Sympos. Pure Math*, vol. 41, American Math. Soc., Providence R.I., 1984, 1–7.

[BEL3]   S. Bell, Differentiability of the Bergman kernel and pseudo-local estimates, *Math. Z.* 192(1986), 467–472.

[BEB]   S. Bell and H. Boas, Regularity of the Bergman projection in weakly pseudoconvex domains, *Math. Annalen* 257(1981), 23–30.

[BEC]   S. Bell and D. Catlin, Proper holomorphic mappings extend smoothly to the boundary, *Bull. Amer. Math. Soc.* (N.S.) 7(1982), 269–272.

[BEK]   S. Bell and S. G. Krantz, Smoothness to the boundary of conformal maps, *Rocky Mt. Jour. Math.* 17(1987), 23–40.

[BELL]   S. Bell and E. Ligocka, A simplification and extension of Fefferman's theorem on biholomorphic mappings, *Invent. Math.* 57(1980), 283–289.

S.G. Krantz, *Geometric Analysis of the Bergman Kernel and Metric*,
Graduate Texts in Mathematics 268, DOI 10.1007/978-1-4614-7924-6,
© Springer Science+Business Media New York 2013

[BERE]   F. A. Berezin, Quantization in complex symmetric spaces, *Math. USSR Izvestia* 9(1975), 341–379.

[BER1]   S. Bergman, Über die Entwicklung der harmonischen Funktionen der Ebene und des Raumes nach Orthogonal funktionen, *Math. Annalen* 86(1922), 238–271.

[BER2]   S. Bergman, *The Kernel Function and Conformal Mapping*, Am. Math. Soc., Providence, RI, 1970.

[BES]   S. Bergman and M. Schiffer, *Kernel Functions and Elliptic Differential Equations in Mathematical Physics*, Academic Press, New York, 1953.

[BEC]   B. Berndtsson, P. Charpentier, A Sobolev mapping property of the Bergman kernel, *Math. Z.* **235** (2000), 1–10.

[BERS]   L. Bers, *Introduction to Several Complex Variables*, New York Univ. Press, New York, 1964.

[BLK]   B. E. Blank and S. G. Krantz, *Calculus*, Key Press, Emeryville, CA, 2006.

[BLP]   Z. Blocki and P. Pflug, Hyperconvexity and Bergman completeness, *Nagoya Math. J.* 151(1998), 221–225.

[BLG]   T. Bloom and I. Graham, A geometric characterization of points of type *m* on real submanifolds of $\mathbb{C}^n$, *J. Diff. Geom.* 12(1977), 171–182.

[BOA1]   H. Boas, Counterexample to the Lu Qi-Keng conjecture, *Proc. Am. Math. Soc.* 97(1986), 374–375.

[BOA2]   H. Boas, The Lu Qi-Keng conjecture fails generically, *Proc. Amer. Math. Soc.* 124(1996), 2021–2027.

[BOS1]   H. Boas and E. Straube, Sobolev estimates for the $\bar{\partial}$-Neumann operator on domains in $\mathbb{C}^n$ admitting a defining function that is plurisubharmonic on the boundary, *Math. Z.* **206** (1991), 81–88.

[BOS2]   H. Boas and E. Straube, Equivalence of regularity for the Bergman projection and the $\bar{\partial}$-Neumann operator, *Manuscripta Math.* 67(1990), 25–33.

[BKP]   H. Boas, S. G. Krantz, and M. M. Peloso, unpublished.

[BOC1]   S. Bochner, Orthogonal systems of analytic functions, *Math. Z.* 14(1922), 180–207.

[BOU]   L. Boutet de Monvel, Le noyau de Bergman en dimension 2, *Séminaire sur les Équations aux Dérivées Partielles* 1987–1988, Exp. no. XXII, École Polytechnique Palaiseau, 1988, p. 13.

[BOS]   L. Boutet de Monvel and J. Sjöstrand, Sur la singularité des noyaux de Bergman et Szegő, *Soc. Mat. de France Asterisque* 34–35(1976), 123–164.

[BRE]   H. J. Bremermann, Holomorphic continuation of the kernel function and the Bergman metric in several complex variables, *Lectures on Functions of a Complex Variable*, Michigan, 1955, 349–383.

[BUN]   L. Bungart, Holomorphic functions with values in locally convex spaces and applications to integral formulas, *Trans. Am. Math. Soc.* 111(1964), 317–344.

[BSW]   D. Burns, S. Shnider, R. O. Wells, On deformations of strictly pseudoconvex domains, *Invent. Math.* 46(1978), 237–253.

[CKNS]   L. Caffarelli, J. J. Kohn, L. Nirenberg, and J. Spruck, The Dirichlet problem for nonlinear second-order elliptic equations. II. Complex Monge–Ampère, and uniformly elliptic, equations, *Comm. Pure Appl. Math.* 38(1985), 209–252.

[CAR]   L. Carleson, *Selected Problems on Exceptional Sets*, Van Nostrand, Princeton, NJ, 1967.

[CCP]   G. Carrier, M. Crook, and C. Pearson, *Functions of a Complex Variable*, McGraw-Hill, New York, 1966.

[CAT1]   D. Catlin, Necessary conditions for subellipticity of the $\bar{\partial}$—Neumann problem, *Ann. Math.* 117(1983), 147–172.

[CAT2]   D. Catlin, Subelliptic estimates for the $\bar{\partial}$Neumann problem, *Ann. Math.* 126(1987), 131–192.

[CAT3]   D. Catlin, Boundary behavior of holomorphic functions on pseudoconvex domains, *J. Differential Geom.* 15(1980), 605–625.

[CNS]   D.-C. Chang, A. Nagel, and E. M. Stein, Estimates for the $\bar{\partial}$-Neumann problem in pseudoconvex domains of finite type in $\mathbb{C}^2$, *Acta Math.* 169(1992), 153–228.

[CHE]   S.-C. Chen, A counterexample to the differentiability of the Bergman kernel function, *Proc. AMS* 124(1996), 1807–1810.

[CHF]   B.-Y. Chen and S. Fu, Comparison of the Bergman and Szegö kernels, *Advances in Math.* 228(2011), 2366–2384.

[CHS]   S.-C. Chen and M. C. Shaw, *Partial Differential Equations in Several Complex Variables*, AMS/IP Studies in Advanced Mathematics, 19. American Mathematical Society, Providence, RI; International Press, Boston, MA, 2001.

[CHENG]   S.-Y. Cheng, Open problems, *Conference on Nonlinear Problems in Geometry Held in Katata*, September, 1979, Tohoku University, Dept. of Mathematics, Sendai, 1979, p. 2.

[CHM]   S. S. Chern and J. Moser, Real hypersurfaces in complex manifolds, *Acta Math.* 133(1974), 219–271.

[CHR1]   M. Christ, Global C^∞ irregularity of the $\bar{\partial}$?–Neumann problem for worm domains, *J. Amer. Math. Soc.* 9(1996), 1171–1185.

[CHR2]   M. Christ, Remarks on global irregularity in the $\bar{\partial}$-Neumann problem, *Several complex variables* (Berkeley, CA, 1995–1996), 161–198, Math. Sci. Res. Inst. Publ. 37, Cambridge Univ. Press, Cambridge, 1999.

[COL]   E. Coddington and N. Levinson, *Theory of Ordinary Differential Equations*, McGraw-Hill, New York, 1955.

[COW]   R. R. Coifman and G. Weiss, *Analyse Harmonique Non-Commutative sur Certains Espaces Homogenes*, Springer Lecture Notes vol. 242, Springer Verlag, Berlin, 1971.

[COH]   R. Courant and D. Hilbert, *Methods of Mathematical Physics*, 2nd ed., Interscience, New York, 1966.

[DAN1]   J. P. D'Angelo, Real hypersurfaces, orders of contact, and applications, *Annals of Math.* 115(1982), 615–637.

[DAN2]   J. P. D'Angelo, Intersection theory and the $\bar{\partial}$- Neumann problem, *Proc. Symp. Pure Math.* 41(1984), 51–58.

[DAN3]   J. P. D'Angelo, Finite type conditions for real hypersurfaces in $\mathbb{C}^n$, in *Complex Analysis Seminar*, Springer Lecture Notes vol. 1268, Springer Verlag, 1987, 83–102.

[DAN4]   J. P. D'Angelo, *Several Complex Variables and the Geometry of Real Hypersurfaces*, CRC Press, Boca Raton, FL, 1993.

[DIE1]   K. Diederich, Das Randverhalten der Bergmanschen Kernfunktion und Metrik in streng pseudo-konvexen Gebieten, *Math. Ann.* 187(1970), 9–36.

[DIE2]   K. Diederich, Über die 1. and 2. Ableitungen der Bergmanschen Kernfunktion und ihr Randverhalten, *Math. Ann.* 203(1973), 129–170.

[DIF1]   K. Diederich and J. E. Fornæss, Pseudoconvex domains: An example with nontrivial Nebenhülle, *Math. Ann.* 225(1977), 275–292.

[DIF2]   K. Diederich and J. E. Fornæss, Pseudoconvex domains with real-analytic boundary, *Annals of Math.* 107(1978), 371–384.

[DIF3]   K. Diederich and J. E. Fornæss, Pseudoconvex domains: Bounded strictly plurisubharmonic exhaustion functions, *Invent. Math.* **39** (1977), 129–141.

[DIF4]   K. Diederich and J. E. Fornæss, Smooth extendability of proper holomorphic mappings, *Bull. Amer. Math. Soc.* (N.S.) 7(1982), 264–268.

[EBI1]   Ebin, D. G., On the space of Riemannian metrics, *Bull. Amer. Math. Soc.* 74(1968), 1001–1003.

[EBI2]   Ebin, D. G., The manifold of Riemannian metrics, 1970 Global Analysis (*Proc. Sympos. Pure Math.*, Vol. XV, Berkeley, Calif., 1968), pp. 11–40, Amer. Math. Soc., Providence, R.I.

[ENG1]   M. Engliš, Functions invariant under the Berezin transform, *J. Funct. Anal.* 121(1994), 233–254.

[ENG2]   M. Engliš, Asymptotics of the Berezin transform and quantization on planar domains, *Duke Math. J.* 79(1995), 57–76.

[EPS]   B. Epstein, *Orthogonal Families of Functions*, Macmillan, New York, 1965.

[ERD]   A. Erdelyi, et al, *Higher Transcendental Functions*, McGraw-Hill, New York, 1953.

[FED]   H. Federer, *Geometric Measure Theory*, Springer-Verlag, New York, 1969.

[FEF1]   C. Fefferman, The Bergman kernel and biholomorphic mappings of pseudoconvex domains, *Invent. Math.* 26(1974), 1–65.

[FEF2]   C. Fefferman, Parabolic invariant theory in complex analysis, *Adv. Math.* 31(1979), 131–262.

[FOK]   G. B. Folland and J. J. Kohn, *The Neumann Problem for the Cauchy-Riemann Complex*, Princeton University Press, Princeton, NJ, 1972.

[FOL]   G. B. Folland, Spherical harmonic expansion of the Poisson–Szegő kernel for the ball, *Proc. Am. Math. Soc.* 47(1975), 401–408.

[FOM]   J. E. Fornæss and J. McNeal, A construction of peak functions on some finite type domains. *Amer. J. Math.* 116(1994), no. 3, 737–755.

[FOR]   F. Forstneric, An elementary proof of Fefferman's theorem, *Expositiones Math.*, 10(1992), 136–149.

[FRI]   B. Fridman, A universal exhausting domain, *Proc. Am. Math. Soc.* 98(1986), 267–270.

[FUW]   S. Fu and B. Wong, On strictly pseudoconvex domains with Kähler–Einstein Bergman metrics, *Math. Res. Letters* 4(1997), 697–703.

[GAM]   T. Gamelin, *Uniform Algebras*, Prentice-Hall, Englewood Cliffs, NJ, 1969.

[GAS]   T. Gamelin and N. Sibony, Subharmonicity for uniform algebras. *J. Funct. Anal.* 35 (1980), 64–108.

[GAR]   J. B. Garnett, *Analytic Capacity and Measure*, Lecture Notes in Math. 297, Springer, New York, 1972.

[GARA]   P. R. Garabedian, A Green's function in the theory of functions of several complex variables, *Ann. of Math.* 55(1952). 19–33.

[GAR]   J. Garnett, *Bounded Analytic Functions*, Academic Press, New York, 1981.

[GLE]   A. Gleason, The abstract theorem of Cauchy-Weil, *Pac. J. Math.* 12(1962), 511–525.

[GOL]   Goluzin, *Geometric Theory of Functions of a Complex Variable*, American Mathematical Society, Providence, 1969.

[GRA1]   C. R. Graham, The Dirichlet problem for the Bergman Laplacian I, *Comm. Partial Diff. Eqs.* 8(1983), 433–476.

[GRA2]   C. R. Graham, The Dirichlet problem for the Bergman Laplacian II, *Comm. Partial Diff. Eqs.* 8(1983), 563–641.

[GRA3]   C. R. Graham, Scalar boundary invariants and the Bergman kernel, *Complex analysis, II* (College Park, Md., 1985–86), 108–135, Lecture Notes in Math. 1276, Springer, Berlin, 1987.

[GRL]   C. R. Graham and J. M. Lee, Smooth solutions of degenerate Laplacians on strictly pseudoconvex domains, *Duke Jour. Math.* 57(1988), 697–720.

[GRA]   I. Graham, Boundary behavior of the Carathéodory and Kobayashi metrics on strongly pseudoconvex domains in $\mathbb{C}^n$ with smooth boundary, *Trans. Am. Math. Soc.* 207(1975), 219–240.

[GRL]   H. Grauert and I. Lieb, Das Ramirezsche Integral und die Gleichung $\bar{\partial}u = \alpha$ im Bereich der beschränkten Formen, *Rice University Studies* 56(1970), 29–50.

[GKK]   R. E. Greene, K.-T. Kim, and S. G. Krantz, *The Geometry of Complex Domains*, Birkhäuser Publishing, Boston, MA, 2011.

[GRK1]   R. E. Greene and S. G. Krantz, Stability properties of the Bergman kernel and curvature properties of bounded domains, *Recent Progress in Several Complex Variables*, Princeton University Press, Princeton, 1982.

[GRK2]   R. E. Greene and S. G. Krantz, Deformation of complex structures, estimates for the $\bar{\partial}$ equation, and stability of the Bergman kernel, *Adv. Math.* 43(1982), 1–86.

[GRK3]   R. E. Greene and S. G. Krantz, The automorphism groups of strongly pseudoconvex domains, *Math. Annalen* 261(1982), 425–446.

[GRK4]   R. E. Greene and S. G. Krantz, The stability of the Bergman kernel and the geometry of the Bergman metric, *Bull. Am. Math. Soc.* 4(1981), 111–115.

[GRK5]   R. E. Greene and S. G. Krantz, Stability of the Carathéodory and Kobayashi metrics and applications to biholomorphic mappings, *Proc. Symp. in Pure Math.*, Vol. 41 (1984), 77–93.

[GRK6]   R. E. Greene and S. G. Krantz, Normal families and the semicontinuity of isometry and automorphism groups, *Math. Zeitschrift* 190(1985), 455–467.

[GRK7]   R. E. Greene and S. G. Krantz, Characterizations of certain weakly pseudo-convex domains with non-compact automorphism groups, in *Complex Analysis Seminar*, Springer Lecture Notes 1268(1987), 121–157.

[GRK8]   R. E. Greene and S. G. Krantz, Characterization of complex manifolds by the isotropy subgroups of their automorphism groups, *Indiana Univ. Math. J.* 34(1985), 865–879.

[GRK9]   R. E. Greene and S. G. Krantz, Biholomorphic self-maps of domains, *Complex Analysis II* (C. Berenstein, ed.), Springer Lecture Notes, vol. 1276, 1987, 136–207.

[GRK10]   R. E. Greene and S. G. Krantz, Techniques for Studying the automorphism Groups of Weakly Pseudoconvex Domains, *Several Complex Variables* (Stockholm, 1987/1988), 389–410, Math. Notes, 38, Princeton Univ. Press, Princeton, NJ, 1993.

[GRK11]   R. E. Greene and S. G. Krantz, Invariants of Bergman geometry and results concerning the automorphism groups of domains in $\mathbb{C}^n$, Proceedings of the 1989 Conference in Cetraro (D. Struppa, ed.), to appear.

[GRK12]   R. E. Greene and S. G. Krantz, *Function Theory of One Complex Variable*, 3rd ed., American Mathematical Society, Providence, RI, 2006.

[HAP1]   R. Harvey and J. Polking, Fundamental solutions in complex analysis. I. The Cauchy-Riemann operator, *Duke Math. J.* 46(1979), 253–300.

[HAP2]   R. Harvey and J. Polking, Fundamental solutions in complex analysis. II. The induced Cauchy-Riemann operator, *Duke Math. J.* 46(1979), 301–340.

[HEL]   S. Helgason, *Differential Geometry and Symmetric Spaces*, Academic Press, New York, 1962.

[HEN]   G. M. Henkin, Integral representations of functions holomorphic in strictly pseudo-convex domains and some applications, *Mat. Sb.* 78(120)(1969), 611–632.

[HIL]   E. Hille, *Analytic Function Theory*, 2nd ed., Ginn and Co., Boston, 1973.

[HIR1]   K. Hirachi, The second variation of the Bergman kernel of ellipsoids, *Osaka J. Math.* 30(1993), 457–473.

[HIR2]   K. Hirachi, Scalar pseudo-Hermitian invariants and the Szegő kernel on three-dimensional CR manifolds, *Complex Geometry* (Osaka, 1990), Lecture Notes Pure Appl. Math., v. 143, Marcel Dekker, New York, 1993, 67–76.

[HIR3]   K. Hirachi, Construction of boundary invariants and the logarithmic singularity in the Bergman kernel, *Annals of Math.* 151(2000), 151–190.

[HIR]   M. Hirsch, *Differential Topology*, Springer-Verlag, New York, 1976.

[HOR1]   L. Hörmander, L^2 estimates and existence theorems for the $\bar{\partial}$ operator, *Acta Math.* 113(1965), 89–152.

[HOR2]   L. Hörmander, *Linear Partial Differential Operators*, Springer-Verlag, New York, 1963.

[HOR3]   L. Hörmander, Pseudo-differential operators and non-elliptic boundary problems, *Ann. Math.* 83(1966), 129–209.

[HOR5]   L. Hörmander, "Fourier integral operators," *The Analysis of Linear Partial Differential Operators* IV, Reprint of the 1994 ed., Springer, Berlin, Heidelberg, New York, 2009.

[HUA]   L. K. Hua, *Harmonic Analysis of Functions of Several Complex Variables in the Classical Domains*, American Mathematical Society, Providence, 1963.

[ISK]   A. Isaev and S. G. Krantz, Domains with non-compact automorphism group: A Survey, *Advances in Math.* **146** (1999), 1–38.

[JAK]   S. Jakobsson, Weighted Bergman kernels and biharmonic Green functions, Ph.D.
           thesis, Lunds Universitet, 2000, 134 pages.
[KAT]   Y. Katznelson, *Introduction to Harmonic Analysis*, John Wiley and Sons, New York,
           1968.
[KEL]   O. Kellogg, *Foundations of Potential Theory*, Dover, New York, 1953.
[KER1]   N. Kerzman, Hölder and L^p estimates for solutions of $\bar{\partial}u = f$ on strongly
           pseudoconvex domains, *Comm. Pure Appl. Math.* 24(1971), 301–380.
[KER2]   N. Kerzman, The Bergman kernel function. Differentiability at the boundary, *Math.
           Ann.* 195(1972), 149–158.
[KER3]   N. Kerzman, A Monge–Ampre equation in complex analysis. Several Complex Vari-
           ables (Proc. Sympos. Pure Math., Vol. XXX, Part 1, Williams Coll., Williamstown,
           Mass., 1975), pp. 161–167. Amer. Math. Soc., Providence, R.I., 1977.
[KIMYW]   Y. W. Kim, Semicontinuity of compact group actions on com- pact dierentiable
           manifolds, *Arch. Math.* 49(1987), 450–455.
[KIS]   C. Kiselman, A study of the Bergman projection in certain Hartogs domains, *Proc.
           Symposia Pure Math.*, vol. 52 (E. Bedford, J. D'Angelo, R. Greene, and S. Krantz
           eds.), American Mathematical Society, Providence, 1991.
[KLE]   P. Klembeck, Kähler metrics of negative curvature, the Bergman metric near the
           boundary and the Kobayashi metric on smooth bounded strictly pseudoconvex sets,
           Indiana Univ. Math. J. 27(1978), 275–282.
[KOB1]   S. Kobayashi, Geometry of bounded domains, *Trans. AMS* 92(1959), 267–290.
[KOB2]   S. Kobayashi, *Hyperbolic Manifolds and Holomorphic Mappings*, Dekker, New York,
           1970.
[KON]   S. Kobayashi and K. Nomizu, *Foundations of Differential Geometry*, Vols. I and II,
           Interscience, New York, 1963, 1969.
[KOH1]   J. J. Kohn, Quantitative estimates for global regularity, *Analysis and geometry in
           several complex variables* (Katata, 1997), 97–128, Trends Math., Birkhäuser Boston,
           Boston, MA, 1999.
[KOH2]   J. J. Kohn, Boundary behavior of $\bar{\partial}$ on weakly pseudoconvex manifolds of dimension
           two, *J. Diff. Geom.* 6(1972), 523–542.
[KOR1]   A. Koranyi, Harmonic functions on Hermitian hyperbolic space, *Trans. A. M. S.*
           135(1969), 507–516.
[KOR2]   A. Koranyi, Boundary behavior of Poisson integrals on symmetric spaces, *Trans.
           A.M.S.* 140(1969), 393–409.
[KRA1]   S. G. Krantz, *Function Theory of Several Complex Variables*, 2nd ed., American
           Mathematical Society, Providence, RI, 2001.
[KRA2]   S. G. Krantz, On a construction of L. Hua for positive reproducing kernels, *Michigan
           Journal of Mathematics* 59(2010), 211–230.
[KRA3]   S. G. Krantz, Boundary decomposition of the Bergman kernel, *Rocky Mountain
           Journal of Math.*, to appear.
[KRA4]   S. G. Krantz, *Partial Differential Equations and Complex Analysis*, CRC Press, Boca
           Raton, FL, 1992.
[KRA5]   S. G. Krantz, *Cornerstones of Geometric Function Theory: Explorations in Complex
           Analysis*, Birkhäuser Publishing, Boston, 2006.
[KRA6]   S. G. Krantz, Invariant metrics and the boundary behavior of holomorphic functions
           on domains in $\mathbb{C}^n$, *Jour. Geometric. Anal.* 1(1991), 71–98.
[KRA7]   S. G. Krantz, Calculation and estimation of the Poisson kernel, *J. Math. Anal. Appl.*
           302(2005)143–148.
[KRA8]   S. G. Krantz, A new proof and a generalization of Ramadanov's theorem, *Complex
           Variables and Elliptic Eq.* 51(2006), 1125–1128.
[KRA9]   S. G. Krantz, *Complex Analysis: The Geometric Viewpoint*, 2nd ed., Mathematical
           Association of America, Washington, D.C., 2004.
[KRA11]   S. G. Krantz, Canonical kernels versus constructible kernels, preprint.

[KRA12]   S. G. Krantz, Lipschitz spaces, smoothness of functions, and approximation theory, *Expositiones Math.* 3(1983), 193–260.

[KRA13]   S. G. Krantz, Characterizations of smooth domains in $\mathbb{C}$ by their biholomorphic self maps, *Am. Math. Monthly* 90(1983), 555–557.

[KRA14]   S. G. Krantz, *A Guide to Functional Analysis*, Mathematical Association of America, Washington, D.C., 2013, to appear.

[KRA15]   S. G. Krantz, A direct connection between the Bergman and Szegő projections, *Complex Analysis and Operator Theory*, to appear.

[KRPA1]   S. G. Krantz and H. R. Parks, *The Geometry of Domains in Space*, Birkhäuser Publishing, Boston, MA, 1996.

[KRPA2]   S. G. Krantz and H. R. Parks, *The Implicit Function Theorem*, Birkhäuser, Boston, 2002.

[KRP1]   S. G. Krantz and M. M. Peloso, The Bergman kernel and projection on non-smooth worm domains, *Houston J. Math.* 34 (2008), 9.3.-950.

[KRP2]   S. G. Krantz and M. M. Peloso, Analysis and geometry on worm domains, *J. Geom. Anal.* 18(2008), 478–510.

[LEM1]   L. Lempert, La metrique Kobayashi et las representation des domains sur la boule, *Bull. Soc. Math. France* 109(1981), 427–474.

[LI]   S.-Y. Li, S-Y. Li, Neumann problems for complex Monge–Ampère equations, *Indiana University J. of Math*, 43(1994), 1099–1122.

[LIG1]   E. Ligocka, The Sobolev spaces of harmonic functions, *Studia Math.* 54(1986). 79–87.

[LIG2]   E. Ligocka, Remarks on the Bergman kernel function of a worm domain, *Studia Mathematica* 130(1998), 109–113.

[MIN]   B.-L. Min, Domains with prescribed automorphism group, *J. Geom. Anal.* 19 (2009), 911–928.

[NAR]   R. Narasimhan, *Several Complex Variables*, University of Chicago Press, Chicago, 1971.

[NWY]   L. Nirenberg, S. Webster, and P. Yang, Local boundary regularity of holomorphic mappings. *Comm. Pure Appl. Math.* 33(1980), 305–338.

[OHS]   T. Ohsawa, A remark on the completeness of the Bergman metric, *Proc. Japan Acad. Ser. A Math. Sci.*, 57(1981), 238–240.

[PAI]   Painlevé, Sur les lignes singulières des functions analytiques, *Thèse*, Gauthier-Villars, Paris, 1887.

[PEE]   J. Peetre, The Berezin transform and Ha-Plitz operators, *J. Operator Theory* 24(1990), 165–186.

[PET]   P. Petersen, *Riemannian Geometry*, Springer, New York, 2009.

[PHS]   D. H. Phong and E. M. Stein, Hilbert integrals, singular integrals, and Radon transforms, *Acta Math.* 157(1986), 99–157.

[PHS1]   D. H. Phong and E. M. Stein, Hilbert integrals, singular integrals, and Radon transforms. I. *Acta Math.* 157(1986), 99–157.

[PHS2]   D. H. Phong and E. M. Stein, Hilbert integrals, singular integrals, and Radon transforms. II. *Invent. Math.* 86(1986), 75–113.

[PIN]   S. Pinchuk, The scaling method and holomorphic mappings, *Several Complex Variables and Complex Ggeometry*, Part 1 (Santa Cruz, CA, 1989), 151–161, Proc. Sympos. Pure Math., 52, Part 1, Amer. Math. Soc., Providence, RI, 1991.

[PIH]   S. Pinchuk and S. V. Hasanov, Asymptotically holomorphic functions (Russian), *Mat. Sb.* 134(176) (1987), 546–555.

[PIT]   S. Pinchuk and S. I. Tsyganov, Smoothness of CR-mappings between strictly pseudoconvex hypersurfaces. (Russian) *Izv. Akad. Nauk SSSR* Ser. Mat. 53(1989), 1120–1129, 1136; translation in *Math. USSR-Izv.* 35(1990), 457–467.

[RAM1]   I. Ramadanov, Sur une propriété de la fonction de Bergman. (French) *C. R. Acad. Bulgare Sci.* 20(1967), 759–762.

[RAM2]   I. Ramadanov, A characterization of the balls in $\mathbb{C}^n$ by means of the Bergman kernel, *C. R. Acad. Bulgare Sci.* 34(1981), 927–929.

[RAMI]   E. Ramirez, Divisions problem in der komplexen analysis mit einer Anwendung auf Rand integral darstellung, *Math. Ann.* 184(1970), 172–187.

[RAN]   R. M. Range, A remark on bounded strictly plurisubharmonic exhaustion functions, *Proc. A.M.S.* 81(1981), 220–222.

[ROM]   S. Roman, The formula of Faà di Bruno, *Am. Math. Monthly* 87(1980), 805–809.

[ROW]   B. Rodin and S. Warschawski, Estimates of the Riemann mapping function near a boundary point, in *Romanian-Finnish Seminar on Complex Analysis*, Springer Lecture Notes, vol. 743, 1979, 349–366.

[ROS]   J.-P. Rosay, Sur une characterization de la boule parmi les domains de $\mathbb{C}^n$ par son groupe d'automorphismes, *Ann. Inst. Four. Grenoble* XXIX(1979), 91–97.

[ROSE]   P. Rosenthal, On the zeroes of the Bergman function in doubly-connected domains, *Proc. Amer. Math. Soc.* 21(1969), 33–35.

[RUD1]   W. Rudin, *Principles of Mathematical Analysis*, 3rd ed., McGraw-Hill, New York, 1976.

[RUD2]   W. Rudin, *Function Theory in the Unit Ball of* $\mathbb{C}^n$, Grundlehren der Mathematischen Wissenschaften in Einzeldarstellungen, Springer, Berlin, 1980.

[SEM]   S. Semmes, A generalization of Riemann mappings and geometric structures on a space of domains in $\mathbb{C}^n$, *Memoirs of the American Mathematical Society*, 1991.

[SIB]   N. Sibony, A class of hyperbolic manifolds, *Ann. of Math. Stud.* 100(1981), 357–372.

[SIU]   Y.-T. Siu, Non Hölder property of Bergman projection of smooth worm domain, *Aspects of Mathematics—Algebra, Geometry, and Several Complex Variables*, N. Mok (ed.), University of Hong Kong, 1996, 264–304.

[SKW]   M. Skwarczynski, The distance in the theory of pseudo-conformal transformations and the Lu Qi-King conjecture, *Proc. A.M.S.* 22(1969), 305–310.

[STE1]   E. M. Stein, *Singular Integrals and Differentiability Properties of Functions*, Princeton University Press, Princeton, NJ. 1970.

[STE2]   E. M. Stein, *Boundary Behavior of Holomorphic Functions of Several Complex Variables*, Princeton University Press, Princeton, 1972.

[STW]   E. M. Stein and G. Weiss, *Introduction to Fourier Analysis on Euclidean Space*, Princeton University Press, Princeton, NJ, 1971.

[STR]   K. Stromberg, *An Introduction to Classical Real Analysis*, Wadsworth, Belmont, 1981.

[SUY]   N. Suita and A. Yamada On the Lu Qi-Keng conjecture, *Proc. A.M.S.* 59(1976), 222–224.

[SZE]   G. Szegő, Über Orthogonalsysteme von Polynomen, *Math. Z.* 4(1919), 139–151.

[TAN]   N. Tanaka, On generalized graded Lie algebras and geometric structures, I, *J. Math. Soc. Japan* 19(1967), 215–254.

[THO]   G. B. Thomas, *Calculus*, 7th ed., Addison-Wesley, Reading, MA, 1999.

[TRE]   F. Treves, *Introduction to Pseudodifferential and Fourier Integral Operators*, Vol. II, Plenum Press, New York 1982.

[WAR1]   S. Warschawski, On the boundary behavior of conformal maps, *Nagoya Math. J.* 30(1967), 83–101.

[WAR2]   S. Warschawski, On boundary derivatives in conformal mapping, *Ann. Acad. Sci. Fenn.* Ser. A I no. 420(1968), 22 pp.

[WAR3]   S. Warschawski, Hölder continuity at the boundary in conformal maps, *J. Math. Mech.* 18(1968/9), 423–7.

[WEB1]   S. Webster, Biholomorphic mappings and the Bergman kernel off the diagonal, *Invent. Math.* 51(1979), 155–169.

[WEB2]   S. Webster, On the reflection principle in several complex variables, *Proc. Amer. Math. Soc.* 71(1978), 26–28.

[WHW]   E. Whittaker and G. Watson, *A Course of Modern Analysis*, 4th ed., Cambridge Univ. Press, London, 1935.

[WIE]   J. Wiegerinck, Domains with finite dimensional Bergman space, *Math. Z.* 187(1984), 559–562.

[WON]   B. Wong, Characterization of the ball in $\mathbb{C}^n$ by its automorphism group, *Invent. Math.* 41(1977), 253–257.

[YAU]   S.-T. Yau, Problem section, *Seminar on Differential Geometry*, S.-T. Yau ed., *Annals of Math. Studies*, vol. 102, Princeton University Press, 1982, 669–706.

[ZHU]   K. Zhu, *Spaces of Holomorphic Functions in the Unit Ball*, Springer, New York, 2005.

Index

A

admissible
    approach region, 98
    limits, 98
Ahlfors map, 83
almost-the-shortest connector, 261
analytic
    polyhedron, 83
    type, 158
annulus, 43
Arazy, J., 95
Aronszajn, Nachman, 25, 35
automorphism, 14, 54, 99
    uniform bounds on derivatives, 153
automorphism group, 14, 69, 147
    compact, 151
    is a real Lie group, 184
    semicontinuity of, 151
    topologies on, 163
    transitive action, 100
automorphism-invariant metric, 153

B

ball and polydisc are biholomorphically
    inequivalent, 83
balls
    nonisotropic, 111
Banach-Alaoglou theorem, 57
Barrett's counterexample, 236
Barrett's theorem, 191, 202, 204, 236
Barrett, David, 74, 191
Bell's theorem
    localization of, 191
Bell, Steven R., 73, 141
Bell–Boas condition for mappings, 14

Bell–Krantz proof of the Fefferman's theorem, 73
Berezin
    kernel, 91
    transform, 90, 91
Bergman
    transformation law, 89
Bergman basis
    new, 36
Bergman distance, 11
Bergman kernel, 4
    as a Hilbert integral, 246
    asymptotic expansion for, 205
    boundary asymptotics, 205
    boundary behavior, 208
    boundary localization of, 59
    boundary singularity, 41
    calculation of, 14
    constructed with partial differential
        equations, 20
    for a Sobolev space, 54, 221
    for the annulus, 12
    for the annulus, approximate formula, 44
    for the annulus, special basis, 44
    for the disc, 18
    for the disc by conformal invariance, 23
    for the polydisc, 19
    for the unit disc, 12
    in an increasing sequence of domains, 57
    invariance of, 2, 92
    is conjugate symmetric, 4
    on a domain in several complex variables,
        64
    on a smooth, finitely connected, planar
        domain, 62
    on multiply connected domains, 54

S.G. Krantz, *Geometric Analysis of the Bergman Kernel and Metric*,
Graduate Texts in Mathematics 268, DOI 10.1007/978-1-4614-7924-6,
© Springer Science+Business Media New York 2013

Printed in the United States
By Bookmasters

Printed in the United States
By Bookmasters